MW01125085

TRADING ON THE EDGE

Wiley Finance Editions

FINANCIAL STATEMENT ANALYSIS, Martin S. Fridson
DYNAMIC ASSET ALLOCATION, David A. Hammer
INTERMARKET TECHNICAL ANALYSIS, John J. Murphy
INVESTING IN INTANGIBLE ASSETS, Russell L. Parr
FORECASTING FINANCIAL MARKETS, Tony Plummer
PORTFOLIO MANAGEMENT FORMULAS, Ralph Vince
TRADING AND INVESTING IN BOND OPTIONS, M. Anthony Wong
THE COMPLETE GUIDE TO CONVERTIBLE SECURITIES WORLDWIDE, Laura A. Zubulake
MANAGED FUTURES IN THE INSTITUTIONAL PORTFOLIO, Charles B. Epstein, Editor
ANALYZING AND FORECASTING FUTURES PRICES, Anthony F. Herbst
CHAOS AND ORDER IN THE CAPITAL MARKETS, Edgar E. Peters
INSIDE THE FINANCIAL FUTURES MARKETS, 3RD EDITION, Mark J. Powers and Mark G. Castelino
RELATIVE DIVIDEND YIELD, Anthony E. Spare
SELLING SHORT, Joseph A. Walker
TREASURY OPERATIONS AND THE FOREIGN EXCHANGE CHALLENGE, Dimitris N. Chorafas
THE FOREIGN EXCHANGE AND MONEY MARKETS GUIDE, Julian Walmsley
CORPORATE FINANCIAL RISK MANAGEMENT, Diane B. Wunnicke, David R. Wilson, Brooke Wunnicke
MONEY MANAGEMENT STRATEGIES FOR FUTURES TRADERS, Nauzer J. Balsara
THE MATHEMATICS OF MONEY MANAGEMENT, Ralph Vince
THE NEW TECHNOLOGY OF FINANCIAL MANAGEMENT, Dimitris N. Chorafas
THE DAY TRADER'S MANUAL, William F. Eng
OPTION MARKET MAKING, Allen J. Baird
TRADING FOR A LIVING, Dr. Alexander Elder
CORPORATE FINANCIAL DISTRESS AND BANKRUPTCY, 2ND EDITION, Edward I. Altman
FIXED-INCOME ARBITRAGE, M. Anthony Wong
TRADING APPLICATIONS OF JAPANESE CANDLESTICK CHARTING, Gary S. Wagner and Brad L. Matheny
FRACTAL MARKET ANALYSIS: APPLYING CHAOS THEORY TO INVESTMENT AND ECONOMICS, Edgar E. Peters
UNDERSTANDING SWAPS, John F. Marshall and Kenneth R. Kapner
GENETIC ALGORITHMS AND INVESTMENT STRATEGIES, Richard J. Bauer, Jr.
TRADER VIC II: ANALYTIC PRINCIPLES OF PROFESSIONAL SPECULATION, Victor Sperandeo
FORECASTING FINANCIAL AND ECONOMIC CYCLES, Mike P. Niemira and Philip A. Klein
TRADING ON THE EDGE: NEURAL, GENETIC, AND FUZZY SYSTEMS FOR CHAOTIC FINANCIAL MARKETS, Guido J. Deboeck

TRADING ON THE EDGE

Neural, Genetic, and Fuzzy Systems for Chaotic Financial Markets

Guido J. Deboeck, Editor

John Wiley & Sons, Inc.

New York • Chichester • Brisbane • Toronto • Singapore

To Nina, Pascal, and Toni
who through their high school science projects
introduced me to robotics, machine learning,
and neural networks.

To my mother and the memory of my father.

This text is printed on acid-free paper.

Copyright © 1994 by John Wiley & Sons, Inc.

All rights reserved. Published simultaneously in Canada.

Reproduction or translation of any part of this work beyond that permitted by Section 107 or 108 of the 1976 United States Copyright Act without the permission of the copyright owner is unlawful. Requests for permission or further information should be addressed to the Permissions Department, John Wiley & Sons, Inc.

This publication is designed to provide accurate and authoritative information in regard to the subject matter covered. It is sold with the understanding that the publisher is not engaged in rendering legal, accounting, or other professional services. If legal advice or other expert assistance is required, the services of a competent professional person should be sought.

Library of Congress Cataloging-in-Publication Data:

Trading on the edge : neural, genetic, and fuzzy systems for chaotic and financial markets / Guido J. Deboeck, editor.
p. cm. — (Wiley Finance editions)
Includes bibliographical references and index.
ISBN 0-471-31100-6 (acid-free paper)
1. Finance—Decision making—Data processing. 2. Program trading (Securities)—Data processing. 3. Neural networks (Computer science) 4. Genetic algorithms. 5. Fuzzy logic. I. Deboeck, Guido. II. Series.
HG4012.5.T7 1994
332.6′0285—dc20 94-2364

Printed in the United States of America

10 9 8 7 6 5 4

Contents

Foreword

This book adds significantly to the field of neural networks. Its major contribution is to bring the reader up to a working understanding of state-of-the-art applications of neural nets and related technologies to financial decision making. What sets this book apart from other publications in this field is its level of detail and realism in describing real-world financial applications. Without the unique capabilities and commitment of Guido Deboeck, who is an advisor on advanced technology at the World Bank in Washington, and the tangible achievements of the many other authors represented here, this book would not have been possible.

The neural network concepts described here actually originated in the social sciences—in psychology and economics, which are central to understanding the behavior of financial markets. In 1974 I created a learning paradigm for the design of neural networks. This paradigm, known as "backpropagation," is a gradient descent method for finding the best set of weights for capturing nonlinear relationships between independent and dependent data sets. I showed that backpropagation could be generalized so as to work with econometric-style models, and that it could be integrated with "Box-Jenkins" concepts.

The first working application of back-propagation dealt with long-range political forecasting (see *The Roots of Backpropagation*, John Wiley & Sons, 1993). The second application involved forecasting GNP growth in Latin America. Two other key applications that followed produced practical insights into the natural gas industry.

In the early 1980s, cognitive scientists received funding to popularize back-propagation. In 1988, the National Science Foundation funded advanced extensions of back-propagation in engineering. This led to breakthroughs in real-world control and production problems.

Neural networks are very useful in pattern recognition. Several banks have begun using these techniques successfully in real-time applications. The working examples described in this book testify to the development of these advanced, integrated capabilities in the financial sector.

This book also describes how tools such as fuzzy logic and genetic algorithms can be used to build systems to improve upon the judgment of traders, to derive automated trading strategies, and/or to optimize portfolio management.

When trying to find the best blend of various techniques, we are really trying to combine expert know-how, which is fuzzy, with findings from empirical time series, which are patterns that can be detected by neural networks. This is an issue that economists have studied for decades! In technical terms, the true probability distribution for possible future outcomes is based on "convoluting" together the probability distributions from experts and from empirical knowledge. In practical terms, this simply means that we can do better if we make full use of both kinds of knowledge—as this book encourages us to do.

The empirical insights provided in these pages demonstrate that it is possible to trade financial markets *more* intelligently, to account for a wider range of information, more patterns, differences in time horizons, and mental models of market participants. More intelligent trading of financial markets could benefit the economy as a whole—while profiting those who implement these innovations. This book contains fascinating insights.

Dr. Paul J. Werbos
Program Director, National Science Foundation
Past President, International Neural Network Society

Preface

Seven years ago I toyed with trading models in spreadsheets. These were mainly mechanical systems based on technical and fundamental indicators. Systematic application of technical indicators produces poor results, high turnover, and only a modest margin over human trading. As Peter Lynch has pointed out: one cannot predict the future on the basis of rearview mirrors!

After seeing many trading floors around the world, and working with a team of experts on the renovation of the trading floor of the World Bank, I realized that there ought to be a better way to trade financial securities than to watch computer screens, read charts, and follow news developments. I also knew that fuzzy thinking applied to trading does not lend itself to symbolic processing: discretionary investment decisions, often influenced by the news and weekend reading, require less discipline than is necessary to formulate expert systems!

I learned about neural networks and machine-learning techniques in 1988. At a high school science fair I saw a project on how a neural net could simulate the functioning of the brain. Over the next two years I coached my kids on several science projects about robotics, machine learning, and neural networks. Their success had a lot to do with what follows.

In the spring of 1990 I organized a symposium on neural networks and fuzzy logic for financial managers at the World Bank. Neural networks had been around for several years but remained foreign to finance. The prototype neural nets I demonstrated at that time were naive. However, John Loofbourrow showed evidence that several financial institutions on Wall Street were using these techniques. Proprietary trading models had booked successes since 1987. Japanese banks were using fuzzy portfolio management in 1989.

In 1991, with the help of Dalila Benachenhou and Masud Cader, I designed and implemented neural net models, undertook genetic optimizations, and designed several fuzzy portfolio management systems. Edgar Peters, Ted Frison, and Mark Embrechts played crucial roles in furthering my understanding of nonlinear dynamics and chaos theory. Their work led to better understanding of the fractal nature of market signals and, consequently, why some neural net models work, why some do not, and how fuzzy logic can improve trading performance.

This book is the product of more than a dozen authors who over the past three to five years have accumulated hands-on experience in using advanced technologies for trading, risk, and portfolio management in financial institutions. The book was assembled via the global network of electronic communications, Internet—the network of all networks. It brings together the technical and financial expertise of a group of innovators from three continents who possess diverse backgrounds in physics, engineering, computer science, statistics, and economics and who are familiar with a wide variety of financial markets and instruments.

This book reflects the commitment of many to developing and applying advanced analytical techniques to financial markets. It offers:

- An introduction to a wide range of advanced analytical techniques
- Several case studies on the use of these techniques for trading, risk, and portfolio management
- Specific guidance on the design of trading systems
- Methods for pre- and postprocessing of financial data for the development of advanced systems
- A glossary, bibliography, and index to easily trace information on these topics
- A companion CD-ROM that includes support tools for learning about advanced technologies.

Three years ago in a presentation in New York I asked, "How many smoke signals does it really take. . . ?" The metaphor involved a lonely Indian on top of a mountain sending smoke signals. To climb a mountain is difficult; to send smoke signals requires patience. One of the several beautiful calligraphies in this book, which were painted by Kathy Xiao, shows the Chinese idiom *shui di shi chuán*, from *Forest of Cranes and Dews of Jade*, a collection of reading notes and quotations from Taoists of the Southern Song Dynasty. It translates as *constant dripping of water wears away the stone*. According to this idiom, even an infinitesimal force can accomplish a seemingly impossible feat. Little strokes fell great oaks. While many more "smoke signals" may be required to influence traditional ways, persistence eventually pays off. There is no magic, no free lunch, in the techniques described in this book. They do, however, provide *a trading edge* to those who have the courage to stop guessing the next move of the financial markets.

Guido J. Deboeck

Arlington, Virginia
August 25, 1993

Acknowledgments

John Loofbourrow and Casimir "Casey" Klimasauskas contributed to my initial understanding of neural networks. Laurence Davis helped me with genetic algorithms; Lotfi Zadeh and Bart Kasko with fuzzy logic. Edgar Peters stimulated my interest in chaos and fractal market analysis; Mark Embrechts and Ted Frison furthered my understanding of nonlinear dynamics. One never forgets those who taught well!

Many people have contributed to the production of this book. I wish to thank in particular Andrew Colin (Australia), Henry Green (U.K.), Gia-Shuh Jang (Taiwan), Zhongxing Ye (People's Republic of China), Francis Wong (Singapore), Morio Yoda (Japan), and Laurence Davis, Casey Klimasauskas, and James Hall in the U.S., all of whom have performed pioneering work on financial applications of advanced technologies. Their work encouraged me to gather these experiences into one tome. This project could not have been accomplished without them. Several others have contributed to chapters in this book or have reviewed drafts of the manuscript: John Guiver, Feipei Lai, Liting Gu, Robert Mark, Michael Pearson, Lee Slutz, Clarence Tan, and Paul Werbos.

Thanks is also due to Harold Szu, president of the International Neural Net Society, for providing the stimulus for a book about financial applications; to Jane Klimasauskas, editor of *Advanced Technology for Developers*, for encouraging me to write a few articles; to Myles Thompson, editor of the Finance series of John Wiley & Sons; and to Mary Boss and Kathleen Mills for a superb job in preprint processing. Needless to say, the views expressed here are those of the authors and do not necessarily represent those of their current employers. All remaining errors or misrepresentations are mine.

Two dedicated coworkers, Dalila Benachenhou and Masud Cader, who have done wonders in making software work, utilities perform, and models outperform, deserve more than thanks; may they remember the past three years as the best of their college days.

A word of thanks also to Shinji Asanuma, Hywel Davies, Stephen Eccles, Bernard Holland, Jean-François Richard, and Donald Roth for encouraging interpreneurship and innovation. Others who supported this project indirectly are Luc De Wulf, Bill Kinsey, Ronald Ng, Lester Seigel, Emmitt Summers, Paul Staes, Thanit Thangpijaigul, and Avon Thomas. Their friendship and moral support have been invaluable over many years.

Finally, this project could not have come to fruition without the understanding and support of my family. I am blessed with a wonderful family who have put up with many evenings and weekends of computer work, writing, and editing. Thanks again Hennie, Toni, Pascal, Nina, and Charing for having taken care of "all the rest that needed to be done" while I was pursuing this ambitious project.

Calligraphy Art

The Chinese calligraphy on the cover means "scaling new heights," or getting an edge, the central theme of this book. Chinese characters are rich in shapes and patterns. Pattern recognition is as old as the creation of Chinese script itself. According to legend, Chinese script was created by a minister of the Emperor Huáng Dì, Cang, who observed the footprints of birds and beasts and then drew pictures of the objects in accordance with their shapes and forms. Over time the shapes and forms were reduced and stylized. These picture characters, called *xiáng xìng* or "image shapes" by Han lexicographer Xŭ Shèn, are expressed here in cursive or "grass" calligraphy with the bold and expressive style of Kathy Xiao.

Kathy Xiao is the granddaughter of Xiao Lao, a well-known contemporary calligrapher and poet in China. Kathy, who was born in Beijing, began to study calligraphy at the age of nine. Her work has been exhibited in China, Japan, Hong Kong, and the United States. The expressive movement of the lines and shapes in Kathy's grass calligraphy are used on the cover and part opening pages to give spirit to the verbal messages. Kathy Xiao currently lives in Shanghai, China.

—G.J.D.

Main Contributing Authors

Dalila Benachenhou

Ph.D. candidate, American University, Washington, DC

Dalila Benachenhou is currently pursuing a Ph.D. in statistics at American University; she obtained an M.S. degree in computer science in 1990. Her thesis topic was on hybrid expert-neural systems for DNA analysis. Over the past three years she has been involved in advanced technology research at the World Bank. Her research interests include fuzzy logic, Bayesian theory, probability theory, object-oriented programming, and dataflow programming.

Masud Cader

Ph.D. candidate, American University, Washington, DC

Masud Cader is working toward a Ph.D. degree in statistics at American University. He holds B.S. and M.S. degrees in computer science. Over the past three years he has been involved in advanced technology research at the World Bank. His research interests include time-series analysis, pattern recognition, signal processing, neural networks, nonparametric estimators, temporal databases, and nonlinear dynamics.

Andrew M. Colin

President, A.M. Colin & Associates, Toowoomba, Queensland

Andrew Colin holds a B.S. degree in mathematics from Sussex University and a Ph.D. in mathematics from St. Andrews University. At the time of this writing he was a manager within the Treasury of Citibank, with responsibility for developing advanced computer trading models to forecast foreign currency exchange rates. Before joining Citibank, Dr. Colin worked on a variety of computing projects for the British Army. His research interests include machine intelligence and automated strategy acquisition. He is a member of the British Computer Society and a Chartered Engineer.

Laurence Davis

President, Tica Associates, Cambridge, Massachusetts

Dr. Davis is the founder of Tica Associates, a consulting firm specializing in applications of genetic algorithms. Dr. Davis has been applying genetic algorithms to real problems for 11 years. He is the author/editor of *Genetic Algorithms and Simulated Annealing* and *Handbook of Genetic Algorithms*, and developed OOGA, an Object-Oriented Genetic Algorithm.

Guido J. Deboeck

Advisor, Investment Department, World Bank, Washington, DC

Guido Deboeck is an advisor on advanced technology for the World Bank in Washington. Over the past three years he has directed applied research on advanced technology for trading financial markets. Prior to this he managed the renovation of the trading floor of the World Bank. He is chairman of a Special Interest Group on Financial & Economic Applications of Neural Nets established by the International Neural Net Society. Dr. Deboeck holds a Ph.D. in economics from Clark University and a "Licentie-Doctorandus" in economics from the Catholic University of Leuven in Belgium.

Mark Embrechts

Professor, Rensselaer Polytechnic Institute, New York

Mark Embrechts is a professor of nuclear engineering and teaches neural networks and chaos theory in the School of Engineering at Rensselaer. He is the coauthor of *Exchange Rate Theory: Chaotic Models of Foreign Exchange Markets* and is interested in time-series analysis, fuzzy logic, and biocomputing.

Ted Frison

President, Randle Inc., Great Falls, Virginia

Ted Frison is president of Randle Inc., which provides data analysis and software systems based on chaos theory. Mr. Frison has degrees in engineering science and industrial engineering from the Georgia Institute of Technology, as well as an M.S. in ocean engineering from the University of Connecticut and an M.B.A. from George Washington University. He has been involved in nuclear submarine design, satellite orbit analysis, remote sensor systems design, and the development of products based on novel scientific concepts.

Henry Green

Managing Director, Financial Technology Systems Development Ltd., London

Henry Green designs and integrates advanced analytical technology for use by FTSD's client base. He is an executive director of the Kobler Unit at the Centre for Cognitive Systems of the Imperial College in London and is involved in research on artificial intelligence for trading, decision support, and forecasting. He has served as head of research and quantitative trading systems for several international investment banks and has worked for NASA on advanced rocket propulsion systems. Dr. Green holds a Ph.D. in fluid mechanics from Imperial College and an M.S. in engineering physics from the University of California at Berkeley.

James W. Hall

Manager, Investment Analysis, John Deere & Company; Moline, Illinois

At John Deere & Company, Jim Hall has been charged with developing a neural-network-based stock selection tool. He received a Ph.D. in mechanical engineering from the University of Illinois in 1992, specializing in modeling complex systems with neural networks. Earlier degrees in mechanical engineering were awarded by Brigham Young University. Prior to working on financial applications, Dr. Hall was responsible for advanced engineering projects at John Deere, which resulted in several significant new products and nine U.S. patents.

Gia-Shuh Jang

Springfield Financial Advisory Services Ltd., Taipei

Gia-Shuh Jang earned a Ph.D. in electrical engineering from the National Taiwan University in 1993. Dr. Jang serves on the Program Committee of the International Conferences on *Artificial Intelligence Applications on Wall Street*. He received the Taiwan Fuji Xerox Research Award in 1991 and is member of Phi Tau Phi and IEEE Computer Society. In August 1993 he joined the Taiwan Branch of Springfield Financial Consulting, a member of the Morningside Group. His research interests include artificial neural networks, genetic algorithms, fuzzy logic, intelligent decision-support systems for trading and portfolio management, computer architecture, and VLSI design.

Casimir C. Klimasauskas

President, NeuralWare Inc., Pittsburgh, Pennsylvania

Casimir C. "Casey" Klimasauskas is cofounder of NeuralWare Inc., one of the leading suppliers of neural network development tools and advanced technology training courses. He is a well-known speaker on the application of advanced technologies to financial problems. His publications include *Accuracy and Profit in Trading Systems, Hybrid Fuzzy Encoding for Improved Back-propagation Performance*, and *An Excel Macro for Genetic Optimization of a Portfolio*. He holds a bachelor's degree in mathematics from the California Institute of Technology.

Francis Wong

NIBS Pte Ltd., Singapore

Francis Wong, who holds a Ph.D. in electrical and computer engineering, is currently principal investigator of a supercomputer project based on INtelliVEST, developed by NEC and NUS. He is also the developer of NeuroForecaster, a neural network package distributed by NIBS Pte Ltd. He formerly was a research staff member of the Institute of Systems Science (ISS) at the National University of Singapore and has worked as an assistant professor and technical consultant. His research interests include parallel processing, pattern recognition, neural networks, and fuzzy engineering. Dr. Wong has published numerous technical papers and has been an invited speaker and panelist at major international conferences.

Morio Yoda

Manager, Research and Trading Division, Nikko Securities, Tokyo

Over the past several years Morio Yoda has been engaged in the development of trading models. Prior to his involvement with Nikko Securities, he worked for Mitsubishi Chemical Industry in Tokyo. Mr. Yoda has B.A. and M.S. degrees in applied mathematics from the University of Tokyo and was a Rotary Foundation research student at the University of Leeds.

Zhongxing Ye

Professor and Vice Chairman, Shanghai Jiao Tong University, Shanghai

Professor Zhongxing is vice chairman of the Department of Applied Mathematics at Shanghai Jiao Tong University. He received a Ph.D. degree in applied mathematics from Cornell University in 1988 and has bachelor's and master's degrees in mathematics from Nankai University. His research interests include information theory, neural networks, applied probability, and statistics.

Introduction: More Interesting Times

Guido J. Deboeck

Any sufficiently advanced technology is indistinguishable from magic.
Arthur C. Clarke

The Hammer Principle

"When the only tool you have is a hammer, everything begins to look like a nail." This "hammer principle," formulated by Lotfi Zadeh, demonstrates the importance of continuous adaptation to technological change.

Perhaps nothing changes faster than technology in financial markets. Each new gizmo offers a potential "edge." Over the last few years, trading technology has evolved from video-displayed information to digital information feeds, from stand-alone PCs to integrated networks of workstations, from simple graphics to 3-D images of the behavior of financial markets. The next logical step is machine learning, which along with related techniques will be used to recognize patterns, to automate trading decisions, and to optimize portfolios. Such techniques enhance performance while reducing the risks of transacting in financial markets.

This book focuses on advanced technologies for improving trading, risk, and portfolio management—including neural networks, genetic algorithms, and fuzzy logic. It reviews the strengths and weaknesses of each and provides practical advice on the design of trading, risk, and portfolio systems. And it shows how the use of these technologies can provide a "trading edge" to those willing to learn the jargon and apply the techniques.

The advent of advanced technology is always frightening. When "horseless carriages" first appeared around the turn of the century, many people wanted them banned from the streets. When large computers emerged in the forties, some could not imagine them used for anything beyond census surveys. When microcomputers came on the market in the seventies, they were advertised for checkbook balancing. In the early nineties, the mere mention of such terms as "neural networks," "genetic optimization," and "fuzzy control systems" fills most traders, portfolio managers, and systems specialists with a mixture of apprehension, fear, and alarm.

Many traders feel threatened by the idea that a model can achieve better results than they can. Portfolio managers, including those who believe in the value of trading models, must struggle to integrate computer- and human-based trading strategies. Systems managers responsible for implementing trading technology often prefer to remain on familiar ground, focusing their attention on selection of workstations, networks, and operating systems. These reactions to new technology are common to every industry, and every industry generally manages to survive them. Airline pilots, for example, were initially very much opposed to automatic-pilot technology. The integration of automatic pilots in airplanes, and particularly the appropriate role of human versus automatic pilots, has been the subject of many studies. After years of investigation, little doubt remains about the respective roles of human pilots and their electronic counterparts.

Likewise, applications of neural networks, genetic algorithms, and fuzzy systems—the autopilots for traders—are penetrating trading floors. And the financial industry will no doubt survive. These machine-learning techniques are no longer the hobbyhorses of rocket scientists. Applications of these techniques appear in respectable journals, in the popular press, even in the most conservative financial institutions! Furthermore, tools have become available that allow individual users to apply these techniques themselves from within common spreadsheet environments.

As Francis Crick, the discoverer of the structure of DNA, pointed out:

"The path of discovery is often difficult. You've got to take a series of steps, three or four steps, which if you don't make . . . you don't get there, and if you go wrong in any one of them you won't get there. It isn't a matter of one jump—that would be too easy. You've got to make several successive jumps. And usually the pennies drop one after another until eventually it all clicks."

The origin of this book can be traced to a long-term interest in exploring, through play, better uses of computing power. The value of exploration via play cannot be overestimated: through play children as well as adults learn abstract concepts, develop behaviors, and begin to understand the world. Adults often forget *how* to play, dismissing as trivial the concept of "having fun." The last chapter of this book discusses *pezonomics*, the study of the laws of play.

In the late seventies, we "played" with microcomputers first by building (soldering) one and then by teaching others how to use them. From there we introduced microcomputers as a means of processing vast quantities of data, to monitor and evaluate the results of complex development projects. The alternatives were to process the data at expensive mainframe centers or to process the data overseas, with all of the attendant possibilities of losing track of project results. This use of microcomputers for development purposes was highly controversial in the late seventies, as labor-intensive approaches then taken to mean development, did not seem to match with what was considered to be high-tech, capital-intensive computing. Fifteen years later, few work on development projects without at minimum a laptop computer!

The emergence of spreadsheet software in the early eighties allowed us to "play" with databases, and economic models. In those days, simple bar charts were considered an innovation! Through play, we established new standards which would affect design of databases, budgets, and economic models for many years. As increasingly complex models were envisaged, the need to test hypotheses became increasingly important.

In the mid-1980s we obtained the capability of building expert systems or intelligent assistants. We used artificial intelligence (AI) for loan approvals, credit ratings, and network management.

The main lesson learned from "play" problems was that the design of expert systems is most appropriately applied to problems where there is a consensus of expertise and where there is value of retaining such expertise in a system. When the same AI approach was applied to trading, risk, and portfolio management, it turned out that there was no consensus as to what are accepted rules for trading, how to measure performance, risk, or what constitutes optimal portfolio management.

As a result, we began in the late eighties to use neural networks and related technologies. Machine learning and data-driven approaches to extract patterns from financial data, to formulate rules, or to find unknown relationships seemed a fair bet, or better play, than the previous constrained AI approaches.

Many others around the world were making the same bet, and applying neural networks to trading became the "in" thing to do. Under the banner of research and development, small teams in large financial institutions and/or academic environments developed the collection of experiences and case studies elaborated in this book. The papers that started to appear in the early nineties, the conferences and panel discussions on financial applications of neural networks, and the electronic communications between workers in this field have all contributed to the present effort.

As classical thinking about the behavior of financial markets came under increasing scrutiny, and as initial results of trading systems based on neural networks, genetic algorithms, and fuzzy logic added fuel to the fire, a storm of criticism arose. Research on nonlinear dynamics, chaos and complexity theory, rescaled range analysis, and the fractal market hypothesis (by E. Peters in 1991) led to the questioning of traditional approaches to trading based on fundamental and technical analysis. In the next section, we review some of the major flaws of these traditional approaches.

Flaws of Technical Analysis

An overview of technical analysis can be found in many sources. For example, William F. Eng's book, *Technical Analysis of Stocks, Options, and Futures: Advanced Trading Systems and Techniques*, details each of these techniques and contains an excellent table on the behavior of technical indicators in various market cycles. Eng's most recent book, *The Day Trader's Manual*, shows how to apply most of these techniques. Another pleasant introduction to the subject can be found in *Trading for a Living*, by Alexander Elder, who expounds the three M's of trading: mind, method, and money. We will assume that the reader is familiar with the basics of technical analysis.

Traditional technical analysis and trading models based on technical analysis use simplified assumptions to make the design of models less complex and more manageable. The following are ten examples of simple trading rules, based primarily on technical analysis techniques:

Rule 1: Single Moving Trend
IF MOV (t. . .t − n) ≥ MOV(t. . .t-m)
THEN GO LONG
ELSE GO SHORT
where MOV is a moving average and $n < m$ (e.g., $n = 10$ and $m = 40$).

Rule 2: Composite Moving Trend
IF MOV (t. . .t − n) ≥ MOV(t. . .t − m) days AND MOV (t. . .t − k) > MOV (t − 1. . .t − k − 1)
THEN GO LONG
ELSE GO SHORT
where $k < n < m$ (e.g., $k = 5$, $n = 10$, and $m = 40$).

Rule 3: Channel Breakout
IF C(t) ≥ Maximum [H (t. . .t − m)]
THEN GO LONG
ELSE
IF C(t < Minimum [L(t. . .t − m)]
THEN GO SHORT
where C(t) is closing price of a security on day t; H is the high of the day; and L is the low of the day.

Rule 4: Relative Strength Index
IF RSI (1) > v
THEN GO SHORT
ELSE
IF RSI (l) < w
THEN GO LONG
where RSI is the relative strength index; and $k < l < n$ and $v > w$ (e.g., $l = 14$ and $v = 70$, $w = 30$).

Rule 5: Relative Strength with Retrace
IF RSI (k) > v and RSI retraced to v'
THEN GO SHORT
ELSE
IF RSI (k) < w and RSI retraced to w'
THEN GO LONG
where $v > v' > w' > w$ (e.g., $v = 70$, $w = 30$, $v' = 60$, and $w' = 40$).

Rule 6: Lane Stochastic Statistic
IF STO%$K < a$ and STO%$D < a$ AND STO%$D \geq$ STO%K
THEN GO LONG
ELSE
IF STO%$K > b$ and STO%$D > b$ AND STO%$D \leq$ STO%K
THEN GO SHORT
where STO%K is the %K Lane stochastic indicator; STO%D is the %D Lane stochastic indicator; and $a < b$ (e.g., $a = 0.1$ and $b = 0.9$).

Rule 7: Directional Movement and Relative Strength
IF Maximum $[H(t) - H(t - m)], L(t) - L(t - m)]$ AND RSI $(Y) =$ LONG
THEN GO LONG
ELSE
GO SHORT

Rule 8: Volatility
IF $C(t) \geq L + z$ *TR$(t. . .t - 2)$
THEN GO LONG
ELSE
IF $C(t) \leq L - z$ *TR$(t. . .t - 2)$
THEN GO SHORT
where TR is the true range; and $0 < z < 1$ (e.g., $z = 0.67$).

Rule 9: N Day Filter
IF $L(t) =$ Minimum $(L(t. . .t - t - n)$
THEN GO LONG
ELSE
IF $L(t) =$ Maximum $(H(t. . .t - t - n)$
THEN GO SHORT

Rule 10: All of the Above Combined

Many traditional trading models are simple rule-based systems that use "what if" scenarios. Some of these rule-based systems have fewer than ten rules. The simpler ones use various kinds of moving averages, technical analysis indicators, or other pattern descriptors.

Rule-based systems are static and deal only with symbolic information where inputs, thresholds, and decision-rules change in discrete steps. Such systems have a difficult time dealing with nonlinearity. As a consequence, rule-based systems must be frequently fine-tuned according to changing market circumstances. Continual fine-tuning prohibits accurate performance measurements; relative performance versus relative risk is seldom computed.

Rule-based systems assume that the influences of various factors on one another can be separated and thus that all factors are independent. They filter out dependencies, orthogonalize inputs, or are based on underlying assumptions of a Gaussian distribution of price changes.

The problem with these assumptions is that many factors that affect the behavior of prices are eliminated because they "interfere" with a deeper understanding of actual behavior of the underlying system. Furthermore, it is traditional to treat time as unimportant. It is often assumed that when an event perturbs the system, it will revert to equilibrium after a short time span. In other words, one assumes that markets have short-term memory.

It is also assumed that traders react to the same information in a linear fashion, that groups of traders are not prone to fashion, and that in the aggregate all traders are rational (even if they do not behave rationally individually). In fact, the main assumption is that trader skills and experience are equally distributed. If all information is already discounted in the price, no trader or group of traders can outperform the markets—hence the justification for indexing.

The next section provides an overview of advanced technologies which when applied to trading, risk, and portfolio management provide a radically different approach to trading in financial markets.

Overview of Advanced Trading Technology

The traditional technical analysis approach is only one of several that can be deployed for trading in financial markets. Figure I.1 uses machine intelligence as the core concept for grouping various advanced technologies that can be used to develop more advanced systems for trading. Machine intelligence is the level of exploitation of computing power, from the elementary level of calculating with spreadsheets, to computer reasoning with rule-based AI tools, to rule induction and data-driven approaches for pattern recognition with neural networks, to natural ways of evolving problem solutions using genetic algorithms, to decision-making and control using fuzzy logic. Figure I.1

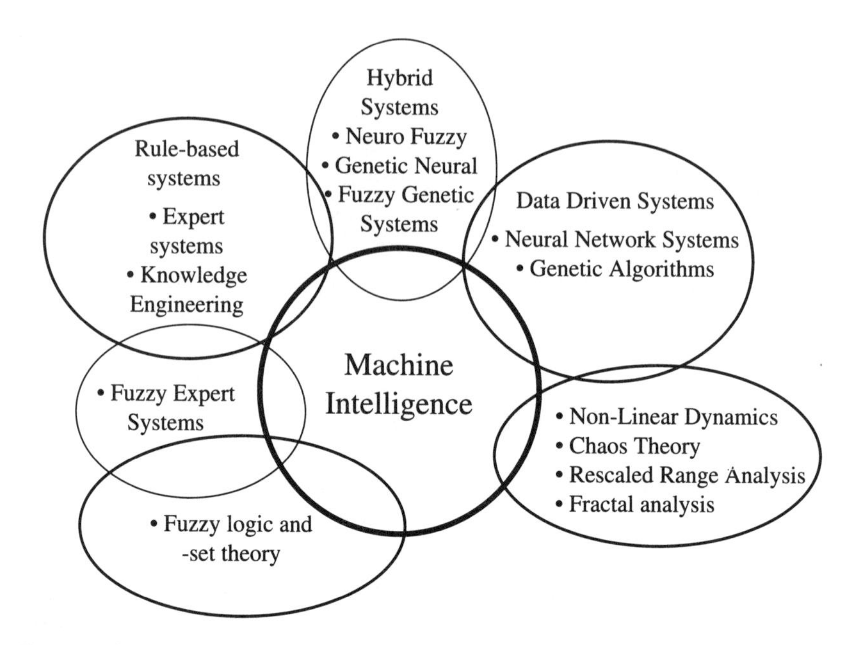

FIGURE I.1
Machine intelligence: a core concept for grouping various advanced technologies

also shows various combinations of rule- and data-driven approaches, or hybrid systems, which will be discussed in Chapter 14.

Figure I.2 translates this abstract bubble diagram into a more specific diagram of trading techniques. It shows an overview of what we have defined as *advanced trading analytics*. The remainder of this section discusses each part of this diagram in detail.

Mechanical Trading Systems

Mechanical trading systems are systems based on one or more rules, usually embedded in a spreadsheet. As indicated above, these systems are characterized by simple "if-then" rules. These are executed in a linear fashion–for example, top row and column to bottom row and column, or some other linear processing approach. A simple example is a spreadsheet that contains daily closing prices for the Standard & Poor's 500 and that uses a moving average of 5 and 20 days for determining buy and sell points. The buy and sell points expressed as +1 and −1 then provide a basis for computing *daily profit or loss* as

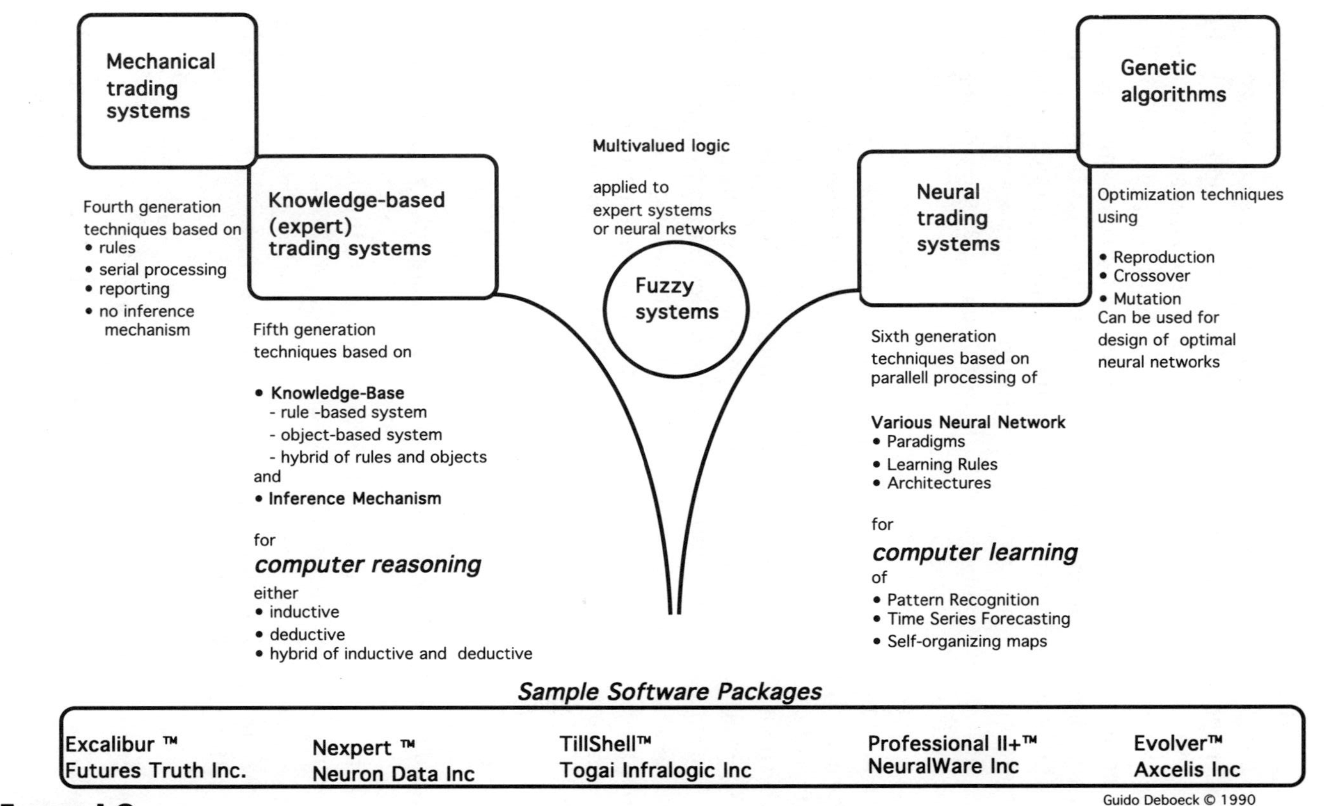

FIGURE I.2
Advanced trading analytics

well as *profit and loss for each trade.* The profit and loss of multiple trades can be aggregated to provide the *total profit and loss* (P&L) over a given period. A very basic measure of *risk* is the standard deviation of the daily returns. This basic trading system can be constructed in spreadsheet form so as to maximize profitability or optimize profit subject to risk. Through an iterative process one could find the best combination of moving averages so as to maximize profit, minimize risk, or achieve both over a given historical horizon. Such an elementary approach often results in "curve fitting"; that is, the system finds the best parameters for a particular historical period. These parameters may not be useful in the future. Excellent ways to avoid curve fitting are discussed in R. Pardo's book, *Design, Testing and Optimization of Trading Systems.*

The disadvantages of curve fitting are well known. The better the fit to a historical time period, the poorer the performance of the system in the near future. In the *Dow Jones-Irwin Guide to Trading Systems* (1989), Bruce Babcock described numerous mechanical trading rules and systems. Two-thirds of the way through the book, Babcock provides performance reports of various systems for trading the S&P 500, U.S. Treasury bonds, Eurodollars, Swiss francs, Japanese yen, gold, oil, soybeans, sugar, live cattle, and the like. The average performance figures for all systems reveal that mechanical systems can achieve between 34% and 42% winning trades. Some systems in each group do better. For example, among 65 systems for trading the S&P 500, four achieve more than 50% winning trades. Among these four, two are profitable, with 54 to 55% profitable trades; the other two achieve 69% to 82% profitable trades but have a negative overall P&L.

Much has been written about the Bruce Babcock systems. These and many other trading systems have been independently evaluated by Futures Truth Inc. The CD-ROM companion to this book provides summary tables of the evaluations of many commercial trading systems performed by Futures Truth Inc. The historical records in these tables show that the majority of commercial systems do not provide high ratios of profitability to risk.

Knowledge-based or Expert Trading Systems

Knowledge-based or expert trading systems, which we prefer to call "intelligent trader assistants," are based on fifth-generation software techniques that use a knowledge base with an inference mechanism. The knowledge base can include rules, objects, or a combination of rules and objects. The inference mechanism provides for computer reasoning through inductive, deductive, or hybrid inductive and deductive reasoning. Very simple as well as very complex applications of these types of systems to trading are in use in many financial institutions. An example is an "expert" system that sorts through all possible options in the markets and provides, on the basis of a set of rules and inheritances, recommendations concerning the best bets for trading options. This expert system leaves the selection of the actual options to trade to the trader, thereby acting as a true *intelligent assistant.*

The main shortcoming or weakness of knowledge-based systems is that they are based on symbolic processing and thus require the expression of rules and/or objects in computer-readable English sentences. To arrive at these rules it is often necessary to undertake *knowledge engineering,* that is, to interview traders as to the rules that they apply. One firm which undertook such an effort several years ago collected 600 trading rules by pulling together the rules from each trader. One imperative for a knowledge base of 600 rules is to check for consistency and validity. Even simple systems with only a dozen rules are complex and difficult to maintain.

Neural Network-based Trading Systems

A neural net trading system is an automated way of trading a financial security or financial market. They can be used as decision-support or decision-making systems. Neural net models differ substantially from mechanical models as well as from expert systems. The relationships and rules in neural networks are implicitly generated and are ideally sufficiently general to interpolate accurately in high-dimensional spaces. Unlike mechanical models, neural networks can generalize or interpolate between known values in a very high-dimensional input/output space. To generalize or interpolate between values implies that neural networks provide more than recall from memorization. Actually, the better a neural network has memorized the sample data, the worse it is likely to perform on future data. The success of a neural net design thus depends on its capability to perform against unknown data or data the system has never seen during the design stage.

Genetic Algorithms

Genetic algorithms (GAs) are problem-solving techniques that possess an astonishing property: they solve problems by evolving solutions as nature does, rather than by looking for solutions in a more principled way. Genetic algorithms, sometimes hybridized with other optimization algorithms, are the best type of optimization algorithm available across a wide range of problem types. Genetic algorithms can be applied in several ways in the financial field: to optimize the inputs to a trading system; for rule induction; to construct rules that can be used for trading; or to combine models from a given set of plausible models.

Fuzzy Systems

Trading systems designed on the basis of fuzzy logic allow the inclusion of trading rules provided by traders. They also contain an explanatory capability of the trading recommendations provided by the system. They avoid overreliance on quantitative data. A simple model of a fuzzy trading system may consist of one or more inputs (e.g., trend and volatility measures), one output variable (e.g., desired trading pattern), and a few fuzzy rules expressing the relationships between them. Such a model requires that inputs be "fuzzified." Membership functions are created; associations between inputs and outputs are defined in a fuzzy rule base; and fuzzy outputs are translated into crisp trading recommendations. All of these concepts will be explained in later chapters.

A fuzzy approach to the design of a trading system requires only a partial fill of a rule matrix. Fuzzy rule bases can be designed by experts or can be estimated from trading samples. A fuzzy set encodes a structure that can be mapped as a minimal fuzzy association of part of the output space with part of the input space.

The advanced techniques outlined in Figure I.2 can be applied to many financial areas. The most common applications are listed below; examples of several will be described in this book:

- Stock market forecasting
- Foreign exchange trading
- Forecasting economic turning points
- Selection and trading of stocks
- Trading of commodity futures
- Bond rating
- Design of bond trading systems
- Loan approvals

- Risk management
- Credit card fraud detection
- Bankruptcy projections
- Portfolio selection and management
- Targeted marketing
- Production job scheduling

What Makes This Book Unique?

Numerous books have been written on neural networks, genetic algorithms, fuzzy logic, and nonlinear dynamics. Most of them are rather mathematical. Few contain financial applications. Few demonstrate the use of these techniques for trading, risk, or portfolio management.

This book is intended primarily for financial professionals, portfolio managers, investors, and anyone interested in learning about technological advances in finance. Many universities offer courses on neural nets, genetic algorithms, and chaos theory, but few of these courses are aimed at business, finance, or economics students. This book offers:

1. An introduction to a wide range of advanced analytical techniques
2. Case studies of the use of these techniques for trading, risk, and portfolio management
3. Specific guidance on the design of trading systems
4. Methods for the pre- and postprocessing of financial data for the development of advanced systems
5. A glossary, bibliography, and index to easily trace information on these topics
6. A companion CD-ROM containing a vast amount of information on advanced technology, including learning tools.

The main themes addressed are:

1. More intelligent inferences gathered using machine-learning techniques and nonlinear dynamic analyses can provide a trading edge. Machine-learning techniques, such as neural networks, are powerful tools for pattern recognition and forecasting. They can enhance the design of profitable trading systems that outperform standard market indexes.

2. Neural net trading systems, when implemented without human interference, can outperform active trading strategies pursued by traders.
3. Genetic algorithms, which evolve solutions in a natural manner in a computer, can facilitate the design of neural net trading systems. Genetic algorithms allow the extraction of trading rules, the selection of inputs and investment strategies, and the optimization of trading systems aiming at multiple financial objectives under hard and/or soft constraints.
4. Applications of fuzzy logic, including the fuzzification of data, automated extraction of fuzzy rules, and use of fuzzy expert systems, are proven technologies that produce superior results in trading as well as in portfolio management.
5. Financial risk management of any size portfolio should take into account overall volatility as well as the correlation between the volatilities of various instruments in a portfolio. Simple risk management approaches, such as using betas or duration-based risk measures, are insufficient and inadequate for measuring the actual financial risk exposure of a portfolio.
6. An alternative asset pricing theory based on rational beliefs and the fractal market hypothesis provides a new framework for better understanding the behavior of financial markets.
7. The rapid evolution of technology and the increased use of virtual reality and robotics continue to challenge trading operations. Without adequate research, development, and experimentation with new technology, you are bound to lose out on the future.

This book is structured in five parts. Part One introduces financial applications of neural networks and provides guidelines for the design of neural net models for forecasting and pattern recognition. Part Two introduces the basic concepts of genetic algorithms and shows how they can be applied to optimize investment strategies and improve neural network design. Part Three shows why fuzzy logic is important, how to design a fuzzy model and how to combine fuzzy models with other techniques in hybrid systems. Part Four offers more advanced analyses of market signals. Concepts of nonlinear dynamic analysis and chaos theory are applied to financial time series. Part Five outlines ways to improve risk management and discusses the impact of technology on financial markets. This book is not meant to be read from cover to cover. The best starting point is to use your own neural net to detect the patterns! It is very important that after reading some you commit to building an application. Recipes for nonlinear learning are provided below.

Recipes for Beginners

If you are a total newcomer to advanced technologies, you may want to start by reading the first chapter of each part (Chapter 1 on neural net concepts, Chapter 8 on genetic algorithms, Chapter 11 on fuzzy logic, and Chapter 15 on nonlinear dynamics and chaos theory), then glance over some of the application chapters. Managers of investment portfolios, pension plans, as well as anyone supporting system developments in these environments, need a basic understanding of advanced technologies. *Fiduciary responsibilities can no longer be met without it.*

Recipes for Intermediate Learning

If you already know about advanced technologies, you may want to proceed directly to the application chapters. You can also start by reading Chapters 19 and 20 to get a sense of why it is important to augment your knowledge and skills regarding these subjects.

Recipes for Advanced Learning

If you already have hands-on experience with one or more advanced techniques, you may want to look up specific applications or tackle the more advanced techniques described in Part Four.

Recipes for Those Who Want to Get the Most Out of This Book

You can get the most out of this book by building applications yourself. A good way to start is to order the companion CD-ROM. Many software packages are currently available in this area. You can explore several of them using the vendor-supplied demos contained on the CD-ROM. Once you have identified a prototype application and have selected a software package, you can use the practical advice provided by the book's case studies and apply these techniques in your own environment. If you do, you may find yourself living in "more interesting times."

PART ONE

Trading with Neural Networks

dẽng pũn:

"Scaling new heights" or attaining new insights via pattern recognition can be achieved with neural networks.

CHAPTER 1

Neural Network Techniques

Casimir C. "Casey" Klimasauskas

The only thing new about back-propagation is that it is an efficient way of calculating derivatives.

Paul Werbos, 1974

This chapter provides a brief introduction to neural networks and their use for pattern recognition and/or financial time-series prediction; it also offers guidance on how to successfully develop financial applications of neural nets. Casimir C. "Casey" Klimasauskas, president of NeuralWare Inc., shares his years of experience and expertise in applying these techniques to various financial problems. His keen interest in making neural net technology easier to use is well known and has been documented in various articles.

What Are Neural Networks?

Neural networks are a collection of mathematical techniques for curve fitting, clustering, and signal processing. As an engineering discipline, they bear little resemblance to the biological systems that inspired them. As a matter of historical interest, although many of these techniques had originally been invented by economists and statisticians, their importance became lost in a maze of technical reports and their potential was never realized. Many of these techniques were rediscovered by cognitive scientists and neurobiologists, and their importance to commercial problems was recognized and promoted by individuals familiar with their work.

The class of techniques described in this chapter are known as *nonparametric* models. This is in contrast to *parametric* models. A parametric model is a formula with a form derived from some external theory which describes the dynamics of a market or security. For example, the moving average convergence divergence (MACD) method of trading uses the difference between two moving averages as a trading signal. The length of the fast-moving average is the first parameter and the length of the slow-moving average is the second parameter in this parametric function. Nonparametric models use a very general formula. Typically, they are capable of *approximating* a wide variety of relationships between the input and output variables. In particular, nonlinear nonparametric models developed using the back-propagation neural network are capable of approximating almost any relationship between input and output variables. This is the key to their power. Building a neural network is equivalent to constructing a mathematical formula.

Although the most popular instantiations of the technology were developed by cognitive scientists, most neural network algorithms have statistical analogies. These analogies go beyond surface similarities. Many of the methodologies used to develop linear statistical models apply equally well to the development of neural network models. A recent book, *Neural Networks for Statistical Modeling* (Smith 1993), highlights this relationship in its title.

Neural networks are graphical formulas. The formulas or nonparametric models developed using the back-propagation neural network algorithm are quite complex. A graphic notation has been developed to make it easy to describe the formulas. Figure 1.1 shows the graphic equivalent of a linear regression model along with the formula it represents. Much more complex formulas can be constructed using this graphic notation. Figure 1.2 shows a more typical formula used in developing models for market timing. The graphic depictions of neural networks can always be reduced to a mathematical formula or algorithm.

In the past five years, technological forces have converged to make neural networks a viable technology. Today the typical desktop personal computer with an INTEL 486DX-2/66 microprocessor is 140 to 200 times faster than the PC/XT introduced by IBM a decade ago, and costs the same price or less (adjusted for inflation). Workstations such as the IBM RS/6000 and HP 9000/700 offer the power of a circa 1980 supercomputer at a fraction of the price—$40,000 or less! In the

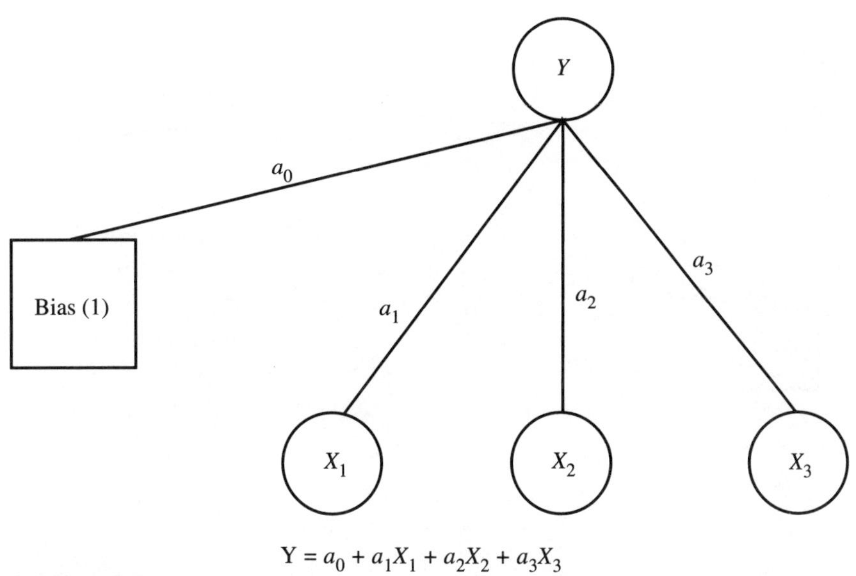

FIGURE 1.1
Example of a linear regression formula. The variables X_1, X_2, . . . X_n are the inputs (explanatory variables) for the linear regression formula. Y is the output result (dependent variable). The coefficients of the linear regression formula are a_0, a_1, . . . a_n. The special box marked bias is a convenient way of constructing a constant input required for a_0.

next few years, the Pentium chip produced by Intel and the PowerPC chip produced by Motorola will increase the power of the average workstation by ten times. Over the next decade, workstation manufacturers expect to increase their power by 100 times. The computationally intensive algorithms used in neural networks that required a supercomputer five years ago can be cost-effectively implemented on desktop computers today. This revolution has made neural network technology cost-effective for use in trading systems.

Neural network techniques have evolved as well. Many of the emerging algorithms provide better solutions than their predecessors in one hundredth the time or less.

Neural networks are math, not magic. Although inspired by studies of the brain, they have strong ties to statistics. Many of the methods used to construct statistical models apply equally to the development of neural models. The diagrams used to describe neural networks are graphical formulas. Improvements in neural technology combined with advances in desktop computing have reduced the time needed

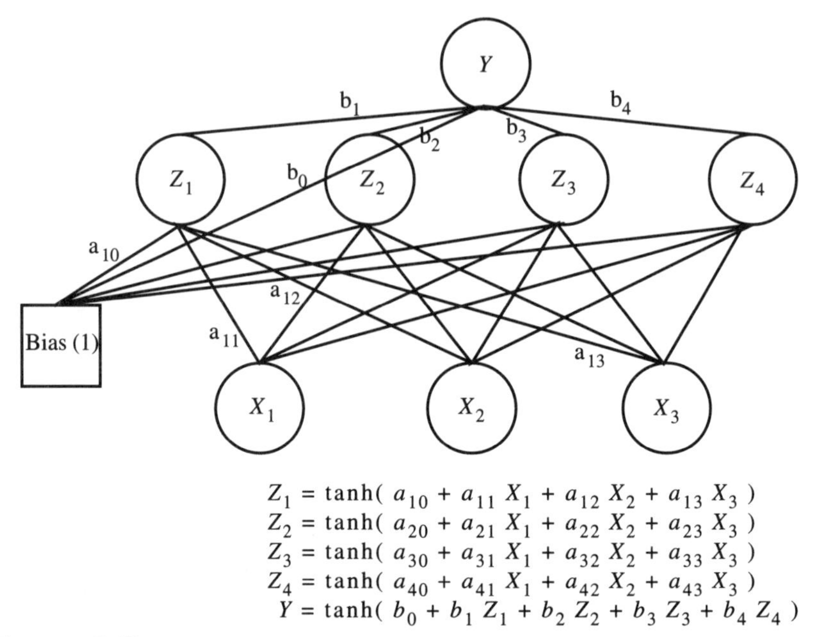

FIGURE 1.2
Example of a neural network formula. Y is the output (dependent variable) and x_1 to x_3 are the inputs. The coefficients a_{10} to a_{43} and b_0 to b_4 are the neural network weights. A nonlinear transformation is provided by the hyperbolic tangent (tanh).

to solve a problem from days in 1986 to minutes or at most hours in 1994. Neural networks have become a practical technology.

Time-Series Prediction

Setting up a time-series prediction problem is the first step toward its solution. The problem is essentially to use data from the recent past to construct one or more *indicators* which can be used as inputs to a profitable trading system. In time-series prediction, neural networks are used to construct a new class of *indicators* which have predictive power. Most technical indicators, such as moving averages (MA), relative strength indicators (RSI), and directional indicators (+DI, −DI, ADX, ADXR), are parametric models. They are formulas that have been developed to measure some effect which is believed present in the data. Various parameters, such as a smoothing period, are adjusted to maximize profit when incorporated into a larger system. The output of these indicators is typically used in conjunction with other indicators as one component of a trading strategy.

Neural network indicators are very similar. As with technical indicators, one or more components of a financial time series is sampled at several points in the past. This forms the input to the neural network formula (neural network indicator). The output may have any of a variety of meanings, depending on how the neural network indicator was developed. The neural network output is also used as one of the inputs to a larger trading strategy. As with certain technical indicators, such as ADX, neural network indicators may use other indicators as input. In contrast to most technical indicators, neural network indicators have the ability to integrate input from several financial time series simultaneously. This makes it possible to develop an indicator which encapsulates relationships that span several markets. For example, a neural network directional indicator for the S&P 500 may use as inputs the effective interest rate on the long bond for the past few days, the cash price of gold, open interest in New York light crude, and the past few days of the S&P 500 itself. Using inputs from multiple markets may increase the predictive power of the neural network indicator. Other chapters in this book describe methods for determining sampling periods from an analysis of the data itself. Often, applying some form of transformation to the inputs improves performance.

Neural network indicators are developed from data. The most commonly used neural network technology for time-series forecasting is a curve-fitting technology known as *back-propagation* (BP). A BP model is developed from data in much the same way as a linear regression model. Several inputs and the related output are assembled in a data set. In the case of a neural network indicator, the target output, for model development purposes, is created using 20-20 hindsight. Examples of target outputs described in the literature and at conferences are:

- The magnitude and direction of a particular security's price five days hence.
- Whether a large move (>2%) is about to occur and its direction (up or down).
- The number of days until the next turning point in the market.
- Market direction: trending up, down, or sideways.

- The degree to which a market is predictable: fractal dimension.
- Trading signal: long, short, out.

Neural networks in time-series prediction are used as special-purpose indicators. Just like technical indicators, they use data from the recent past as inputs and provide an output that may be useful in trading. One of the challenges in developing such an indicator is construction of an example data set. The BP neural network is the most popular algorithm for developing neural network indicators.

Measuring Performance

When using neural networks or any mechanical trading system, it is essential to define how performance of the total system as well as of the individual components will be measured. Criteria typically used in neural network development, such as root mean square (RMS) error or linear correlation coefficient, may have little validity in terms of the system's overall objectives.

Neural network indicators are only part of a trading system. When building a system that uses neural network indicators, a trading strategy and an evaluation system to measure the performance of the combined system must also be developed. Some excellent articles have been written on these topics (see [Deboeck 1992] and [Pardo 1992]).

A neural network indicator was developed that correctly predicted the next-day direction of the long bond (30 year) 85% of the time. Was it tradable? In this particular instance, it was not! The system correctly predicted market direction when the moves were small, but it was wrong on almost every large move. Moreover, transaction costs made the small moves unprofitable. The system consistently lost money.

This is not an isolated phenomenon. To investigate this systematically, a simple trading strategy was investigated using a rapidly trained neural network. The relationship between accuracy and profit was measured. The tests were done on the Deutsche mark from 11/85 through 9/88, with the last 100 days of this period used for testing. Prior data were used to train the network, and a neural network cousin to a nearest-neighbor classifier with linear regression was used to make the predictions. The output of the network was the predicted price some number of days into the future. If the predicted price was up, the trading system bought a contract (long). If the price was down, the trading strategy sold a contract (short). At the end of each prediction period, the contract was liquidated. No transaction costs were considered in the evaluation. Figure 1.3 shows the relationship between accuracy and profit. One might expect a linear or at least a proportional relationship and, in general, this is true. However, an intermediate level of accuracy (about 65%) produced the greatest profit. Figure 1.4 shows the relationship between RMS error and profit. Again, one might expect that the smaller the RMS error on the test set, the higher the profit—and again, generally this is true. Notice though that at the lowest levels of RMS error, the difference in profit between various networks is quite substantial!

One plausible explanation is this: Because of noise in the system, it is never possible to make perfect predictions (100% accuracy, or zero RMS error). Performance then depends on the kinds of mistakes that are made. High accuracy (low RMS error) may be achieved by a network that correctly predicts only small moves, which occur the majority of the time. However, this may be at the expense of missing the small proportion (<2%) of larger moves. On the other hand, it may be possible to correctly predict all of the relatively large market moves while missing many of the smaller ones. This would result in

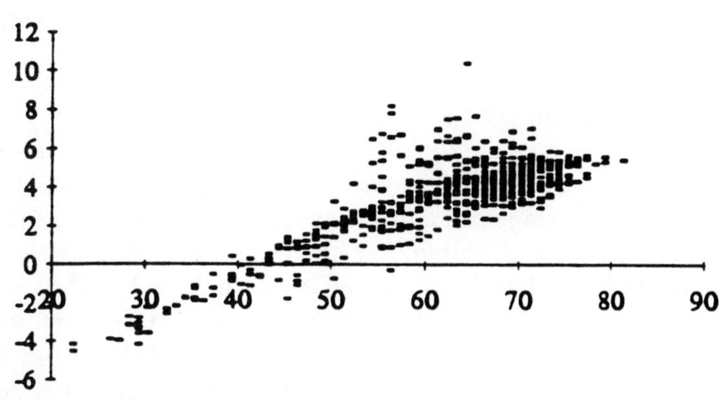

FIGURE 1.3

Profit as a function of accuracy. Accuracy is a measure of the percentage of the time that the direction of the forward prediction was correct. In general, the more accurate, the more profitable. However, the most profitable network was only about 65% accurate. (From [Klimasauskas 1992]. Used by permission of High-Tech Communications.)

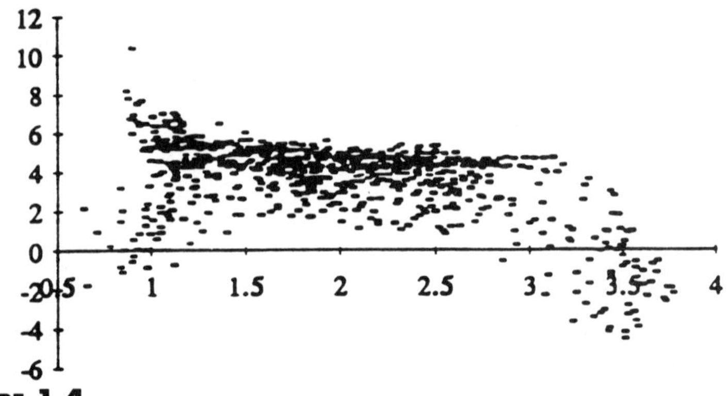

FIGURE 1.4

Profit as a function of RMS error of the difference between actual price and predicted price. Notice that as accuracy improves (smaller RMS error), profit generally improves. However, more specifically, the most accurate networks were the least profitable. Slightly less accurate networks were most profitable. (From [Klimasauskas 1992]. Used by permission of High-Tech Communications.)

lower accuracy (higher RMS error) than the prior network but could realize much higher profit. This can only happen if market moves are not normally distributed (do not have a Gaussian distribution). In fact, most markets have very long tails (for example, see Figure 1.5). These long tails are at least partially responsible for the discrepancy between accuracy (or RMS error) and profit.

There are three lessons to be learned from this. First, standard neural network criteria, such as RMS error, linear correlation coefficient, or accuracy, are usually poor predictors of performance when integrated into a deployed system. Second, it is essential to measure performance of the complete system, including the neural network and trading strategy associated with it. Third, appropriate system objectives and measuring mechanisms must be developed. In this regard, profit is not the only objective. A system may be very profitable, but trade too often or suffer unacceptable levels of drawdowns. One way to apply these lessons is by using a train-test approach to neural network development. The process is as follows. The training process is periodically interrupted, and the current network is tested in the deployed environment on a validation data set. Performance is measured in terms of system objectives (profit, sharp ratio, maximum drawdowns, etc.). If the current network scores higher than prior ones, it is saved as the best network. Training stops when measured system performance is repeatedly less than the current best performance.

Neural network performance can be evaluated only in terms of the performance of the entire system. Standard technical measures such as RMS error may be quite poor predictors of system performance. Defining system objectives and then integrating them into the neural network development process is essential.

Neural Network Development Methodology

NeuralWare Inc. has developed a seven-step methodology for constructing neural network indicators:

1. Data collection
2. Data analysis and transformation
3. Train, test, validation sets
4. Variable selection
5. Model development and optimization
6. Model verification
7. Model deployment

Highlights of each of these steps will be discussed in the next several sections. This process is not a single-pass procedure but an iterative one. At any time, it may be necessary to reexamine the information from a prior step or even to collect new or different data.

Data Collection

When collecting data for developing a neural network indicator, three issues must be carefully addressed: (1) availability of data in the deployed environment, (2) handling missing data, and (3) scrubbing data.

When building a real-time trading system that is expected to provide short-term predictions of the future, volume information is often difficult if not impossible to obtain. As such, though possibly very predictive in retrospect, it cannot be used in the development of the neural network indicator. Another example, somewhat more subtle, is that historical data, particularly fundamental data found in historical databases, are often updated substantially after initial publication. For example, if a system to predict which stocks will appreciate substantially in the following quarter is developed using quarterly historical fundamental data, it will fail. Why? Because the quarterly information is often not available for weeks or months after the end of the quarter. When the reports are received, the database is updated retroactively. To effectively develop a system using such data requires that the developer collect data on a snapshot basis at the instant it would be required in the deployed system.

Handling missing data is particularly challenging when developing neural network indicators that span several markets. The challenge is that certain markets may trade when others are closed for the day. Another key decision is how to handle the missing data during the development process and then in the live trading environment. One approach is to eliminate all of the data if any is missing. This is often important when using indicators such as ADX. The alternative is to assume that the missing data remained the same as the prior datum. Other methods of handling missing data are described in the next section. The choice of how to handle missing data depends on the system design.

Scrubbing data helps to reduce noise during the development process. It also provides an essential clue as to the quality of the data that will be used to make decisions in the live trading environment. Equally important, bad data can substantially skew evaluation procedures. Quality data pays for itself.

Data Analysis and Transformation

In many ways, this is the most important step in the process of developing a neural network indicator. Again, neural networks are math, not magic. There is no substitute for understanding the data. In this regard, the most successful neural network indicator projects have both an experienced trader and an analytically oriented individual on the development team. There are two types of neural network indicators: technical indicators and synthetic indicators. Development of either type must also address issues of missing data and scaling.

Neural network technical indicators usually use raw data for input and process it through a formula. One example was provided by Phil Erlanger (1993), a technical analyst who specializes in estimating the *strength* of a stock. As an experiment, he went through one stock and rated its strength over a period of several weeks. The same data he usually looks at for estimating the strength of a security were used as input. His estimates of strength were used as the target for the indicator to predict. This is an example of developing a neural network technical indicator that mimics the estimates made by an expert. Erlanger tested the effectiveness of the neural network indicator on several other securities. Surprisingly, it was quite effective in estimating the strength of many, though not all, of those tested.

Synthetic neural network indicators typically attempt to predict long-term market trends and turning points. For example, a particular trader may look at Lane's stochastics, relative strength indicators, directional movement indicators (+DI, −DI, ADX, ADXR) and combine them through a set of ad hoc rules to decide what the appropriate position should be: long, short, or out. With hindsight, this trader might go back through several months of historical data and mark the best position (long, short, out). The same indicators used in rules together with this *ideal* trading signal are used to develop a synthetic neural network indicator. The neural network finds the relationships between the various input indicators and the ideal position in the market. It should be noted that just as ad hoc rules sometimes fail, synthetic neural network indicators will also fail at times.

Much of the data selected for input to a network is time-series data. Although several papers have been written about using recurrent neural networks for modeling, in practice this technique is fraught with problems and challenges. The alternative is to capture time-varying information by using a sliding window on each of the data inputs. This method usually works. Other chapters of this book describe methods for capturing time history by sparse sampling of past data.

Unlike most technical indicators, neural network indicators usually cannot deal with the wide range of values found in raw data. The data must be scaled into a usable range, usually [0 . . . +1] or [−1 . . . +1]. When scaling data, there are two considerations. (These may be overridden by the issue of outliers described later.) First, if the data are expected to exceed prior limits, such as new highs or lows, this should be addressed in the scaling process so that the new highs and lows become the extreme. The second issue is specific to neural network indicator output. The nature of the functions used to synthesize the output is such that it is difficult to get outputs very close to 1 or 0 (−1 in some cases). The best way to address this is to scale the output into the range [+0.2 ... +0.8] or [−0.8 ... +0.8]. The basic rescaling formula is:

$$X_s = \text{Scale} \times X_u + \text{Offset}$$

where

X_s is the rescaled data value
X_u is the raw or unscaled data value
T_{min} = target minimum (0)
T_{max} = target maximum (1)
R_{min} = raw minimum
R_{max} = raw maximum
Scale = $(T_{max} - T_{min}) / (R_{max} - R_{min})$
Offset = $T_{min} - \text{Scale} \times R_{min}$

Although many products offer automatic or semiautomatic facilities for rescaling, it is important to use them correctly. Scaling parameters should be computed once using all of the data, including future expectations on the range of the data. In some cases, outliers should be ignored when scaling data, and truncated to the effective minimum or maximum when performing the transformation.

There are several approaches to handling missing data, a few of which were described in the previous section. Another is to simply set the network input to zero if the data are missing. The reasoning behind this follows from the way in which parameters are adjusted in a BP model. The weight update formula is:

$$\Delta W_{ij} = \alpha \delta_i X_j$$

where α is the learing rate, δ_i is the error internal to the current processing element, and X_j is the input to the connection. If X_j is zero, no changes will occur to ΔW_{ij}.

So, setting the network input to zero for missing data prevents learning from occurring in the weights associated with that input. If desired, a second neural network input can be provided to detect the presence or absence of missing data. Another approach to missing data is to replace the value with a zero-mean random number. When a substantial portion of data is missing, a linear or neural network model may be developed to approximate the value as a function of the other inputs. This approach has been successfully used in a variety of targeted marketing, insurance, and other large database applications.

Outliers can substantially affect the ability of a neural network to use the input data. One of the characteristics of the numeric algorithms used in back-propagation is that they tend to smooth out noise. This is good in that it allows effective modeling of noisy systems. However, if all of the data is concentrated in a very small portion of the input range, that input to the network may have little effect on the resulting model. Figure 1.5 shows a histogram of day-to-day volatility of the closing price of the S&P 500. Typical of volatility for many time series, most of the data is concentrated around zero. However, as shown, the series has large, infrequent events. Rescaling [−0.1 . . . +0.05] into the range [−1 . . . +1] creates a neural network input that may be completely ignored! Seventy-seven percent of the data lies within 6% of the rescaled input range of the processing element. Eliminating 1.6% of the data in the tails produces the histogram in Figure 1.6. Rescaling [−0.015 . . . +0.015] to the range [−1 . . . +1] spreads out the data to a degree where it becomes useful. As a general rule, a BP network will not distinguish between differences in input of less than 0.01. (Note: The actual magnitude is determined by the number of inputs and the average magnitude of the weights in the network.) This means that a BP network will effectively ignore an input where most of the data is concentrated in a range of less than 0.1 (5% of the range [−1 . . . +1]). (Note: This does not mean that the weight associated with this input will be zero, but rather that if this input were set to its average value and fixed at that value, the performance of the network would not change.) If the outlier data are important for the indicator under development, two additional network inputs can be derived: one to detect those to the left of the tail, and the other to detect those to the right of the tail. Either zero-one indicators or fuzzy membership sets can be used.

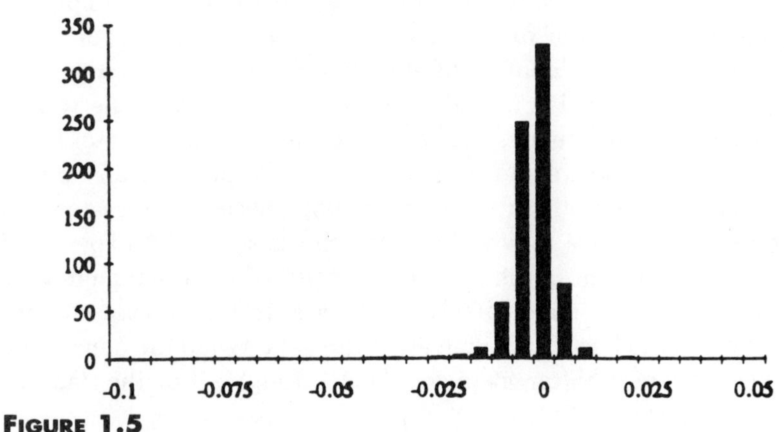

FIGURE 1.5

Histogram of the volatility of the S&P 500 from 11/11/85 through 9/23/88. Volatility was computed as log (S&P close today) − log (S&P close yesterday).

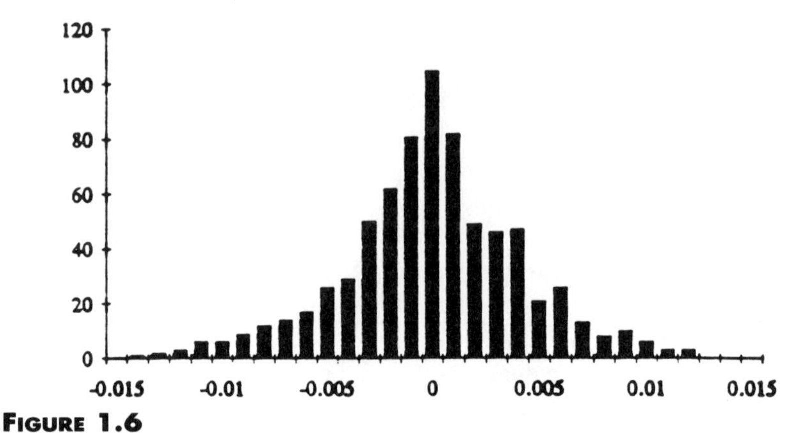

FIGURE 1.6
Rescaled histogram of the volatility of the S&P 500 from 11/11/85 through 9/23/88. Approximately 1.6% of the data in the tails was omitted when computing the histogram.

A variety of other techniques for data analysis and transformation has been developed for neural networks. However, most of those techniques are more pertinent to developing indicators based on fundamental analysis rather than time-series analysis.

Train, Test, and Validation Sets

Training sets are used to develop the model. The test set measures how well the model interpolates over the training set. The validation set is used to estimate actual performance of the model in a deployed environment. Selection of the training and test sets is particularly important to the development of effective models.

As mentioned earlier, neural networks, in particular BP networks, build a formula that relates the inputs to the outputs. In actual use, the formula interpolates on prior experience to produce an output. The numerical algorithms used to develop these formulas tend to optimize performance where the data are dense, at the expense of sparse areas. Figure 1.7 shows an example of data sampled very densely in one area and sparsely elsewhere. In this case, a straight line provided the best fit through all of the data. When the dense data are subsampled, a better fit is achieved through all of the data, as shown in Figure 1.8.

The moral of this story is that it is often possible to build better models with fewer data. When developing systems to predict changes

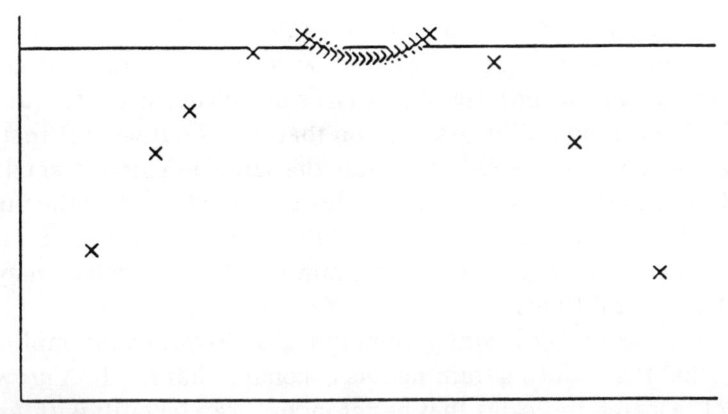

FIGURE 1.7
Results of fitting a curve through raw, unsampled data. The curve fits best where the data are dense, at the expense of sparser regions. In this case, a straight line provided the best fit through all of the data.

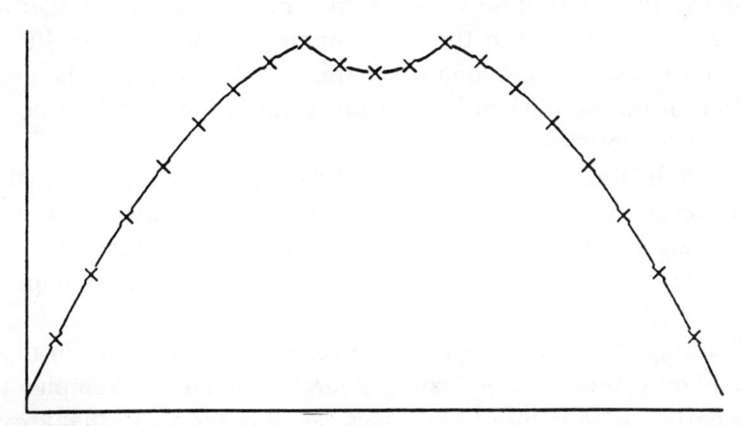

FIGURE 1.8
When dense data are subsampled, a better fit is achieved through all of the data. This curve was fit with 19 points (versus 126 points for Figure 1.7). Contrary to first intuition, it is often possible to achieve better results with fewer training examples.

in prices, this means that most of the days when prices change only a small amount should be eliminated from the training set. On the other hand, all of the data from large moves in the market should be included. When testing the power of the system to interpolate, a hold-out subsample of the set used for training should be used. This can help during model development but, as described earlier, is not appropriate for measuring the deployed performance of the system.

Another way of approaching the issue of data density is to *multiply* data. This is based on the assumption that two input vectors that are very close should have approximately the same outcome. A small (< 0.05) Gaussian distributed random value is added to each of the inputs for each of the sparse data vectors, and the same output is used. These new data vectors are added to the training set. This process may be repeated several times.

One of the concerns which sometimes arises with subsampling of data is that the resulting training sets become rather small. A growing body of research indicates that better models can be built with fewer data rather than more (see [Owens and Mocella 1991] and [Plutowski and White 1990, 1992]). In one test, a network was trained on the Mackey-Glass time series (fractal dimension approximately 2.1) using only a dozen data points. The performance of this trained network was better than that of networks trained on substantially larger data sets. Other research into the fundamental mathematics of BP networks indicates that the amount of training data required is more a function of the number of hidden units than of the total number of weights in a network.

The validation set, which is usually contiguous data taken at the beginning or end of the train-test set, is used to measure performance in a deployed system. It is also used to determine which of several possible models are most appropriate for integration into the deployed system.

Training and test sets should be selected to cover uniformly all possible outcomes. Concentrating a large number of examples in a small portion of the output range may result in networks that provide lazy solutions to minimizing RMS error. Since the purpose of the test set is to measure the effectiveness of how well the model interpolates, it should have the same distribution as the training set. The validation set should include all of the data for a contiguous period of time. This is used to estimate the performance of the deployed system.

Variable Selection

Variable selection is the process of reducing the number of inputs to a neural network indicator. Often, models developed with fewer inputs are more effective (read: profitable) than those constructed with many inputs. When a *good* set of inputs has been found for a specific market or a particular class of securities, those inputs usually do not change. There are three approaches to the selection of which inputs to use and how often those inputs should be sampled. The first approach is to use analytic techniques. Chapters 15 to 17 discuss the use of nonlinear dynamic systems analysis (chaos) for choosing how often to sample data. Statistical methods are also used. The other two approaches use the neural network to decide which inputs are appropriate.

When a neural network has been trained, a process borrowed from statistics known as *sensitivity analysis* can be used to determine which inputs are important. Ranking each input based on importance and eliminating the bottom of the list can be quite effective. Statistical or chaos-based methods may miss inputs that are important only when they interact with other variables. Sensitivity analysis can detect these. The essential process is to select a set of data for input, train a network until an acceptable level of performance is achieved, apply sensitivity analysis, and eliminate all variables below a threshold. When several neural indicators are developed on a single training set, they may hone in on different features. If this happens, some of the variables selected may differ between models. Those which are consistently important should be used in further model development.

An alternative to sensitivity analysis is to determine the impact that setting an input to its average value has on overall performance. If the network is not using an input to discriminate between possible outcomes, fixing that input to its average value should actually *improve* overall performance.

Either or both of these methods may be used to reduce the number of inputs to a network, which will often improve the performance of the neural network model. (As a side benefit, it also reduces the time needed to train the network.) One caution: Any one network may use a select group of specific features in the input space and ignore other features. As a result, rather than training a single network and eliminating variables based on its performance, several networks should be trained and variable elimination performed on each one individually. After several trials, a set of inputs can be selected for further model development.

Model Development and Optimization

The process of developing a neural network is still somewhat of an art. This section describes a set of heuristics that usually results in effective neural models.

A number of possible issues might be addressed in developing and optimizing a neural network model. Rather than describing all of the dead ends or near-dead ends, this section focuses on some of the more successful approaches. These approaches, on the whole, substantially reduce the complexity of the neural network development problem.

For financial modeling, the most commonly used neural network technique is back-propagation. Several techniques exist to maximize performance of back-propagation and its variants, including train-test methodology, weight initialization, selection of the transfer function, epoch size, derivative offset, nonstandard error functions, and network architecture. Each of these will be addressed separately.

Train-Test Methodology

As alluded to in prior sections, effective model development requires a train-test methodology. Depending on the objective of the application, the best network may be based on how well it interpolates (as measured by performance on the test data set), or how well it performs in a deployed environment (using the validation data set). To evaluate how well a network interpolates, the network training process is periodically interrupted and the network is tested. If performance on the test set improved over the prior best, the new network is saved. If it fails to improve for several trials, training is terminated. As a variant on this, rather than testing the interpolation capabilities of the network on the test set, it is tested on the validation set integrated into a complete trading system. The network that best meets the system objectives is saved.

Weight Initialization

Each time a new network is developed, it should be randomized with a different set of initial weights. This can usually be accomplished by serially randomizing each network to be trained, or by manually setting the random number seed to a different value prior to each initialization. Typically, it is better to initialize a network with small (1/number of inputs) zero-mean Gaussian distributed weights rather than with larger uniformly distributed values (as described in earlier literature).

Transfer Function

Both from theoretical considerations as well as practical experience, BP networks built from blocks using the hyperbolic tangent function typically work better than those built from units using the sigmoidal function. There are exceptions, but few. Although Lapedes and Farber (1987, 1988) have suggested that linear output units are better than sigmoid or hyperbolic tangent for time-series prediction, getting them to work properly is somewhat tricky. The key to success is setting the learning rates on the output layer very low (e.g., 0.001). In practice, the author has found that linear outputs are often more trouble than can be justified in terms of performance improvement.

Epoch Size

Setting the training epoch or accumulation period can have an impact on how well a neural network performs. Bottou (1991) provides a convergence proof for back-propagation which requires that (1) the epoch size be greater than 1 and less than the entire data set, and (2) the learning rate be bounded by $1/n$ (n = number of epochs). This means that standard back-propagation without momentum is not guaranteed to converge. It also means that epoch size for cumulative back-propagation should not be set to the entire training set size. Paul DuBose (Chapel Hill, NC) has developed a method for picking the optimal epoch size for back-propagation learning. The procedure is based on the empirical observation that for a fixed number of epochs, as the epoch size is increased, performance on the training set will improve to a maximum value (as measured by linear correlation coefficient) and then decrease and level off. The optimum epoch size is the one where the performance peaks. To implement this, build a network and save it. Reload the network and set the epoch size to 2. Train the network for 10 to 50 epochs until the RMS error starts to drop and level off. Record the number of epochs and the R^2 value. Reload the network and set the epoch size to 4. Train for the same number of epochs as before. Repeat this process for epoch sizes of 8, 12, 16, 20, and so on. Plot them on a graph and pick the epoch size that produces the highest R^2 value. Once selected, the epoch size tends to work well for a data set, even when that data set is augmented with more recent data. The intuition behind this method is that the noise in a data set has a certain frequency. At the optimum epoch size, the noise is averaged to zero and the best gradient information can be computed from the data.

Derivative Offset

Another technique for ensuring that all or many of the hidden units in a network are effectively used is to adjust the derivative of the transfer function. This technique was first described by Fahlman (1988). The most common practice is to set F' offset to 0.1 for sigmoidal units and 0.2 for hyperbolic tangent units. This offset is added to the derivative of the error function prior to multiplying it by the error. It has the effect of always forcing a small amount of error back into the processing element, thus preventing it from becoming stuck.

Nonstandard Error Functions

Nonstandard error functions are an advanced technique for incorporating the costs of different kinds of errors into the network training process. For example, previous sections of this chapter have shown that making mistakes on small moves is less important than making mistakes on large moves. This can be incorporated into the training process by defining an error function which is not just the difference between the target and actual output, but which multiplies this by a penalty function. The penalty function can be defined as a matrix, as shown in the table below. The target output would range from minimum to maximum along the columns. The actual network output would range from minimum to maximum from the bottom to the top row of the table. Intermediate values can be interpolated. This technique usually works best with a single output network.

10.0	7.0	1.0	0.5	0.5
7.0	1.0	0.5	0.5	0.5
1.0	0.5	0.5	0.5	1.0
0.5	0.5	0.5	1.0	1.0
0.5	0.5	1.0	7.0	10.0

Network Architecture

Regardless of which variant of back-propagation is used, it is *essential* to recognize that the first network trained is probably not the right one to use in the deployed environment. The difficulty of the modeling problem, the noisiness of the data, and the idiosyncracies of the numerical algorithms involved in back-propagation all require that several models be developed and compared. In one successful system, thousands of networks were trained before deciding on which one to deploy. Effective measurement of the performance of a network in a deployed system is crucial.

A single hidden-layer BP model has certain advantages, including simplicity. It will typically train quickly. When using this approach, variations of back-propagation that automatically adjust learning rates, such as extended-delta-bar-delta (EDBD) and quick-prop, are rapid and make good use of the hidden units. There are three approaches to optimizing the size of the hidden layers: fixed, constructive, and destructive. Optimizing the size of the hidden layer is important, because networks with the smallest hidden layer which solve the problem tend to interpolate (generalize) better.

In the *fixed approach*, several networks are trained with increasing numbers of hidden units. Each of these is evaluated, and the smallest number of hidden units that provides the best deployed performance is used for future development. This typically works well but, due to the nature of the numeric algorithms, requires that a substantial number of networks with different initial randomization be developed to optimize the performance of the system.

The most popular *constructive approach* was proposed by Fahlman and Lebiere (1990). In this approach, hidden units are added one at a time and trained until performance stabilizes. To avoid some of the local minima in this approach, several candidate hidden units are tried for each possible addition. NeuralWare has developed a variant on this in which more than one hidden unit may be added at a time. As with the original algorithm, these are taken from a pool of possible candidates. The process of adding additional units continues until a point of diminishing returns. Alternatively, the train-test methodology can be used to establish a stopping point. Several networks should be developed and one or more selected for deployment based on performance.

Another approach to optimizing the size of the hidden layer is to start with a large hidden layer and to prune out units during the training process (NeuralWare 1993). The rationale behind this *destructive approach* is the empirical observation that networks with many hidden units on average perform better than those with fewer hidden units, and that one or more of the hidden units contributes little to the solution. Hidden units are ranked in terms of the degree to which they affect the performance of the network. Those with little or negative impact are eliminated. In a test of two randomizations of each of 24 different networks with two to 24 hidden units in one or two layers, this method reduced all of them to two or three hidden units.

Sometimes it is not easy to decide which network to deploy. One may work better in upward trending, another in downward trending, and a third in sideways markets. An approach to solving this problem is to use all three. A fourth neural network can be trained to decide which of the other three to use at any given time. This is the genesis for the idea of the modular neural network (Jacobs et al. 1991), which consists of a gating network and several local experts—both of which are standard BP neural networks. The gating network learns which local experts are good at solving a particular part of the problem. In this way, the modular neural network learns not only how to divide the problem into smaller pieces but also the functional relationships in each of those smaller pieces. When applied to time-series forecasting, empirical results show that the local experts actually specialize in different types of markets. All of the caveats and suggestions for simple BP networks also apply to modular neural networks.

Genetic optimization of neural networks is becoming a hot topic. One of the challenges of using genetic algorithms to optimize the weights in a neural network is the problem of overtraining. Just as BP networks can be overtrained on a particular segment of data, so can genetically optimized neural networks. One solution is to use a train-test procedure similar to that used in back-propagation and to devise a fitness function that takes into account performance on both the training set and test set, as well as the discrepancy between them.

A variety of techniques are available to facilitate the development of a neural network. Many of the introductory products are quite easy to use. More sophisticated products, such as NeuralWorks Professional II/PLUS, though more complex, provide a broader range of options that under many instances can provide an incremental edge in performance.

Model Verification

The validation set is used to test the model in a simulated deployed environment. This step is essential to establishing how the system will perform when actually deployed. Other chapters in this book provide an excellent description of methods for evaluating a trained network. However, performance on the validation set does not necessarily guarantee that the network will work as well in actual trading. An easy trap is to assume that if an integrated system trades well on historical data or even on a validation data set, it will continue to work well when put into a live trading environment. The means used to evaluate a network indicator also need to be thought through carefully.

One of the pitfalls in using data-derived indicators is that they may sometimes hone in on relationships of a certain type of market, failing to model other types. A neural network indicator may work quite effectively in a trending market and fail in a trading market. If the development process focused on a trading market and the market has recently shifted, the system may not work. For this reason, it is essential to artificially divide development data from testing data. It is typical to use six to nine months of data to develop an indicator, and to use the most recent one or two months to verify that continued performance. Even then, many systems are paper-traded for a period of time prior to going "live." Another lesson is that a live system must contain built-in monitoring systems to measure the quality of the performance. When performance begins to degrade, immediate steps need to be taken to understand the causes.

Model Deployment

The objective of developing these models is to deploy them in trading systems that make money. Most neural network vendors provide the technical mechanisms for integrating trained neural networks into deployed systems. Beyond this, several important issues must be addressed. The three key issues are measuring on-going performance, determining when the model is reliable, and trusting the model.

There is no substitute for knowing your model and understanding how it works under different conditions. To the degree that it is possible to learn under which conditions a model works well and under which it works poorly, it is possible to build a more effective trading system.

Another danger often overlooked when developing trading systems is the human factor. Dean Barr (LBS Capital Management, Safety Harbor, Florida) noted that "it is easier to fire a computer than a fund manager." The traders and developers of successful systems have learned that once you believe in your system, do not try to second-guess it. Second-guessing often leads to losses. The exception is in the event of extraordinary market-shaping forces, such as the assassination of a president, an oil embargo, outbreak of war in the Persian Gulf, the collapse of a particular securities market, or a major scandal. Neural network indicators almost never capture such events. This is the time for human intervention, never otherwise.

Summary

Neural networks are powerful tools for developing formulas useful in building trading systems. They are *model-free estimators* that provide input/output functions based on sample data which capture the dependence of outputs on inputs. They are *dynamic systems* that can be trained to learn the principle features contained in sample data. Following a disciplined methodology to train neural networks improves the likelihood of success. A variety of insights into the model development process have been described in this chapter, along with key issues that must be addressed during the process.

References

Bottou. L. 1991. "Stochastic Gradient Learning in Neural Networks." *Neuro Nimes* (Paris).

Deboeck, G. 1992. "Pre-processing and Evaluation of Neural Nets for Trading Stocks." *Advanced Technology for Developers* 1 (Aug.).

Erlanger, P. 1993. Unpublished oral presentation. *AI on Wall Street.*

Fahlman, S. E. 1988. "An Empirical Study of Learning Speed in Back-Propagation Networks." Pittsburgh: Carnegie-Mellon University, Computer Science Dept.

Fahlman, S. E., and C. Lebiere. 1990. "The Cascade-Correlation Learning Algorithm." In *Advances in Neural Information Processing Systems*. Vol. 2. D. S. Touretzky, ed. Palo Alto, CA: Morgan Kaufman.

Jacobs, R. A., M. I. Jordan, S. J. Nowlan, and G. E. Hinton. 1991. "Adaptive Mixtures of Local Experts." *Neural Computation* 3.

Klimasauskas, C. 1992. "Accuracy and Profit in Trading Systems." *Advanced Technology for Developers* 1 (June).

Lapedes, A., and R. Farber. 1987. "Nonlinear Signal Processing Using Neural Networks." In *Prediction and System Modeling*. Los Alamos, NM: Los Alamos National Laboratories.

Lapedes, A., and R. Farber. 1988. "How Neural Nets Work." In *Neural Information Processing Systems*. D. Z. Anderson, ed. New York: American Institute of Physics.

NeuralWare. 1993. *Neural Computing Reference Manual*. Pittsburgh: NeuralWare Inc.

Owens, A. J., and M. T. Mocella. 1991. "An Experimental Design Advisor and Neural Network Analysis Package." In *Proceedings of the IWANN.*

Pardo, R. 1992. *Design, Testing and Optimization of Trading Systems*. New York: John Wiley & Sons.

Plutowski, M., and H. White. 1990. "Active Selection of Training Examples for Network Learning in Noiseless Environments." In *Technical Report* CS91-180. San Diego: University of California.

Plutowski, M., and H. White. 1992. *Selecting Concise Training Sets from Clean Data*. San Diego: University of California.

Smith, M. 1993. *Neural Networks for Statistical Modeling*. New York: Van Nostrand Reinhold.

CHAPTER 2

Pre- and Postprocessing of Financial Data

Guido J. Deboeck and Masud Cader

Detecting tendencies and patterns is a crucial skill. The more accurately we can get the picture from minimal information, the better equipped we are to survive.

Marilyn Ferguson

Chapter 1 defined neural networks as model-free, nonparametric estimators similar to statistics and other mathematical estimation techniques. Neural networks are not magic! To use neural nets for trading requires a good understanding of financial markets, trading, performance measures, risk, and money management practices—indeed, such an understanding is more important than understanding neural network paradigms themselves. Financial applications of neural nets are best implemented by a team with financial acumen, neural net expertise, and programming skills. Obtaining neural networks that perform according to desired financial criteria is difficult. This chapter provides guidelines on preprocessing and postprocessing of financial data for the design of profitable neural trading systems. The range of issues covered includes preparation of financial data, selection of training and test data sets, training and testing of neural nets, and evaluation of neural trading systems.

Setting Objectives

The purpose of any financial application of neural nets is to use machine learning for time-series forecasting, pattern recognition, or organization of data so as to obtain a system that can produce recommendations that are valuable for decision-making. A neural net trading system is a completely automated way of trading a financial security or financial market. It can be used as a decision-support or a decision-making system.

Neural net models differ substantially from mechanical models (Babcock 1989) in that relationships and rules are implicitly generated and are ideally general enough to interpolate accurately in high-dimensional spaces. Thus, unlike mechanical models, neural networks can generalize or interpolate between known values in a very high-dimensional input/output space. To generalize or interpolate between values implies that neural networks provide more than recall from memorization. Actually, the better a neural network has memorized the sample data, the worse it is likely to perform on future data. The success of a neural net design thus depends on its capability to perform against unknown data or data the system has never seen during the design stage. Optimizing the capability of a neural net to perform against unknown data brings many different factors into play:

- The original data collection
- Data analysis and transformations
- The selection of the neural net architecture
- The learning paradigm
- Training, testing, and evaluation of the system

Development of a neural net trading system requires a lot of experimentation. It is important to start by defining specific objectives (Deboeck 1992). Ultimately the performance of the neural net must be tested, evaluated, and/or integrated into a trading environment. All neural trading systems need a benchmark whereby predicted neural net results are measured against market behavior, the performance of traders, or some other customer-defined parameter.

In order to sell the idea of undertaking a neural net project, the concepts, techniques, and potential benefits must first be understood. In our experience this can best be done by organizing a workshop for management where prototype applications are demonstrated. We specifically suggest focusing on senior managers, because traders may be too preoccupied with day-to-day operations to understand the potential of neural nets. In the end, management is responsible for determining the appropriate level of diversification, that is, the mix of human versus computer-based trading.

Once the idea of undertaking a neural net project has been sold, specific project objectives must be set. A useful way to define these objectives is to brainstorm with experts. One method of doing this is to arrange a meeting with traders and managers. The agenda should be to explore potential areas of interest and priorities. Once the meeting begins, all of the ideas that pop up are recorded on a flip chart. After a dozen ideas are listed, the following questions should be asked:

- Which topics would be easiest?
- What data would be needed?
- Which topics require new data acquisition and/or new database design?
- Which topics may produce the highest payoff?

Moving quickly down the list of topics, a column is created in answer to each of these questions; check marks, question marks, or ratings of the degree of difficulty and payoff are added. Once the table is complete, the list of potential topics is rewritten in order of payoff, degree of difficulty, availability of data, etc. The topics that emerge on top are then discussed in detail and priorities are set. Priority may be given to project objectives that offer significant potential return, that have readily available data, or that are relatively easy to develop. It is important to distinguish among *ideal*, *realizable*, and *salable* objectives. Project objectives that are too ambitious, that will take a lot of time to accomplish, or that may yield unsalable results are best avoided.

When a satisfactory list of project objectives has been established, it should be reviewed in light of the actual allocation of resources. For example, while the brainstorming session may produce a wish list of applications, the actual allocation of project resources may only allow the undertaking of one or two. If this is the case, either pick the top-priority objective or phase in the project so as to accomplish a specific output of several objectives (e.g., by building prototypes without completing the full task).

Data Collection, Analysis, and Transformations

Once the objectives are spelled out and are consistent with allocated resources, the project can actually get underway. The first step is data collection. For most financial applications, plenty of data exist and access is relatively easy. Data on stocks, bonds, foreign exchange, commodities, futures, and options are readily available from various market information services. However, the quality of the data varies widely!

Data collection involves the acquisition of high-quality data from reputable sources that provide complete and consistent information for the desired data fields. If fundamental data are needed, remember that fundamental economic data are frequently revised. This raises the issue of which series to use (the original series that influenced the market, or the revised series available a month or quarter later). An additional complication of using fundamental data is that when the revised data are made available, the original data are overwritten.

Some projects fail because they never get out of the data collection phase. Bernstein (1992), who collected one of the largest historical databases on the futures industry, called it "both a blessing and a curse." It is essential to maintain a pragmatic engineering perspective—a balance between research and production requirements. *This provides a strong argument for developing neural net trading systems with a well-balanced team of financial experts, neural net engineers, and goal-oriented management.*

Data downloaded electronically require validation, that is, checking of formats and ranges, and detection of bad data. If electronic data are used as input to various data transformations, then data errors can have a devastating impact on the neural net results.

In selecting sources for data retrieval, one must ensure that the data obtained will be available for the future production environment. While neural network design is feasible based on historical price series, the implementation of a trained neural network requires regular availability of or access to current data. Thus, if a neural network for trading depends on daily closing prices, it is mandatory to select a data source that provides on-line access to daily closing prices.

Selection of the proper data series or fields depends on what the financial expert needs in order to represent the problem. Typically, a neural net for stock trading requires data on opening, high, low, and closing prices, volume of transactions, and sometimes opening interest. These variables are the raw data from which various data transformations can be computed. Data collected for a neural net ought to represent the entire spectrum or as many dimensions as possible of the data of interest. Key questions that need to be addressed in the selection of input variables are:

- Which key variables will influence the output?
- Which variables do the experts use or suggest?
- What types of data transformations are relevant?
- What do classical statistical analysis techniques suggest?
- What do graphical displays suggest?
- What is the appropriate data frequency?

Selection of the appropriate data frequency is very important—a point demonstrated by the application discussed in Chapter 3. A neural net system for daily market forecasts needs at a minimum daily data, and possibly even intraday data. A neural net system to forecast major economic turning points—for example, in GNP, unemployment, or inflation—needs at a minimum weekly or monthly data. Some systems may require both high-frequency data (e.g., intraday or daily data) and low-frequency data (e.g., weekly or monthly data). If this is the case, special attention should be given to maintaining a proper balance or combination of the high- and low-frequency data inputs.

The selection of data sources, the selection of data fields, and the scope and frequency of the data will all have a bearing on the cost of the data. What is the appropriate amount to spend on data collection? It is impossible to determine this in the abstract, but frequently there is a tendency to spend too much rather than too little! Access to too much historical data or use of too many variables may be a curse. In Part Three we will see that there is a real cost to high precision. Remember, all of the time spent on data collection is time not available for data analysis and model design.

Raw data are seldom adequate for neural net training. Data analysis and transformations are necessary to enhance information that provides a better descriptor of the trends or processes present in the raw data. The objective of data analysis and transformations is to pull the data apart and thus simplify classification by the neural network. Three years ago we thought that neural networks were so powerful that they could find a "needle in a haystack." After a lot of experimentation, we have found that they need help. The neural net designer must know how many needles (i.e., patterns) are in the haystack and where they are actually hiding! Again, don't expect miracles or magic; magic will result only if you design the tricks yourself.

The data transformations should be made to help:

- Classify the data
- Transpose nonlinear problems into linear ones
- Focus on a portion of the input range and/or
- Let the neural net find the needles!

The simplest data transformations are percentage differences and log transformations; more sophisticated ones are statistical transformations and technical analysis. Log transforms and percentage differences are easy to implement in a spreadsheet. For example, the first-order logs on the closing prices can be implemented by creating a column using the equation:

$$X_i = \ln\left(\frac{P_t}{P_{t-1}}\right)$$

where P_t is the current closing price and P_{t-1} is the previous day's closing price.

Examples of statistical transformations are regression slopes that measure the direction and acceleration of trends in the data. More complex transformations include Fourier transformations and wavelet transforms. Fourier and wavelet transforms may be useful; however, their distinct disadvantage is that most traders do not understand them!

More common data transformations for financial applications are derived from technical analysis. Technical analysis is well understood by most traders and thus provides a better base for the development of inputs for a neural net. Technical indicators can be used to enhance the description of market signals (Plummer 1991)—for example, use of moving averages to identify trends, volatility measures to identify trading markets, and oscillators to identify oversold or overbought conditions. The following technical indicators are potential ways to enrich raw data:

- Moving averages: simple, weighted, or exponential
- Average range channel
- Commodity channel index
- Directional movement
- Oscillators
- Relative strength index
- Stochastic indicators
- Parabolic or stop and reversals
- Volatility indicators

- Point and figure boxes
- Market profiles
- Candlestick patterns

An extensive discussion of most of these technical indicators can be found in *Technical Analysis of Stocks, Options & Futures: Advanced Trading Systems and Techniques* (Probus Publishing, Chicago, 1988) by William F. Eng. Actual use of these techniques by traders is discussed in detail in Eng's *The Day Trader's Manual: Theory, Art and Science of Profitable Short-Term Investing* (John Wiley Finance Editions, 1992). The main result of the data transformations is a set of candidate input variables from which actual input can be selected.

Types of Data and Data Scaling

Several other issues emerge with regard to data transformations when different types of data are used. This section discusses common data classifications, time-series data, categorical data, strategies for missing data, log transformations, and data scaling.

In general, there are four major classifications of data: nominal, ordinal, interval, and ratio. *Nominal data* are typically symbolic, with no order relation (for example, accept or deny credit, buy or sell), and are thus represented in neural nets using *1-of-N encoding*. *Ordinal data* permit order relations between data, but not exact differences between data elements. Typically, ordinal data are represented using a *thermometer coding* in which $N - 1$ input neurons are used to represent N data values (for example credit ratings of good, mediocre, and bad would be represented with two neurons). *Interval data* permit ordering and difference relations (for example salaries of loan applicants) and can be represented using a single neuron for each variable. However, the input should be scaled in such a manner that it is consistent with the activation function used in the neuron, and before the data series are divided into test and train sets. The latter ensures that the network can consistently respond to ranges of both training and testing sets. Oftentimes, interval data are represented using a percentage encoding in one neuron. In developing financial models, one often faces the problem of integrating many of these data types into a single model. We will now describe how the nature of the data and the model objective conspire to permit useful networks.

Financial time-series data can be presented to a simple feed-forward neural modeling process in different ways. One method for short-term predictive models uses as inputs the price or yield data at time $t, t - 1, t - 2, t - 3 \ldots t - n$ and trains the neural net to predict the prices or yields at $t + 1, t + 2, t + 3 \ldots t + m$. This is one of the simplest approaches, which may or may not be effective. In essence, one assumes that there is a direct relationship between the prices over the last n days and the future prices over m days. However, this approach does not take into account that many price movements are just noise and that the underlying patterns cannot be detected without eliminating the noise. Part Four of this book demonstrates why this approach to time-series forecasting may not be very effective.

Another much more common method of describing time series is to filter out the noise and use moving averages to represent the trends. As pointed out earlier, the use of multiple moving averages allows representation of short-, medium-, and/or long-term trends. Taking into account the earlier discussion on data scaling, it is often useful to use, instead of moving averages themselves, the difference between the current price (P_t) and one or more moving averages. For example,

$$MOV^*_{5\ days} = P_t - MOV_{5\ days}$$
$$MOV^*_{10\ days} = P_t - MOV_{10\ days}$$
$$MOV^*_{40\ days} = P_t - MOV_{40\ days}$$

Yet another way of representing time series is to calculate the slopes or regression coefficients via sliding windows over 10 or 20 days. For example, compute the beta's from $Y = \alpha + \beta X$, where X is the original price series and Y is a moving average of 5 or 10 days. The second derivatives of these can be added to obtain the rate of change of the slopes or accelerations of the differences. The essence of all these techniques for describing time series is to filter out the noise and achieve a better description of the underlying trends.

Describing categorical data requires other techniques. Categorical data may require a 1-of-N code transformation, depending on whether the order in the data is important. If the data are nominal, for example, a variable on trade recommendations where 1 could stand for "buy" and 0 for "sell," then there are two ways to use them as inputs in a neural net: (1) use a single series with 1 and 0; or (2) use two series where buy transactions are represented as {1}, {0} and sell transactions are represented as {0}, {1}.

If the data are ordinal, that is, if order has importance, a 1-of-*N* code can be translated into a thermometer code. For example, if trading day of the week is an important input variable, then the trading days can be represented as:

Monday	{1,0,0,0,0}
Tuesday	{1,1,0,0,0}
Wednesday	{1,1,1,0,0}
Thursday	{1,1,1,1,0}
Friday	{1,1,1,1,1}

Histogram normalizations are transformations of raw data into a code that distinguishes between upper and lower quartiles. This is similar to a rank-order transformation. Rank information can be used to represent equal conceptual distances. For example, if one wants to provide a relative position indicator, then a buy recommendation could be represented as 5 for very strong, 4 for strong, 3 for medium, 2 for weak, and 1 for very weak.

When the data set contains many diverse data types, it is important to normalize the data such that no particular variable will influence the error estimates based on its order or magnitude. Thus, if the variables range comparably, errors will more likely be a function of misfit rather than of magnitude discrepancies among variables. Some cases may require that some ordinal variables be encoded by the neuron activation function. For example -1 to -0.61 could be the activation for very weak, -0.6 to -0.2 for weak, -0.19 to $+0.19$ for medium, 0.2 to 0.6 for strong, and 0.61 to 1 for very strong buy recommendations.

What if data are missing? If there are gaps in the data, it may be desirable to:

- Try to find the missing data potentially from other sources
- Interpolate or use average values
- Fill in values that would minimally disturb the patterns, or
- Drop the record in question

Without doubt, it is best if one can return to the data source and find the missing data; however, this is often impossible. As most neural network learning algorithms are robust, it is not crucial to filter noisy data or data which contains elements that might be considered outliers. Common activation functions such as the sigmoid and hyperbolic tangent naturally act to clip or saturate outlier data, so processes that remove outliers may serve to accelerate neural net learning at the expense of omitting possibly important information. Outlier definitions are subjective. Omission or inclusion depends invariably on the application area and the nature of the data, including the number and density of outliers. A much more conservative approach would be to use a compressive transform.

If there are records in the data that represent outliers, it may be important to consider dropping the outliers or to use log transformations. For example, the movement of the stock market on October 19, 1987, may not be relevant to your model. In general, when the tails of a distribution of the data are very wide—for example, major market disturbances such as jumps of 100 to 200 points in the Dow Jones or moves of 5% or more on the Nikkei—then log transformations should be considered. A log transformation compacts the large values into smaller values so that more logical distances are created for the neural networks to learn. Log transformations of trading volumes are clearly a case in point.

Interpolation is an alternative that can be used as long as the data are not too sparse. It is also useful to understand that interpolation assumes a functional form between points and then attempts to "fill in" between two apparently adjacent values. Such an approach can be dangerous if one of the values is considered to be an outlier.

So far we have discussed how to obtain raw data and how to perform data transformations. The result of these processes may be a series of data inputs, most of which may have quite different ranges. The next issue is data scaling.

Why is data scaling necessary? Since neural nets use multiple inputs, the data ranges of each (i.e., the minimum and maximum of each series) can often be quite different. For example, a neural net to select stocks may use capitalization, price-earning ratio, price-book value, expected earnings, and the like as input variables. The range of the capitalization series will be substantially different from the range of price-earning ratios, expected earnings, or other series. In consequence, all data series should be scaled. For logical data, scaling should be between 0 and 1; numerical data should be scaled between -1 and $+1$.

All transformations of market data are intended to classify, describe, or extract features in the data that are significant and that describe relatively stable elements as opposed to unimportant or invariant changes. What has been said thus far concerning inputs to a neural net is also true for outputs. There are, however, several additional issues in connection with the selection or identification of output indicators. These will be discussed next.

Identification of Output Indicators

Depending on which learning paradigm is chosen, it may be necessary to select one or more output indicators. When supervised learning is used, the neural work expects that a desired output is produced against which it can compare the actual neural net output.

What are suitable output indicators? If the objective of the system is to forecast prices, yields, or returns, then output indicators could be future prices, yields, or expected return rates. In forecasting systems it is particularly important to choose those future prices, yields, or returns that correspond to the time window required for trading operations. For example, if the system trades on a daily basis, then the desired price, yield, or return forecast should be for that time period.

If the objective of the neural net system is not forecasting but pattern recognition or classification of data, then any number of other output indicators can be considered. For example, patterns in the historical data may be related to patterns in the future data. An important consideration in choosing an output indicator is the manner in which it makes patterns "stand out." Another consideration is the number of patterns the neural net is supposed to learn. Neural nets with limited capacity cannot learn hundreds of different patterns; alternatively, neural nets with a lot of capacity (i.e., processing elements at the hidden layer) can memorize patterns. The selection of one or more appropriate output indicators usually constitutes the proprietary nature of the neural network design.

Useful output indicators can be constructed based on the desired level of profitability, the desired trading strategy (e.g., long, short, and out of the market positions, or long and short positions, or long positions only), or level of turnover. Unless the neural net designer knows what it is the neural net needs to learn, it is unlikely that a neural net can come up with relevant relationships between inputs and outputs.

Preparing Training and Test Data Sets

A common procedure in the design of neural nets is to split the available data into two parts: one set for training and another for testing. Usually, two-thirds of the data is used for training and one-third for testing. The basic assumption here is that of all the data that have been collected, two-thirds is relevant for training and one-third should be retained for testing. We used this procedure of biased cross-validation for quite some time, until we tried sampling.

Sampling is the process of reducing a population of data so that a number of criteria can be satisfied. Traditionally, these criteria have been reducing the cost of analysis, speeding the process of analysis, and facilitating more accurate analysis. However, when sampling for neural networks, it is important that the data be proportionally sampled in relation to the learning difficulty. Traditional sampling satisfies such general criteria by scaling the number of observations (random sample) such that the statistical properties (mean, variance, skewness, etc.) of the population and the sample are similar. Here we focus on sampling of financial time-series data with the express purpose of training a feed-forward neural network.

Much of sampling theory has been devoted to studies involving demographics and, in fact, much of the terminology can be traced to such early applications as the U.S. census. It is also useful to note that standard sampling tomes, such as Cochran (1977), mostly ignore the temporal aspects of data. Recent studies focus on temporal sampling methods, such as sequential sampling, in which a fixed population is sampled at different time periods with knowledge extracted from the previous sample to guide the new sampling strategy. While temporal sampling techniques may be useful for some modeling, we will focus on spatial sampling methods.

Supervised feed-forward neural networks (such as the extended-delta-bar-delta [EDBD] and back-propagation networks) require that the sample data be at least divided into a training set and a test set. A training set is the set over which the neural network iteratively works to reduce the error between the network output and the desired output. When sampling the training set, it is usually required that the test and training sets be mutually exclusive and that the training set be sufficiently large to represent the problem space. In developing sampling strategies to speed the development of neural network modeling, we are principally interested in reducing the training sample to about 10% of the data. Also, the sample must either be statistically representative of the database or, alternatively, accentuate some feature of the data set. Sampling methodologies can be used to bring out interesting facets of the data. An example is a technique that samples all records in the tails of the distribution (usually by a multiple of the standard deviation). Such a sampling method usually assumes that patterns of interest are nearly symmetric, and probably Gaussian.

It is intuitively more appealing to utilize an appropriate standard deviation or exception sampler as the output signal. As a result, the neural network learns more important parts of the space rather than the whole space, of which only small subspaces are relevant. A further step is to analyze the points in the input space when either the error of the model or the signal-to-noise (S/N) ratio is large. In such areas, it is imperative to resample so that the model error or S/N ratio exhibited by the model is reduced.

It is often desirable that the data be proportionally sampled in relation to the learning difficulty or the S/N ratio. However, traditional sampling does not satisfy such criteria, and overly directed resampling strategies produce neural networks that are heavily biased. Some modern resampling strategies, such as the bootstrap and jackknife, can be used to indicate the efficacy of the model and to attempt to minimize sampling bias.

In a bootstrap scheme, data are randomly sampled with replacement from the entire data set, then the network is trained. This process is repeated a number of times (usually more than 50), with each bootstrap sample set to about 10% of the total data size. The mean of the weights or parameters is then computed and assigned to be the "best" model (in that it is theoretically an unbiased estimator). Confidence intervals may be computed on each weight or parameter to provide further information about the reliability of the estimated weights. Unresolved issues include whether the neural networks should all be trained for the same number of iterations, and whether they should all have constant model architectures on each training sample (note that the topology must remain constant in order for this methodology to make sense).

Modern statistical methods such as the jackknife can be used to provide a sample which ensures a minimum sampling bias in the estimated neural model weights. The neural network is trained on a training set which precludes one sample data element selected randomly without replacement from the database. The trained model is then tested on this single element. This process is continued until every sample in the data has been precluded once. The averages of the weights of all the models are then used to provide the weight estimates of the "best" model:

$$S / N = 10 \log_{10} \left(\frac{\sum_i (desired\_output_i)^2}{\sum_i (desired\_output_i - net\_output_i)^2} \right)$$

where S/N = 10 implies that the signal is ten times more evident than noise in the series.

The use of sampling techniques can substantially enhance the efficiency of training neural nets. We have found that training neural nets on a carefully designed sample of about 100 records produces equal if not better results than training a neural net on four or five years of daily data (i.e. 1000 to 1250 data records. Obviously, if one trains a neural net on 100 samples selected out of five years, there is less likelihood that the neural net can memorize the data.

Testing and Evaluation of Neural Trading Systems

The most challenging aspect in the design of a neural network for trading is assessment of the quality of a trained neural net. As pointed out in Chapter 1, statistical measures such as the root mean square error between the actual and the desired neural network output are meaningless for systems designed to trade. Systems designed to trade need to produce financial value. Consequently, neural nets should be tested and evaluated on the basis of how much financial value they achieve. In other words, to what degree do they achieve the financial criteria or objectives of the system?

A useful framework for evaluating the financial results of a neural net is composed of three criteria: profitability, consistency, and robustness.

Profitability criteria for the evaluation of financial neural nets include:

- Total profit and loss (P&L)
- Average P&L per annum
- Total number of profitable trades
- Percentage of profitable trades

- Daily and annualized rates of return
- Pessimistic return ratio (i.e., the return weighted inversely by the number of profitable and unprofitable trades)

Consistency criteria include:

- Average profit (or loss) per profitable (or unprofitable) trade
- Average P&L per trade
- Maximum profit/loss
- Maximum drawdown or the cumulative of successive losses
- Standard deviation of P&Ls
- Sharp ratio, or the average P&L divided by the standard deviation

To test the *robustness criteria*, we suggest measuring the sensitivity of the neural net results to:

- Input variables
- Number of hidden processing elements
- Learning coefficients
- Interconnectivity of the net
- Number of data presentations or training cycles
- Size of data sample when training a neural net

The above framework allows neural nets to be optimized to meet specific financial criteria. For example, neural nets for trading can be trained to:

- Maximize the total return
- Minimize the volatility of the daily returns
- Minimize the number of trades or transaction costs
- Maximize the ratio of the average profitability to the maximum drawdown, or
- Combinations of the above

Adopting one or more of these evaluation criteria can produce neural nets for active trading, infrequent trading, or even relatively passive management.

The frequency of trading, or *alpha cycle*, is the total number of trades divided by the period under investigation, normalized to one year:

$$\alpha = \left(\sum_t T / P\right) \times 250$$

where α is the alpha cycle, T is the total number of trades, and P is the training or test period expressed in trading days. For example, if the total number of trades produced over a period of 450 days (or 1.8 years, using 250 trading days per year) is 60, then the alpha cycle is 33 ($= 60/450 \times 250$), or two to three trades per month. The selection of the output indicator(s) for training of a neural net should take into account the desired trading frequency: very active, moderately active, active, or inactive. Clearly, the alpha cycle is very important for assessing the effectiveness of a neural net. If a neural net is trained to achieve certain financial criteria and does so with a relatively low alpha cycle, then this may be more desirable than a neural net with a high alpha. Chapter 4 provides an illustration of this point.

In testing and evaluation, the performance of neural nets can be captured by various measures: model efficiency, walk-forward efficiency, robustness, and stability. Each of these concepts is elaborated in Pardo (1992) and summarized in the following paragraphs.

Model efficiency refers to the performance of trained neural nets versus performance that could be achieved based on desired or ideal output. To compute the model efficiency, one runs the evaluation based on the desired or ideal output and compares the performance with that achieved on the training data using the neural net output. For example, if the annualized return that could be achieved on the basis of the ideal output is 20% per annum and the neural net actual output produces 16% (on the same data), then the model efficiency is 80%. The model efficiency provides an indicator of how well the neural net has learned the training data; it does not provide an indicator of how well the neural net will perform.

Walk-forward efficiency is a more relevant measure of the potential performance of a trained neural net. The walk-forward efficiency measures the performance of a trained neural net on a test data set that contains only out-of-sample data. It compares this performance with what could have been achieved had the desired or ideal output been used. For example, if on the basis of out-of-sample data, the ideal output would have produced 14.5% and the actual neural net performance on the test period is 11.5%, then the walk-forward efficiency is 72%. The walk-forward efficiency provides an indicator of the potential performance of a trained neural net. However, it does not guarantee that such performance will be sustained. Regime or structural changes may create differences between the training data set, the test data set, and the future real-time implementation of the neural net. Consequently it is important to check the robustness of the neural net over time.

Robustness is the performance of a trained neural net over various periods. These could be overlapping or nonoverlapping windows of varying sizes. For example, if several years of historical data are available, it is possible to test the performance of neural nets for all windows of 3, 6, 9, and 12 months. The use of sliding windows produces a distribution of neural net performance figures. Such distributions lend themselves to classical statistical analyses, whereby averages, standard deviations, and *t*-statistics can be computed to determine the expected values, the variance of the expected values, and the significance of these values. A concrete example of testing the robustness of neural net models is described in Chapter 6.

Stability is the performance of a trained neural net based on variable window sizes. The main idea is to undertake Monte Carlo simulations, whereby performance of the trained neural net is measured for multiple periods, each of which has a random starting date and a random ending date—the only constraint being that a minimum (e.g., 6-month) trading period be maintained. Given neural net model outputs and the financial evaluation routine described above, Monte Carlo simulations on the model results can be implemented in a spreadsheet.

What we have described is actually very similar to the testing of a new car. The first level of tests measures the performance of the engine against technical requirements; the next measures the performance of the car on a closed circuit (usually outside the plant); the next level moves to testing in cities and on highways in various weather conditions (the stuff one finds in consumer reports); finally are the endurance tests, usually on racetracks or safaris.

Testing and evaluation of neural net models should be approached in the same way. There is model efficiency (or testing of engine performance), walk-forward efficiency (or closed-circuit driving), robustness (or testing in all weather and road conditions), and finally stability (or endurance) testing. Only after these tests should you consider to implement a neural net in real-time.

Summary

This chapter discussed preprocessing and postprocessing of financial data for neural trading systems. Financial applications of neural networks remain somewhat of an art because there are no engineering designs, blueprints, or templates. There are many neural network paradigms, transformation functions, and ways to build neural nets, select inputs, produce data transformations, and define desired outputs. Neural nets are not for the unadventurous. They require strong statistical, financial and advanced computer skills. They require different personality traits than those of traders!

Neural networks provide unlimited possibilities for creating trading systems for financial markets. As shown in this chapter, neural nets can be designed to meet one or more financial performance criteria. They are essential tools for designing disciplined trading strategies.

The next five chapters provide illustrations of applications of neural nets to various financial markets. Each contains important lessons on how to put theory into practice and how to build profitable systems to achieve various objectives.

References

Babcock, B. 1989. *Trading Systems*. Homewood, IL: Dow Jones Irwin.

Bernstein, J. 1992. *Timing Signals in the Futures Market*. Chicago: Probus.

Cochran, W. G. 1977. *Sampling Techniques*. 3rd ed. New York: John Wiley & Sons.

Deboeck, G., H. Green, M. Yoda, and G. S. Jang. 1992. "Design Principles for Neural and Fuzzy Trading Systems." In *Proceedings IJCNN* (Beijing).

Pardo, R. 1992. *Design, Testing and Optimization of Trading Systems*. New York: John Wiley & Sons.

Plummer, T. 1991. *Forecasting Financial Markets: Technical Analysis and the Dynamics of Price*. New York: John Wiley & Sons.

William, F. E. 1988. *Technical Analysis of Stocks, Options & Futures: Advanced Trading Systems and Techniques*. Chicago: Probus.

CHAPTER 3

Adaptive Selection of U.S. Stocks with Neural Nets

James W. Hall

The key to making money in stocks is not to get scared out of them.
Peter Lynch
Beating the Street, *1993*

The first application of neural nets concerns building a stock selection system. Jim Hall, who manages a sizable portfolio with such a system at John Deere & Company, explains the techniques required. His system attempts to autonomously discover current market trends and to rediscover new trends as the market evolves to new states. Unlike previous stock selection methods, this system does not contain rules or patterns discovered by humans; its decisions are based entirely on patterns it discovers by itself.

Style-based Stock Portfolios

Style-based selection of a stock portfolio is founded on the belief that market anomalies will, at times, result in groups of stocks that significantly outperform the market average. Several styles that are commonly referenced include *growth stocks* (fast-growing companies with rapid sales and earnings increases), *value stocks* (companies with low price-to-book ratios), and *cyclical stocks* (companies with regular market cycles).

One of the earliest references to value stocks was the formulation of the *dividend discount model* by Williams (1936). The basic premise of these early models was that the price of a stock should equal the discounted stream of its expected dividends. Miller and Modigliani (1961) later showed that similar models of a stock's value could be formed from corporate earnings or cash flow.

More recently, it has become common practice to use the ratio of a stock's price-to-book value to characterize value stocks. BARRA (1992), in collaboration with Standard and Poor's Corporation, has constructed the S&P/BARRA Value Index and the S&P/BARRA Growth Index, based solely on the price-to-book ratio. The value index contains those stocks from the S&P 500 group with the lowest price-to-book ratios. The price-to-book cutoff that divides the two groups is established so that the value index has 50% of the capitalization of the S&P 500 index.

Although the S&P/BARRA Value Index is determined strictly by price-to-book ratio, the stocks in this group tend to have other characteristics generally associated with value stocks: low price-to-earnings ratios, high dividend yields, and relatively low earnings growth. Typical industry sectors that fall into the value category are utilities and financial institutions. Stocks in the growth index tend to have the opposite characteristics: high price-to-earnings ratios, low dividends, high earnings growth, and high sales growth. Typical industry sectors for growth companies are computers, software, and drugs. Over long periods the performances of both indexes roughly equal that of the S&P index. From 1978 to 1992, the growth index had a cumulative absolute return of 966%, while the value index yielded 1009%.

Several investment management firms, including Wilshire Asset Management (1990), attempt to outperform the market average by switching between value or growth portfolios in response to market conditions. A typical period for holding a particular style of stocks might vary from 3 to 24 months. This type of portfolio selection is often referred to as *style-based management*.

There is considerable empirical evidence to support the intent behind style-based stock selection. Figure 3.1 shows the annual returns for growth and value stocks relative to the S&P 500 average for the years 1978 to 1992. Although both the style and value indexes had similar performances over the entire period, Figure 3.1 shows that significant differences occurred from year to year. During these 15 years, the mean absolute value of the difference between the two styles was 9.5%, with a maximum difference of 21% in both 1981 and 1991. If one could correctly predict which style would be in vogue

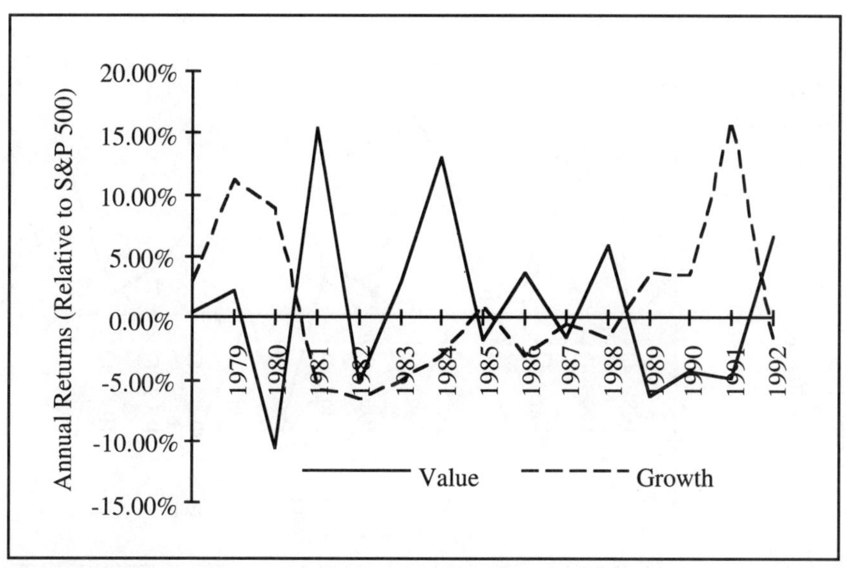

FIGURE 3.1
Annual returns for growth and value stocks relative to the S&P 500 index

for the coming year, it would be possible to obtain average annual gains of 9.5% above the market!

Although the theoretical potential for style-based portfolio selection is well known, few style-based portfolio managers have been able to significantly and consistently outperform the market. One reason is that trends in market styles are not nearly as clear as Figure 3.1 would seem to imply. Surprisingly, it is often quite difficult even to identify the currently favored style. Figure 3.2 shows the quarterly returns for value and growth portfolios based on the Wilshire data and definitions. Note that the best performing style often switches from quarter to quarter. What appeared in Figure 3.1 to be a "style year" or a "growth year" was actually the net effect of several quarterly trends and style reversals from Figure 3.2.

Figures 3.3 and 3.4 show the returns of a typical 25-stock growth portfolio and a 25-stock value portfolio plotted on a weekly and a daily basis, respectively. The best performing style can change from week to week, and even from day to day! Due to this characteristic, as well as the market response to unpredictable news events, it is very difficult to know in real time whether one is currently in a long-term growth period, a value period, or in the middle of a rotation from one

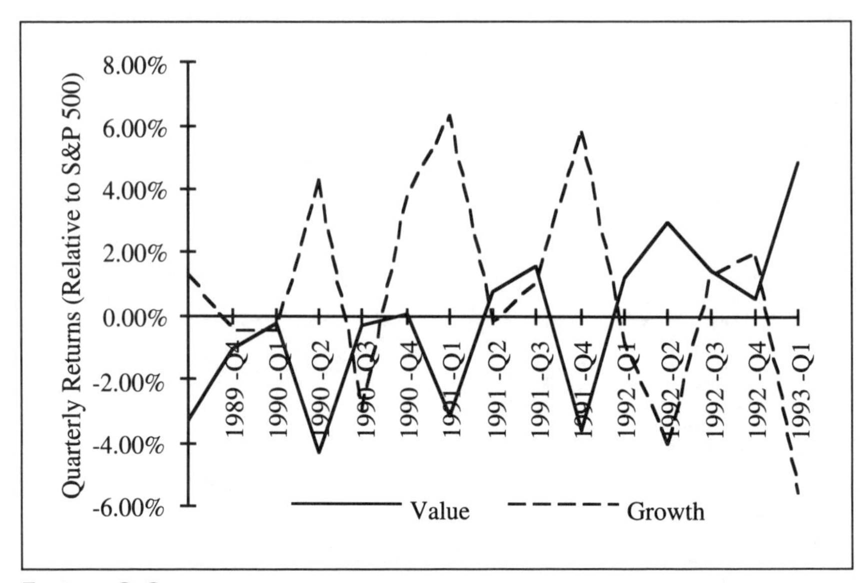

FIGURE 3.2
Quarterly returns for growth and value stocks relative to the S&P 500 index

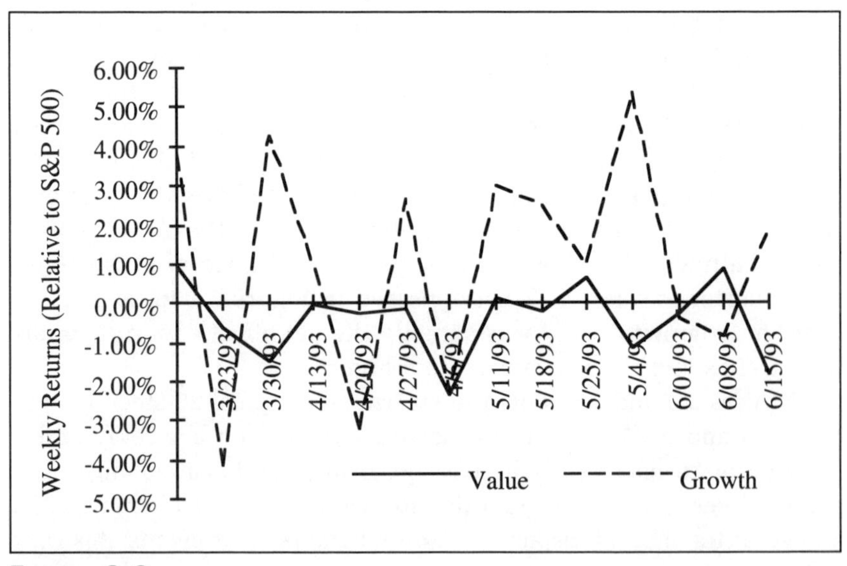

FIGURE 3.3
Weekly returns for growth and value stocks relative to the S&P 500 index

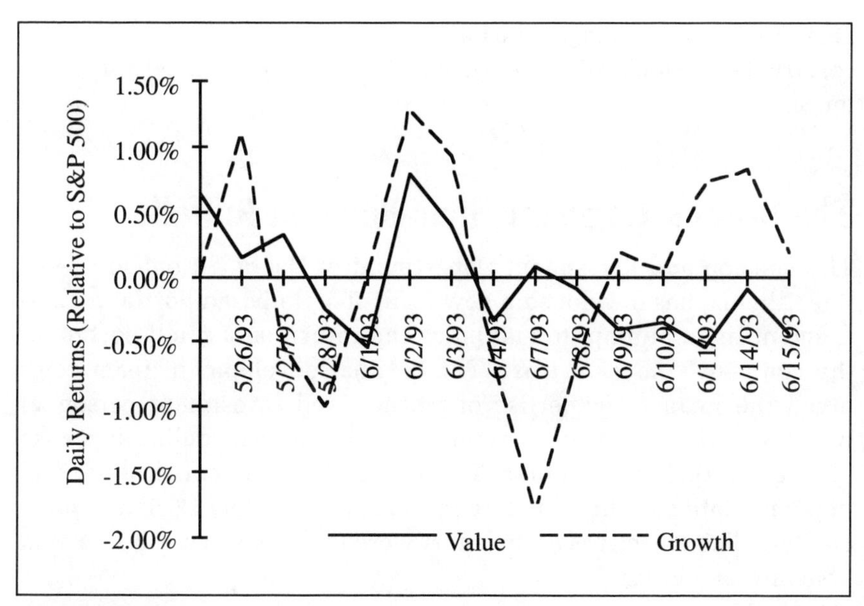

FIGURE 3.4
Daily returns for growth and value stocks relative to the S&P 500 index

style to the other. When it finally becomes clear that a rotation has been completed, most of the opportunity for realizing excess returns is past. For this reason, it is difficult to successfully implement a style-based portfolio.

A second reason for the inability of most style-based money managers to capture excess returns is that the concept of only two states, growth or value, is too simplistic. Figures 3.3 and 3.4 show that at times both growth and value portfolios will underperform the market average. Apparently other groups of stocks were outperforming the market during these periods—but what were their characteristics?

It is very difficult, perhaps impossible, to accurately predict the future. However, a reasonable guess is that tomorrow will likely follow the trends established over the past few days, that next week will follow the trends of last month, and so on. At a minimum, a successful strategy for style-based portfolio selection requires techniques for accurately identifying the characteristics of stocks that are currently outperforming the market. If these characteristics can be identified quickly and accurately, and if some short-term persistence exists in market trends, then it will be possible to capture significant returns above the market average. The following section explains how we have constructed a stock selection system that attempts to meet this challenge.

The Deere & Company Style-Rotation Portfolio

The Pension and Investment Department of Deere & Company (Moline, Illinois) has developed a new form of style-based portfolio management in an attempt to maximize the return on a small portion of the company's pension trust. The original development team combined the financial expertise of pension and investment personnel with the artificial intelligence and complex system modeling expertise of Deere's Technical Center. The initial development, testing, and implementation of the system required approximately 12 staff-months of effort. The objectives of the effort were to develop a stock selection system that would:

1. Identify the style (or common characteristics) of stocks with the greatest recent price increases, and apply this model to today's market to select a portfolio likely to outperform the S&P index in the short term.
2. Provide both buy and sell signals.
3. Automatically learn and continually adapt to changing market trends.

The various styles identified by the system are not limited to value or growth. The system contains no embedded rules or criteria for classifying stocks; instead the characteristics that define the best performing style are inferred directly from the data. This allows the system to discover the best style for the current market and provides the flexibility for it to adapt as market conditions change. We believe this self-learning capability will result in a system able to operate for long periods of time without extensive modifications.

A schematic representation of the process is shown in Figure 3.5. The universe from which stocks are selected is currently limited to the 1,000 largest U.S. corporations. Each week, the latest values of approximately 40 data items for each of these corporations are gathered and added to a historical database. The data items include technical as well as fundamental indicators. The raw inputs in the data-

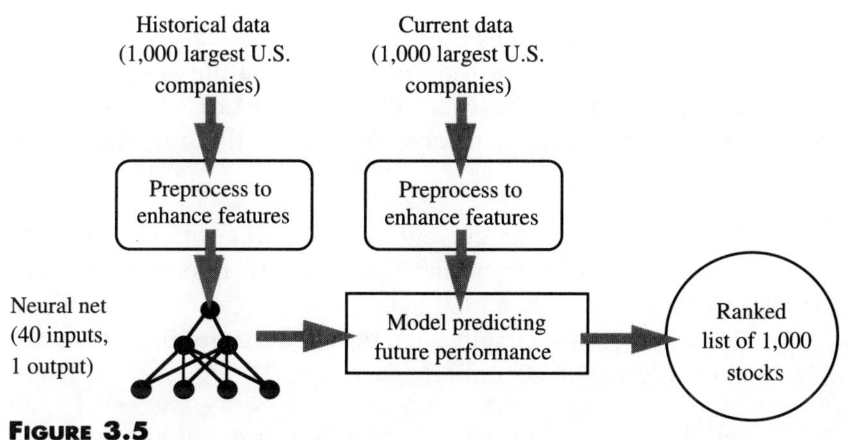

FIGURE 3.5
Schematic representation of the process used in the Deere & Company style-rotation portfolio

base are then preprocessed to yield new variables. This preprocessing enhances selected features while masking some of the noise inherent in the data. Typical preprocessing operators include relative differences and ratios. Only preprocessed variables are used as inputs to the modeling; no raw data are included.

The primary tool used to model the stock market is an artificial neural network. The network training set consists of several weeks of historical data for each of the 1,000 corporations in the database. Each data record in the set contains 37 preprocessed input variables and a single output. The output of the neural network is the predicted change in a stock's price (relative to the S&P average) for some future period. When the network is trained on this historical data, it learns the characteristics of stocks that have recently been outperforming the market. This model provides the criteria for selecting those stocks that are most likely, in the short term, to also outperform the market. In addition, it provides a basis for determining when the market is undergoing a major shift in preferred styles, so that appropriate action can be taken.

We have found that the standard back-propagation techniques for training neural networks do not work well for this problem. Consequently, we have developed a modified form of back-propagation that enhances the network predictive power and gives more stable results.

After training, the network is used to predict the future change in stock price for each of the 1,000 corporations. Current data on each corporation are transformed by the same preprocessing algorithms and presented to the trained network. The network output is a prediction of the future price change for each stock. The stocks are rank-ordered according to predicted performance, and the portfolio is rebalanced using a simple strategy:

1. Sell any stock in the current portfolio when its ranking falls below some predetermined threshold for a given number of weeks.
2. Replace any stocks sold with the highest ranking stocks not currently in the portfolio.

Each week the most recent data on the 1,000 corporations are added to the neural network training set, and the oldest set of training examples is discarded. Weekly retraining of the network using this evolving data set forces the system to continually and automatically adapt to changes in market conditions. The system contains no expert-type rules or embedded knowledge. All decisions concerning stock selection are inferred solely from market patterns contained in the training data. Careful controls on the number of weeks of historical data included in the training set yield a system capable of adapting to changing market trends without overreacting to the noise inherent in the market.

In December 1992, a $100 million stock portfolio was initiated using this technology. Some fine-tuning of the system parameters and the sell criteria has occurred over time, but the basic system remains as originally implemented. The portfolio contains 80 stocks, each equally weighted. To date, no attempt has been made to take long or short positions on any stock. Performance of the portfolio after all transaction costs has exceeded the S&P 500 performance since inception. Actual performance results and specific details of the system are proprietary.

The following sections describe the process, tools, and basic principles used in developing the system. These may prove useful to others attempting to build models of similarly complex systems.

Sizing Up the Problem

Once the true nature of a problem is understood, it is half solved. Based on our work, we believe the stock market to be an *evolutionary*, *complex, nonlinear dynamical system.* This section explains what we mean by these terms and our reasons for this assertion. Clearly understanding these characteristics of the stock selection problem will help us know where to look for solutions.

First, however, our overall approach to this problem should be clarified. The objective was to build a working system for internal use. We addressed the problem from an engineering perspective rather than as researchers. Where basic scientific principles were available, they were used as guides in solving the problem. However, a problem of this complexity involves many areas for which no theoretical foundations exist. When entering unknown territory, our main concern was finding something that worked and that seemed robust over time and under changing conditions. Once empirical evidence indicated that we were successful, we moved on to the next problem.

We feel that the aggregate approach is sound and provides a useful guide—both for practitioners attempting to model very complex systems and for researchers shoring up the underlying scientific foundations.

Nonlinear Dynamical Systems

We assert that modeling the stock market requires an *NLD (nonlinear dynamical)* system. As the name implies, the output of a nonlinear system is determined by nonlinear relationships among the input variables. Other typical characteristics of NLD systems are a strong sensitivity to changes in the initial conditions and self-similarity of the system output. *Self-similarity* means that the variability within the output of the system appears similar when viewed at widely different time grains (or time units). Several researchers have demonstrated that many financial markets exhibit such self-similarity. For example, Peters (1991) demonstrated that a time history of the S&P 500 index looks surprisingly similar whether plotted on a monthly, weekly, or daily basis.

Figures 3.1 through 3.4 plot the returns of growth and value stocks for yearly, quarterly, weekly, and daily time grains. The variability of these returns appears similar for all of the plots, even though the time grain varies by two orders of magnitude. This self-similarity of the system output provides some empirical evidence that building a stock selection model requires an NLD system.

The previous section outlined the basic approach used in the Deere & Company style-rotation portfolio. The core modeling tool is a neural network trained to discover the relationship between 37 different inputs and future changes in stock prices. After training on historical examples, the neural network was able to build a model that predicted the direction of future stock prices in a significant number of cases. When we attempted to apply linear regression techniques to these same training data sets, the resulting linear models had no predictive ability. Experiments such as this provide further empirical evidence that the relationships which predict a stock's performance tend to be nonlinear rather than linear.

In general, neural networks produce what are known as *black-box models*. Such models can accurately map the system inputs to the outputs, but the actual relationships formed cannot be expressed in terms that are comprehensible to humans. However, some hints about the input-output relationships contained within the network can be obtained by sensitivity analysis. To implement such an analysis, an example is selected from the data set and placed on the network inputs. Each input is varied by a fixed percentage, one at a time, while all others are held constant. The change in the system output attributable to each input gives some indication of the first-order, or linear, relationships within the model. Repeating the entire process for many different input examples and comparing the results reveals the existence of nonlinear relationships. More information about these second-order, or nonlinear, relationships can be obtained by conducting a sensitivity analysis while varying the inputs in a pairwise fashion. However, for complex models, the large number of pairwise combinations can make this procedure quite tedious.

Additional information about the nature of the models contained within neural networks can be gained by examining the connection weights linking the inputs and the nodes in the hidden layer of the neural network. When we examined those trained neural networks that showed some success in selecting stocks that would outperform the market, both a sensitivity analysis and an examination of the internal network connections indicated that the neural network had formed highly nonlinear models of the input parameters. Thus our assertion that stock selection requires an NLD system.

This conclusion agrees with those of other researchers, such as Jacobs and Levy (1989), whose research shows that the stock market is not an ordered system that can be explained by simple rules. Neither is it a totally random system for which no predictions are possible. They found the market to be a complex system, in which portions of the system's behavior could be explained and predicted by a set of complex relationships between many inputs.

Complex Systems

Next, we assert that selecting stocks that will outperform the market requires modeling a *complex* NLD system. Most of the research on NLD systems to date has focused on simple systems with only one or two inputs related in nonlinear ways. Such systems can exhibit very complicated behavior, but they are classified as simple NLD systems because of the small number of inputs and relationships that determine system behavior. Peters (1991) used fractal dimension analysis to estimate that only three inputs were required to predict the behavior of the S&P 500 index.

Analysis of the connection weights in our neural network models trained for stock selection indicates that generally four to eight inputs are required to estimate the future price of an individual stock. We do not claim to have found the optimum solution nor the minimum number of required inputs, but this result does give some indication as to the complexity of the problem. Problems involving many variables interacting in nonlinear ways are classed as complex NLD systems. Unfortunately, little research has been done on modeling such systems.

Evolutionary Complex Systems

Many complex NLD systems are stationary; that is, they have a stable existence in only one state. This state, or the relationships between the system inputs and outputs, may be very complex, but it never changes. Successfully modeling such systems is not a simple matter. Generally it requires skill in using a nonlinear modeling tool such as neural networks, expert knowledge to provide insights into possible relationships, and considerable trial and error.

Some complex NLD systems exhibit bistable behavior; that is, they can exist in two completely different states. In the first state, the output can be predicted by fixed relationships among the input variables. However, the system can also have a stable existence in a second state, in which a completely different set of relationships holds true. Modeling such a system requires two separate models—one for each state—as well as a gating model that identifies the current state and calls up the correct model. Since the inputs driving the system may be different for the two states, the combined model must include both sets of inputs, even though at times some of the inputs may be latent.

Some complex NLD systems are neither stationary nor bistable but exhibit *evolutionary* behavior. An evolutionary system switches from time to time between a large (perhaps infinite) number of stable states. At any point in time, the output of such a system can be explained by a nonlinear combination of the inputs. However, such relationships may be stable only for a short period before evolving to a completely different set of relationships. An evolutionary system may reenter previous states, or evolve into totally new states. Modeling such a system requires sensing when the system is moving into a new state, then quickly discovering the new relationships between system inputs and output. Since this happens repeatedly, the task can be quite challenging.

Stock selection requires modeling an evolutionary, complex NLD system. Our experience in building neural network models of the stock market for the period 1990–1993 indicates that the market generally has stable periods lasting 6 to 12 weeks in which the system relationships remain fairly constant. Within these stable periods, daily noise, or oscillations, occurs from time to time, but the system returns to the previous state following these perturbations. When leaving a stable state, the system generally oscillates for several weeks before settling into a new state. These major state changes are sometimes correlated with outside events or sometimes occur for no obvious reason. Due to the local noise and oscillations in the system, it is difficult to discriminate these major state changes in real time. The time constant of the modeling system must be adjusted so that it will adapt quickly when major state changes occur, but the response must be sufficiently damped so that the model does not overrespond to the accompanying local noise and oscillations.

Modeling evolutionary, complex NLD systems requires a technique for accurately identifying the occurrence of state changes and a method for quickly discovering the new system relationships following a state change. The remainder of this chapter outlines the techniques and principles that have proved useful in building such a system.

Guidelines for Modeling Evolutionary Complex Systems

Once the basic characteristics of the problem have been identified, appropriate tools and techniques from the realms of finance, physics, statistics, dynamic system modeling, and artificial intelligence can be applied. These techniques have been used to solve similar problems in other disciplines, and we feel that they constitute the best arsenal for attacking adaptive stock selection. This section outlines the tools and techniques used and discusses why they were selected. It is presented in the form of guidelines for modeling complex systems.

Guideline #1: Be Willing to Trade Short-Term Accuracy for Long-Term Performance

Modeling complex systems requires a special mindset. The relationships are so complex that it is generally not realistic to expect more than rough estimates from the models. However, for many systems, the payback from even rough estimates can be high. The data from complex systems tend to be very noisy. When these facts are forgotten, it is easy to fall into the trap of thinking that models which fit some representative set of data more precisely will also provide more accurate predictions.

In Figure 3.6, the eighth-order polynomial has high precision in fitting the data. However, if these data came from a complex system, one would expect them to be quite noisy. A wise modeler would know that a model with fewer degrees of freedom, such as the second-order polynomial in Figure 3.6, would be more likely to offer useful predictions about the underlying system. For simple systems we intuitively avoid overfitting the data by examining data plots and mentally estimating the amount of noise in the data. For complex systems with large numbers of parameters, it is not possible to graphically view the data or to mentally estimate the noise. This makes it easy to succumb to overfitting the data, in the belief that greater precision in modeling the data will certainly result in better predictions. *In general, for evolutionary, complex NLD systems, the accuracy of a model in explaining some local condition is inversely proportional to its usefulness in discovering and explaining future states.*

Putting this principle into practice means that useful models will generally have very few degrees of freedom. Caution should be used

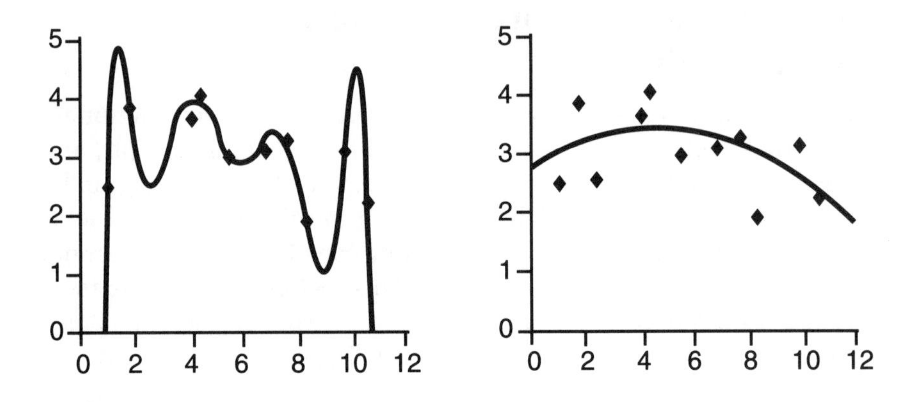

FIGURE 3.6
Noisy data modeled with an eighth-order polynomial (left) and a second-order polynomial (right)

before adding more modeling degrees of freedom to obtain incremental gains in apparent accuracy; often one is merely modeling system noise, and the predictive power of the model will suffer.

Models with a low degree of freedom require careful selection of system inputs. Proper preprocessing of the data can reduce the number of system degrees of freedom by enhancing the pertinent features within the data and reducing the number of model inputs. When done correctly, preprocessing can also remove some of the noise in the data without decreasing information content. Discovering the proper set of model inputs and the appropriate set of preprocessing algorithms requires considerable trial and error and usually consumes more than half of the total modeling effort. In this phase of the effort, expert knowledge of the problem can be especially helpful in providing hints as to which types of preprocessing might be appropriate.

It is critical to remember that with evolutionary systems the relevant inputs of the model will change over time. Although the final model must have few degrees of freedom, the model building tool must have access to a fairly exhaustive set of possible inputs as it attempts to discover the new relationships. This requirement for a large number of potential inputs, yet a model with few degrees of freedom, places severe constraints on the modeling technique. This leads us to the next guideline:

Guideline #2: Select the Proper Modeling Tool

It is critical that the modeling tool selected can meet the requirements of the problem. For a stock selection system, neural networks were selected as the primary modeling tool. Neural networks have demonstrated their ability to discover the underlying relationships in NLD systems in many previous applications. More importantly, *neural networks have unique properties in the way model degrees of freedom are used.*

In regression analysis, the modeling degrees of freedom are fixed by the regression model selected and can never be smaller than the number of system inputs. In contrast, the degree of freedom for neural network models is somewhat nebulous. The exact number of degrees of freedom in a neural network model cannot be determined in the traditional sense, but is related to the network architecture, the training technique, and the characteristics of the training data set. Hence, for properly constructed and trained neural networks, it is possible to build models that consider large numbers of possible inputs, but which have very few degrees of freedom. In this case, the model degrees of freedom within the neural network are assigned only to the most significant relationships. This characteristic of the neural network modeling tool is important to the stock selection problem. However, actually constructing such low-dimensional models requires skill and care. Simply throwing data into a neural network package is not enough.

When building models to predict the behavior of complex NLD systems, excessive degrees of freedom can also be introduced from sources other than the primary modeling tool. One should be cautious when incorporating human intervention or screening into the modeling system, especially for postprocessing the model output. Such systems have the potential for infinite degrees of freedom. We have kept careful records of each time we have performed a human override of our stock selection model. To date, the net result of all human overrides has been a decrease in system performance. Humans are not adept at discovering models of complex systems with many nonlinear relationships, and they tend to cling to models of what happened in the past rather than recognize the evolutionary nature of these systems.

For similar reasons, hybrid systems such as combined neural-network/expert systems should be implemented with caution. When not properly designed, such systems can also have high degrees of freedom. In addition, it is difficult to construct expert rules that recognize and adapt to the evolutionary nature of the problem.

Armed with the proper perspective and the correct modeling tools, one is ready to begin investigating some of the time-dependent characteristics of the system:

Guideline #3: Discover the Proper Predictive Period for the System

The evolutionary nature of the stock selection problem and the probability of uncontrollable external events make long-term predictions unreliable. The large amount of noise inherent in this type of system also makes most short-term predictions unreliable. However, there is some period into the future for which useful predictions can be made. *Discovering the proper predictive period is a critical task in modeling evolutionary, complex NLD systems.*

For simple NLD systems with only one or two inputs, the techniques of rescaled range analysis developed by Hurst (1951) can be useful in estimating the optimum predictive period. However, researchers have not yet shown how these techniques can be applied to systems with more inputs. One difficulty is that traditional rescaled range analysis requires data covering several complete cycles of the system. In evolutionary systems, each system cycle represents a different set of input/output relationships, so they cannot be used for rescaled range analysis. For problems such as stock selection, an estimate of the predictive period must be made by experimentation over a period of several system cycles.

The proper predictive period will be affected by many factors, including the accuracy of the model, the amount of noise in the data, and the underlying dynamics of the problem itself. It is also strongly influenced by the characteristics of the trading system used to implement the model's recommendations. Figure 3.7 illustrates the concept that the accuracy of the model's predictions is generally poor for both very short and very long periods into the future. Trading systems with short predictive periods will have more opportunities to trade than systems with long predictive periods. However, trades based on longer holding periods will generally have larger price movements—both losses and gains. For stock selection, the cost per trade is fairly constant, although trading systems that require very short predictive periods may incur a penalty because of the short time in which the

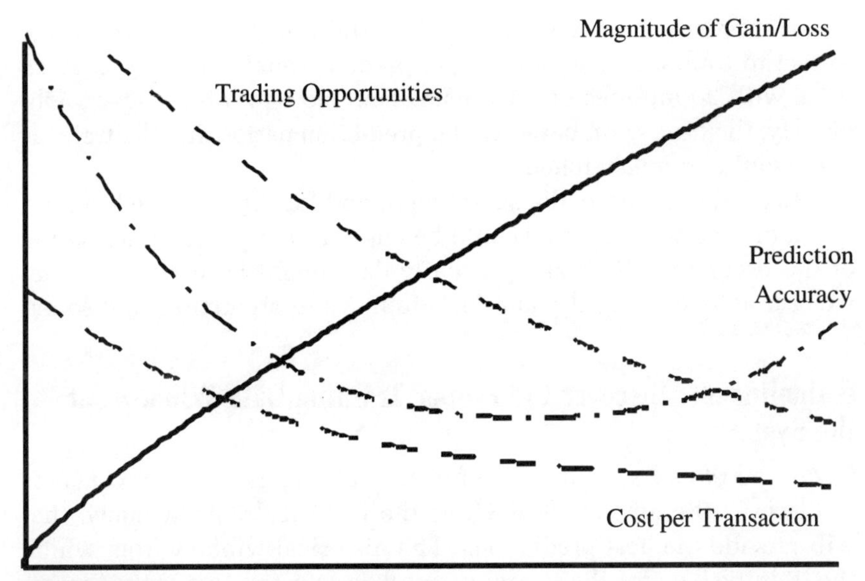

FIGURE 3.7
Factors affecting the proper predictive period for stock selection models

transaction must be completed. These and many other factors interact to determine the proper predictive period for the system model.

For relationships of this complexity, there is no true optimum. The best predictive period will be determined by all of the factors illustrated in Figure 3.7, as well as the investor's risk tolerance, market volatility, accuracy of the model, transaction costs, and amount of noise in the data. The only method for discovering the proper predictive period is through repeated simulation of the entire system, including simulated trading, over an extensive period of time. A historical database of the system inputs and an accurate model of trading costs are required for such a simulation.

Guideline #4: Select a Sampling Rate for the Training Data That Is Compatible with the Prediction Period

Once the proper predictive period has been estimated, data for training the neural network model must be selected. For effective models, the time grain, or the sampling rate, of the training data must be compatible with the time period of the model forecast. Given the desired predictive period, most individuals have an intuitive sense for an appropriate sampling rate for the training data. One would not expect to train a neural network to predict weekly changes in stock price with 30 minutes of price data, sampled one minute apart. Obviously, the time grain between the prediction period and the training data would be mismatched.

As a rule of thumb, the sampling period for the data used to train the neural network model should be one-tenth to one-half the period of the prediction. For example, daily data might be used for a one-week prediction, weekly data for a four-week prediction, and so on.

Guideline #5: Discover the Proper Training Data Window for the System

Given the proper sampling rate for the training data, the next task is to discover the amount of data, or the historical data window, that will provide the best predictions. The historical window from which the training data for the neural network is selected acts as the "memory" of the system. The neural network forms its model based only on the patterns contained within the training data. Typically, the length of the training window remains fixed as it slides through time. Each week the newest data are added to the training set and the oldest data discarded. Weekly retraining of the network with this evolving data set forces the system to continual and automatically adapt to changes in market conditions.

When the training data window is long, the system is better able to differentiate between noise and the true system response. *Thus, the accuracy of the model often improves as the training window is increased, but only if the current state of the system remains constant.*

When the system begins evolving to a new state, the training data forming the system memory should be modified. If the new state is similar to the previous state, some of the historical data might be useful in discovering the new system model. However, if the new state is significantly different from the old state, it may be best to completely clear the data memory and begin with new data.

The minimum length of the historical training window determines how the model will respond during these state changes. If the minimum training window is too long, the model will be slow to respond to state changes. Since much of the opportunity for outperforming the market occurs early in the state changes, the effectiveness of the total trading system will be low.

If the training window is too short, the model may overreact to noise. It may also overreact to the oscillations that normally occur during the state changes. In effect, the model becomes unstable during these periods. An investor who responds to an unstable model will be whipsawed, quickly switching back and forth between two types of portfolios at exactly the wrong time. Again, the effectiveness of the total trading system will be low.

Two strategies can be used in selecting appropriate training windows. One is to find the fixed-length window that provides the best system performance over an extended period of time. Another is to adapt the length of the training window to the current system state. The length of the window might be allowed to grow longer and longer provided that the system remains in a given state. When it is determined that a state change is occurring, the training window must be reset to the minimum size. Such a strategy requires some external indicator of when the system state changes occur.

Guideline #6: Improve Performance by Tuning the Total System

After the initial attempt to build a total trading system following the guidelines outlined above, some tuning may be desirable to improve system performance. Our experience indicates that the greatest performance improvements come from matching the elements of the total system. Tuning the length of the prediction window and the training data window can produce significant improvements, especially as experience is gained with live trading. Experimenting with various strategies for utilizing the predictions from the model in making buy and sell decisions is also beneficial. A third method for producing significant improvements is to experiment with additional techniques for preprocessing the input data.

In our experience, fine-tuning the neural network training process itself produced only minor improvements in total system performance. Attempts to modify the distribution of examples in the neural network training set, such as removing the extreme examples or removing a portion of the samples that occur with high frequency, did not improve system performance.

Summary

Many portfolio managers have developed systems for selecting stocks that they believe will outperform the market. However, all of these techniques rely on patterns or rules that were first discovered by humans, then embedded into the selection system, either as stock screens or as expert-system rules.

We believe that ours is the first autonomous stock selection system. Unlike previous methods, this system has no embedded rules or logic; its decisions are based entirely on patterns it discovers by itself. It is designed not only to discover current market trends but also to rediscover new trends as the market evolves to new states.

A critical task in the development of this system was understanding the basic nature of the stock market itself. Our work has led us to assert that the stock market is an evolutionary, complex, nonlinear dynamical system. Evolutionary refers to the concept that the rules determining which stocks will outperform the market do not remain fixed, but evolve over time. Complex refers to the fact that the rules for stock selection are not simple linear equations with one or two variables but instead have many variables. Nonlinear dynamical systems have complex, nonlinear relationships among the system variables, which makes the system output chaotic. Chaotic systems appear to be random, but in fact at least a portion of the system output can be predicted by complex models.

Once the basic characteristics of the problem had been identified, appropriate tools and techniques from finance, physics, statistics, dynamic system modeling, and artificial intelligence were applied. These techniques had been used to solve similar problems in other disciplines, and it was felt that they constituted the best arsenal for attacking adaptive stock selection. Using these techniques, it is possible to build crude models of the stock market as an evolutionary, complex NLD system. These models allow a portion of the rules that determine which stocks will outperform the market to be discovered.

Rather than describe the proprietary details of the Deere & Company stock selection system, this chapter described the tools and techniques that were used to build this stock market model. In the long term, these guidelines should prove useful both to practitioners who are attempting to build models of similarly complex systems, and to researchers who are attempting to develop a more scientific foundation for such models.

References

BARRA Newsletter. 1992 (May-June).

Hurst, H. E. 1951. "Long-term Storage of Reservoirs." *Transactions of the American Society of Civil Engineers.*

Jacobs, B., and K. Levy. 1989. "The Complexity of the Stock Market." *The Journal of Portfolia Management* 16.

Miller, M., and F. Modigliani. 1961. "Dividend Policy, Growth and the Scale Structure of Dividend Discount Returns." *Journal of Business* (Oct.).

Peters, E. 1991. *Chaos and Order in the Capital Markets*. New York: John Wiley & Sons.

Williams, J. B. 1936. *The Theory of Investment Value*. Cambridge: Harvard University Press.

Wilshire Asset Management. 1990. *Style Portfolios: A New Approach to Equity Investment Management.*

CHAPTER 4

Predicting the Tokyo Stock Market

Morio Yoda

Let the following be one of the unfailing rules by which the individual investor and, needless to say, the pension and other institutional-fund manager are guided: there is the possibility, even the likelihood, of self-approving and extravagantly error-prone behavior on the part of those closely associated with money.

John K. Galbraith
A Short History of Financial Euphoria:
Financial Genius Is before the Fall, 1990

A few years ago a paper was published on the application of neural net technology to forecasting of the Tokyo stock exchange price index (TOPIX). This system has now been operational for more than three and a half years. It has produced 145 predictions, with a correct prediction rate of 62.1% for both rising and falling markets. Morio Yoda, who developed this system, is a manager with the Research and Trading Division at Nikko Securities Company in Tokyo. He has also developed numerous mechanical systems. In this chapter, Yoda first explains his neural net system for forecasting the TOPIX and reviews the experience gained since the system was put into operation. Next, he offers a comparison of the performance of this system and of a trend-following mechanical system. The chapter concludes with a discussion of how tradable information can be captured from noisy price movements using rescaled range analysis and Hurst coefficients.

Can the Tokyo Stock Market Be Predicted?

Let's start by asking a simple question: Can the Tokyo stock market be predicted? There are two possible answers. A simple "no" is based on the assumption that people are perfectly rational and always act in a perfectly rational manner. If this is indeed true, then when news breaks anywhere in the world (and, given current information technology, instantaneously spreads to every market), everyone reacts alike and stocks are immediately priced to reflect that information. As such, although people are believed to be predictable, world events and markets are not.

An alternative answer is "maybe." This is based on the observation that people are not always rational and may act in different ways at different times. As a consequence, it will take some time for market prices to reflect all the available information. Thus, market predictability remains. Chapter 19 elaborates on alternative market hypotheses and demonstrates why markets are not random.

If the randomness of markets is not assumed a priori, then the randomness hypothesis can be tested for each market. The question then is: How can the predictability of the stock market be tested?

Suppose you are asked, "How much will the TOPIX go up or down in a week?" If you are unfamiliar with the Tokyo stock market, the correct answer may be, "I don't know." However, if you know that the average yearly return of the Tokyo stock market has been 11.8% a year for a long period of time, then your answer may be, "It will go up 0.23% in a week." Suppose that you also know that the yield of long-term government bonds is decreasing; then your answer may be, "It will go up 0.53% in a week." Prediction of a market is nothing more than a conditional expectation, influenced by your information about the market and inferences drawn from that information.

What affects the Tokyo stock market? We picked the following factors:

1. A vector curve consisting of regression coefficients over time of changes in the weekly Dow Jones index (DJI)
2. A moving average of the interest rate of the long-term Japanese Government Bond (JGB)
3. A vector curve for the JGB
4. A technical indicator called the IT radar, developed by Nikko.

We normalized each factor between 0 and 1 and calculated conditional weekly returns of the future TOPIX for all deciles of each of these four factors.

Table 4.1 shows the mean of annualized weekly future returns (%) (i.e., the TOPIX of next week divided by the TOPIX of the current week minus 1), the annualized standard deviation (%), and the number of observations for each decile of each factor. For example, the first set of numbers shows that the average annualized weekly return of the TOPIX in a week, after the DJI vector curve was between 0 and 0.1 was 26.83%; that the standard deviation on this return was 9.06%, and that this was derived from 14 cases. The first decile (from 0 to 0.1) indicates the largest decline in the DJI, the lowest level of JGB long-term interest rate, its quickest fall, and the lowest level of IT radar. Based on this table, we can say that if the DJI goes up in a week, then the TOPIX will also go up the next week, and that if the JGB interest rate is low or falls sharply or if IT radar remains low in a week then the TOPIX will likely go up the next week. This may sound like nothing new; the important thing is that there seems to be some predictability in the Tokyo stock market which can be detected by a simple statistical approach.

What we want to do is to develop a system to provide precise predictions in as many cases as possible. For example, what will happen to the TOPIX next week if last week the DJI remained almost unchanged, the level of interest rates stayed low or went up slightly, the IT radar was at an intermediate level, and the yen/dollar rate got stronger?

We used a neural network for this problem because networks provide a flexible way of learning complicated relationships between input data and the "best" predictions.

The Neural Net Prediction System

Figure 4.1 shows the basic architecture of our prediction system. The system consists of several neural network modules that used historical data to learn the relationships between various technical and economic indexes and the timing for when to buy and sell on the TOPIX. Each module predicts the coming week's returns for the TOPIX from current technical and economic indexes. The output of the system is an arithmetic average of the outputs of the modules.

TABLE 4.1 *TOPIX Average Returns and Standard Deviations with Respect to a Single Factor (January 1983 to August 1990)*

	Factors[a]			
Range	DJI vector curve	JGB long-rate moving average	JGB long-rate vector curve	IT radar moving average
0–0.1	*26.83	−26.57	51.93	0.00
	9.06	10.14	6.33	0.00
	14	8	16	0
0.1–0.2	−106.93	31.45	48.44	41.47
	0.00	5.50	9.10	6.15
	1	64	30	65
0.2–0.3	−43.47	25.47	35.19	10.39
	10.26	6.46	6.03	7.53
	12	77	32	46
0.3–0.4	1.74	29.76	−2.44	3.14
	8.95	7.92	6.24	6.57
	40	40	31	20
0.4–0.5	20.85	−34.16	9.14	−0.70
	6.95	14.17	11.14	10.48
	51	17	18	14
0.5–0.6	4.83	−5.17	14.92	−5.24
	7.99	8.04	6.71	9.00
	62	30	34	12
0.6–0.7	18.49	−26.49	18.08	4.59
	6.96	10.97	6.28	7.25
	27	38	21	18
0.7–0.8	14.89	12.85	17.11	−0.06
	8.53	6.88	6.56	7.81
	48	21	53	25
0.8–0.9	17.11	0.00	−16.21	11.23
	9.93	0.00	9.17	8.34
	30	0	33	68
0.9–1.0	57.20	0.00	−47.34	−19.61
	10.52	0.00	10.88	13.76
	10	0	27	27

TOPIX annualized return 11.80%
TOPIX annualized standard deviation 8.63%

(a) Each group of three numbers represents the annualized return, %; the annualized standard deviation, %; and the size of the sample.

Figure 4.2 shows the basic network model used for the prediction system. The network is hierarchical, consisting of three layers: the input layer, a hidden layer, and the output layer. Each unit in the network receives input from the lowest-level units and computes a weighted sum, which is transformed by a sigmoidal function to produce an output. The output is an analog value from 0 to 1.

The error back-propagation method proposed by Rumelhart et al. (1986) is the typical learning method for hierarchical networks. For

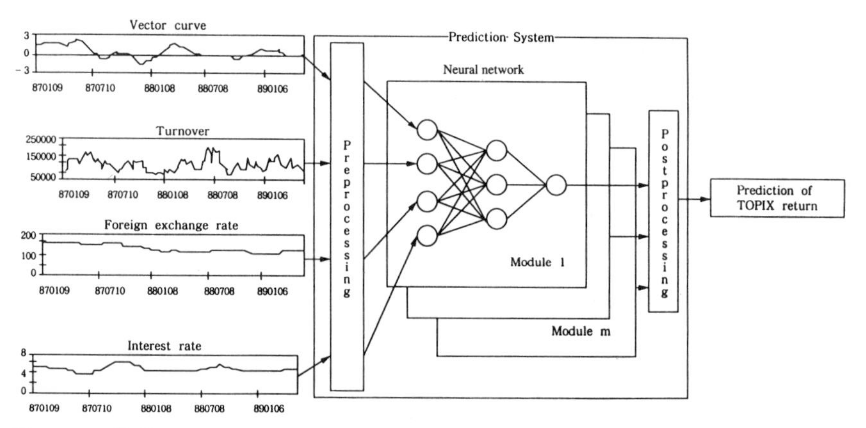

FIGURE 4.1
Basic architecture of the prediction system

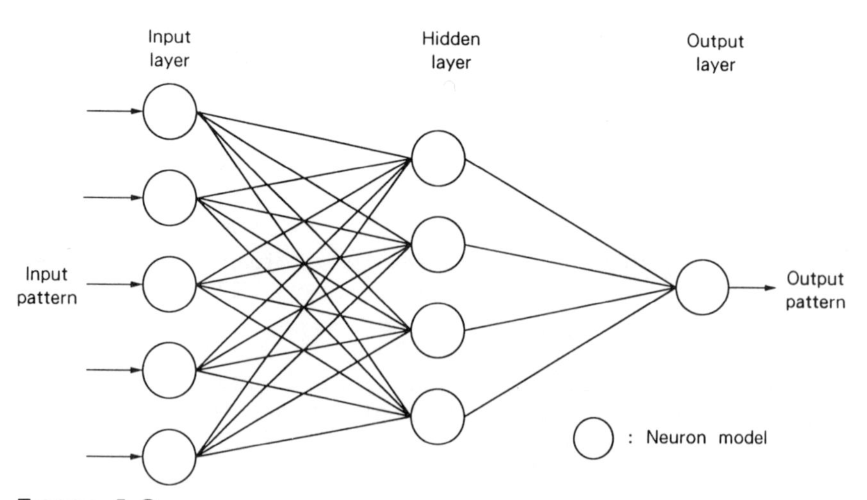

FIGURE 4.2
Neural network model

high-speed learning with large volumes of data, we employ a new high-speed method called *supplementary learning* (Masuoka et al. 1989).[1] Supplementary learning allows the automatic change of learning constants, depending on the amount of learning data. As the amount of learning data changes and learning progresses, the learning constants are automatically updated, thus eliminating the need for changing learning parameters.

The input data for the system assume that stock prices are determined by both macroeconomic and technical data, such as vector curves of the TOPIX, volume, interest rates, foreign exchange rates, and the DJI. These data are chosen so as to cover as large a part of the input space as possible, but not too large. The input variables are then transformed to time-series indexes, such as lags, moving averages, and relative strength indicators. Development of proprietary transformations is highly recommended because it enables effective condensation of sparsely distributed information and development of a unique network. The prediction system uses weekly average data of each index to reduce noise in the data. Each input index is converted into three values: a moving average, a regression coefficient in time, and the difference between current index value and moving average—representing, respectively, level, trend, and relative level indexes. Thus, the time sequence of each index at a particular point in time is replaced with three scalar values. Table 4.2 shows some of the input used. Converted indexes often have either bell-shaped or positively or negatively skewed distributions. These distributions are reshaped as close as possible to a uniform distribution with a range from 0 to 1.

TABLE 4.2 *Selected Input Data*

	Converted value[a]		
Input data	**Level**	**Trend**	**Relative level**
Stock price	NU	U	U
Volume	NU	U	U
Long-term interest rate	U	U	U
Yen/dollar exchange rate	NU	U	U
Volume percentage of 10 most heavily traded stocks	U	U	U
DJI	NU	U	U

(a) U, used; NU, not used

The output of the system is the weighted average of weekly returns of the TOPIX, which reduces noise in the return data. We try to predict a longer future. For example, if we average four weekly returns with equal weights, this is equivalent to a monthly return. The optimal number to use for averaging is therefore equal to the most predictable horizon into the future.

If economic time series or stock indexes are used, a large amount of noise and a small amount of information will result. Since the amplitude of noise is proportional to the square root of time, the signal-to-noise ratio decreases in proportion to the square root of t as t approaches zero. Thus, prediction of the very near future is difficult. On the other hand, trying to forecast far into the future does not provide many clues with which to make inferences for the long term. Therefore, there must be an optimal forecasting horizon where the signal-to-noise ratio is not too small and some information exists in the series with respect to the immediate future.

However, if there is reason to believe that certain economic time series are chaotic, then the situation is completely different. Under such conditions there is a deterministic nonlinear mechanism and, by definition, predictability. When a Lynapunov coefficient is greater than 1, its trajectory is very sensitive to initial conditions. Then, only short-term forecasting is possible. In this case, the shorter the prediction horizon, the more reliable the predictions.

When we developed our prediction system, we regarded the TOPIX as information plus noise and examined forecasting horizons of 8, 10, and 12 weeks. Back-testing showed an average of 10 weeks to be most predictable. In actual operation, an average of 8 weeks has thus far been most predictable. We will discuss later whether short-term trading is possible in certain markets.

Testing and Evaluation of the System

The proposed system was evaluated using adaptive prediction with iterative learning and prediction periods. As shown in Figure 4.3, M months before a given point in time are learned to predict the coming L months. This process is iterated to move with time. For evaluation purposes, a concatenation of prediction periods L months apart are considered a prediction result. Table 4.3 lists the correlation coefficients between teaching data and prediction results. The correlation

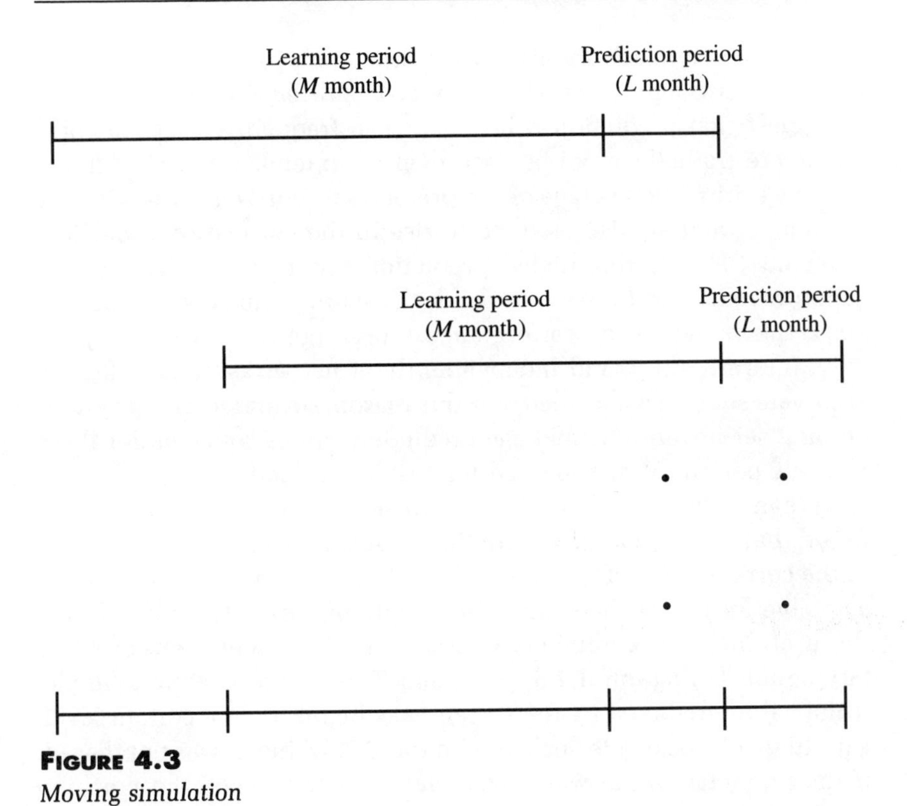

FIGURE 4.3
Moving simulation

TABLE 4.3 *Correlation Coefficients between Teaching Data and Prediction Results*

L	M \ N	8	10	12	Average
1	6	0.442	0.421	0.416	0.435
	12	0.417	0.483	0.446	0.458
	18	0.341	0.386	0.449	0.414
	24	0.415	0.473	0.458	0.457
	avg.	0.498	0.533	0.529	0.528
3	6	0.287	0.236	−0.119	0.175
	12	0.184	0.215	−0.303	0.043
	18	0.080	0.065	0.019	0.060
	24	0.119	0.078	0.047	0.063
	avg.	0.207	0.185	−0.124	0.106

L, prediction period in months; *M*, learning period in months; *N*, number of weekly returns to be averaged coefficients are relatively high when prediction period *L* is set to one month, but are significantly lower for an *L* of three months. This seems to suggest that prediction rules change so frequently that networks which are trained cannot be used over an extended period of time.

The arithmetic average of the predictions, derived by varying the learning period *M*, also produces a rise in the correlation coefficient by about 10%, offering higher prediction accuracy. Further investigation reveals that for each network there are periods when the network shows very poor performance. These periods arise unexpectedly, intermittently, and independently of networks. Every effort to eliminate such periods failed. For this reason, several learning periods are first set up, and the average prediction values collected for these learning periods *M* are adapted for system prediction.

When training the neural net on in-sample or training data, it was found that the system had a prediction accuracy of only 0.5 in terms of the correlation coefficient and that a trained network would not be available for longer than about one month (Figure 4.4). To determine the usefulness of predictions with a correlation coefficient of 0.5, a buying and selling simulation was done. The method used was simple: if the value predicted by the system was higher than a certain level, a portfolio of stocks interlocked with the TOPIX index was purchased; if the predicted value was lower than a certain level, the portfolio was sold. In the upper box in Figure 4.4, the dashed line shows the TOPIX index and the solid line gives the buying and selling perfor-

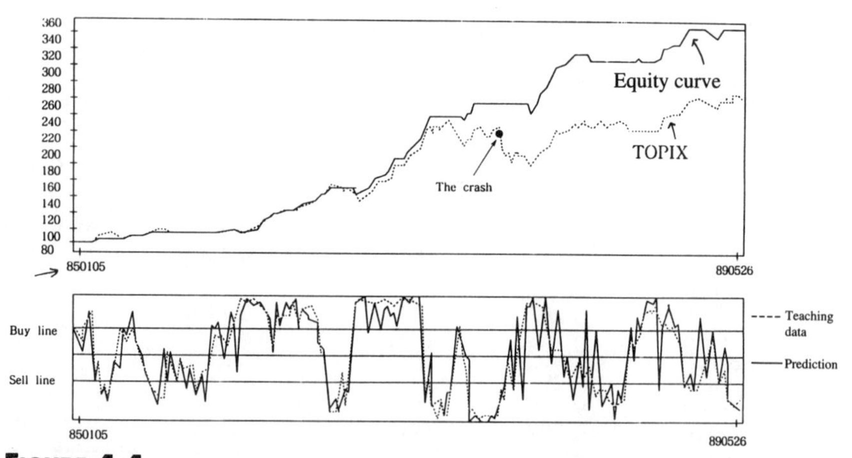

FIGURE 4.4
In-sample trading simulation

mance. In the lower box the dashed line shows teaching data, while the solid line represents system output. Stocks were purchased when the system output rose above the buy line; they were sold when the system output fell below the sell line. The simulation period was from 5 January 1985 to 26 May 1989. Surviving the sharp decline in October 1987, the buying and selling performance was substantially above the TOPIX without significant drawdowns. The system was quite effective on the training data. If it had maintained this level of performance, the result would have been excellent. These findings support the system's usefulness for trading.

Placed in actual service starting in September 1989, the system has yielded 145 predicted values, with a correlation coefficient of 0.293 (R^2 of 8.6%) and a correct prediction rate of 62.1% for rising and falling stock prices. The R^2 swings from 8% to 10% and does not often exceed 10%.

A hedging strategy was coupled with the system predictions for actual trading. Fund managers follow the TOPIX index. They sell futures if they have reason to expect a fall in the market, or buy back the futures if they expect a rise in the market. Because of their liquidity, Nikkei stock futures were used for trading. The results were a per annum return of 1.60% and a per annum standard deviation of 11.14%, far above the per annum return of −20.63% and standard deviation of 23.29% from the TOPIX buy-and-hold strategy for the period from September 1989 to October 1992.

Figure 4.5 shows part of the track record of the actual experiments attributable to the neural network. On the whole, stock prices have been declining throughout this period, in contrast to the strong upward trend when this system was developed. The system recommended sales during the first fall of the market beginning in December 1989 and subsequently avoided losses triggered by the Iraqi invasion of Kuwait in August 1990. However, a considerable loss was incurred because the system failed to recommend sales in a timely manner. During the third fall of the market, which occurred in March 1991, the decline in the absolute return was slower than that in the TOPIX index.

The first fall was generally attributed to internal factors in the Japanese economy that resulted in rising interest rates. The prediction system successfully managed to predict this fall. However, the system was unable to learn and manage the second fall because it stemmed from rising military tension in the Middle East, a factor not easily

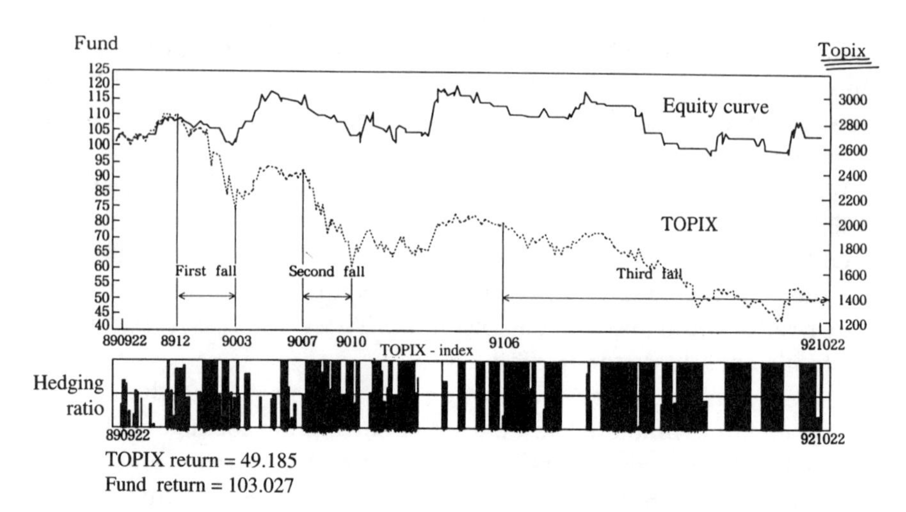

FIGURE 4.5
Out-sample trading

captured numerically. The third fall was a result of the economic slowdown coming to a head. Although economic in nature, such a factor is psychological as well and does not necessarily assume an evident, instant numeric form. The prediction system kept track of falling stock prices in this phase, which resulted in a loss, but managed to direct buybacks of the futures during the subsequent rise. These findings suggest that the system is fairly useful for predicting changes in stock prices associated with factors readily adaptable to numeric representation, but it is not so useful for predicting changes caused by other factors.

Comparison of a Neural versus a Mechanical System

We also developed a mechanical trend-following system using crossover of moving averages. It is well known that crossover techniques work well when markets exhibit trends but work poorly otherwise. So we added some rules to avoid being whipped when the market is choppy.

Compared with the mechanical system, the neural system performed better in the first and third falls and worse in the second fall. Overall performance was close, with an annual return and standard deviation of 1.60% and 11.14% for the neural system and 4.69% and 10.17% for the mechanical system. Since neural nets require market and macroeconomic data that must be carefully selected, as well as sophisticated neural software, these results are rather discouraging.

One important difference between the neural system and the mechanical system is the frequency of trading. There are two issues. One is the total number of trades in a certain period and the other is the average number of trades. The neural system generated a total trade volume of 7,664% with 230 trades, while the mechanical system generated a total volume of 16,800% with 168 trades. Average volume per trade was 33% for the former and 100% for the latter. Thus, trading with a neural system permits position control, whereas trading with the mechanical system is more like speculation. Since the equity curves of both systems are similar, the total number of trades may not seem important. However, some investors are uneasy with highly speculative trading. Therefore, if the performance is roughly the same, the smaller the amount of trading the better!

Trading volume has a substantial impact on the market. The smaller the number of trades, the smaller the market impact. The number of trades is even more important than total volume. Our mechanical system has only two states: full hedge or no hedge. We tried a mechanical system that allowed fractional hedges like the neural system, but the performance was significantly worse. These two factors incorporating macroeconomic data as inputs and fractional infrequent hedges, provide an effective marketing tool. We found clients for the neural system but not for the mechanical system.

Trading in Noisy Markets

Market signals are often considered the sum of drift and stochastic components. Drift components are sometimes regarded as market trends, which trend-following systems can capture. Absence of trend means choppy markets, in which trend-following systems lose out. Usually stochastic components dominate in markets, and clear trends often do not appear. Worse, markets tend to shift from trendy to choppy as they mature; very infrequently do they change from choppy to trendy. Many financial markets that offer substantial liquidity have already become choppy. For example, the S&P 500, U.S. Treasury bonds, FT 100, and sterling pounds are extremely choppy. Japanese Government Bond futures, Nikkei 225 futures, and Japanese yen are still trendy but are becoming choppy (especially Nikkei 225 futures).

Both of our models, neural and mechanical, are trend-following in nature as developed and applied to the Tokyo stock market. Tests of our trend-following mechanical system applied to choppy U.S. markets such as S&P 500 futures have produced very poor results. Neural nets may have some usefulness in choppy markets because they are less vulnerable. Since choppy markets are becoming more dominant, trend-following techniques to catch drift components in markets may decrease in their effectiveness.

Therefore, more attention should probably be paid to stochastic components. By definition, stochastic components have no tradable information. If one regards noisy movements in markets as stochastic, there is no chance to predict them. But are they really stochastic? Markets that appear noisy at first sight are not necessarily stochastic. Nonlinear dynamics provides another way to look at noisy markets—namely as being chaotic. If noisy markets are chaotic, then tradable information exists in them. Correlation dimensions and Hurst coefficients are commonly employed tests to check for nonlinearity in data. These techniques are explained in detail in Part Four.

Summary

This chapter has discussed a neural net-based prediction system for the Tokyo stock market. This neural network predicts the timing of buy-and-sell signals for stocks on the TOPIX. The simulations and actual experiments showed that the prediction system's timing of buying and selling of stocks was fairly good. The internal representation of the neural network was discussed. Our neural net system allowed us to forecast the TOPIX with an R^2 value of less than 0.1. Because of this low value, the forecasting power of the neural net succeeded in outperforming the index in the period from 22 September 1989 to 22 October 1992. Although having failed to outperform a mechanical trend-following system, the neural net system has achieved almost the same result with less frequent trading and fractional hedging. This turned out to be important in terms of marketing the neural net based approach to trading the Tokyo stock market.

References

Masuoka, R., et al. 1989. "A Story on Supplementary Learning Algorithm in Back Propagation." *JSAI.*

Rumelhart, D.E., et al. 1986. *Parallel Distributed Processing.* Vol. 1. London: The MIT Press.

Note

1. This supplementary learning method was developed by FUJITSU. It automatically controls learning iterations by referring to test data errors, thereby preventing overlearning of neural nets. This learning control method uses two-thirds of data in the learning period and uses the remainder as test data in the prediction system. The test data are evaluation data for which only forward processing is done during learning, and thus error is calculated but is not back-propagated. Our learning control method is done in two steps. In the first step, learning is done for 5,000 iterations, and errors against test data are recorded. In the second step, the number of learning iterations where learning had a minimum error against the test data is determined, and then relearning is done for that number of iterations. This prevents overlearning and results in a prediction model involving a moderate number of learning cycles. In the second step, learning is done for at least 1,000 iterations.

CHAPTER 5

Intelligent Trading of an Emerging Market

Gia-Shuh Jang and Feipei Lai

Today the key question for pension sponsors and fund mangers is no longer whether and why to invest in the emerging stock markets but; rather, whether they fall short in their fiduciary duty if they fail to recognize the importance of investing a portion of their assets in these markets.

Antoine W. van Agtmael
The World's Emerging Stock Markets, *1993*

In Chapter 4, Morio Yoda focused on forecasting the Japanese stock market. The majority of the published work on neural net applications in finance is on forecasting market signals. In this chapter, Gia-Shuh Jang and Feipei Lai, who worked together at the National Taiwan University for several years, present one of the most carefully researched approaches to stock market predictions. A dual adaptive neural net that can predict the short-term trends of price movements, as well as recognize reversals, is utilized to develop an intelligent and profitable stock trading system for the Taiwan stock market. The authors discuss at length the transformations used to identify both retrospective and predictive features from raw data. Their portfolio of dual modules of neural nets generalizes better than a single module. The annual rates of return achieved by this system during the test period in 1990 and 1991 are superior to the performance of closed-end funds that are focused on Taiwan and managed by human experts.[1]

Problem Statement

Most stock trading decision processes primarily involve weighing evidence gained from various observations of the market and its environment. For example, a price trend prediction could be made by weighing evidence obtained through analyses of monetary, political, and economic fundamentals, technical indicators, psychological factors, current events, and so on. However, making trading decisions based solely on the weighing of a large number of factors by humans may reduce the quality of decision-making (Felsen 1975). With the aid of neural modeling techniques, we can synthesize trading decisions from much more information than the human mind can handle (Hawley et al. 1990). Decisions derived from those neural trading systems may be superior to the judgment of human analysts.

The volatility of the Taiwan stock market[2] is illustrated by comparing its market statistics with those of other major stock markets (Tables 5.1 to 5.3). The highly speculative and volatile nature of the Taiwan stock market makes it a challenging target to model using neural networks (Jang et al. 1991a, 1991b, 1993).

Recent developments in neural network theory have proven that multilayer feed-forward neural networks with a sufficient number of neurons in one hidden layer can be used to approximate any multidimensional function to any specified degree of accuracy, if correct interconnection weights can be found (Cybento 1989). Back-propagation algorithms are often used to determine the appropriate interconnection weights (Rumelhart et al. 1986). However, back-propagation algorithms that adjust the interconnection weights of fixed-structure neural networks suffer from several problems. First, the learning process sometimes gets trapped in a local minimum point of the error surface and the network cannot produce the required accuracy. Second, the choice of an optimum network structure has remained an art. For a given problem, one usually has no a priori information about the number of hidden units needed, so the structure of the neural network must be determined by trial and error. Third, even when a feasible structure of a neural network can be found in the beginning, if the characteristics of the problem change later, the original network structure may be unable to represent the underlying mapping between the input and output vectors and thus be unable to permit the desired accuracy to be achieved through adaptation of interconnection weights.

TABLE 5.1 *Comparison of the Stock Indices of Major Stock Markets*

Year	Stock indices				
	Taiwan	New York	Tokyo	Korea	London
1987	2,135	1,939	23,176	418	1,373
1988	5,202	2,106	27,001	693	1,455
1989	8,616	2,753	34,043	919	1,917
1990	6,775	2,634	29,475	747	1,674
1991	4,929	3,223	24,298	658	1,892

TABLE 5.2 *Comparison of the Turnover Rate of Major Stock Markets*

Year	Turnover rate, %				
	Taiwan	New York	Tokyo	Korea	London
1987	267	73	96	130	142
1988	333	55	98	154	82
1989	590	52	73	112	77
1990	506	45	38	67	70
1991	322	47	28	82	67

TABLE 5.3 *Comparison of the P/E Ratio of Major Stock Markets*

Year	P/E ratio				
	Taiwan	New York	Tokyo	Korea	London
1987	22.1	15.9	58.3	11.4	13.9
1988	26.6	11.2	58.4	11.6	10.4
1989	55.91	11.7	70.6	14.4	11.3
1990	31.11	12.2	39.8	13.3	10.3
1991	32.1	15.9	37.8	11.3	14.7

Stock Trading Using Adaptive-Structure Neural Networks

Stock prices are often affected by ephemeral influences, including seasonal tax-motivated trading, singular international events, market psychology, and the "madness of crowds." Also, the stock market is not stationary; its structure changes as regulatory, tax, and trade policies change (Elder and Finn 1991). With these complications, the prediction accuracy of a stock trading model is affected by the model's tolerance to ephemeral events influencing the market, its generalization ability, and its capability to continue learning when the market changes (Jang and Lai 1993).

The problems addressed above have led us to further research. We have examined the effectiveness of applying learning algorithms with structure-level adaptation ability to the problems of stock market prediction using neural networks.

In this chapter, we are interested in the generation of stock market predictions utilizing dual adaptive-structure neural networks (DAS net). In the following section, we present the stock data model used in our system. Next, we focus on how stock market prediction is done by the proposed dual adaptive-structure neural networks. We conclude with a discussion of performance evaluation.

Stock Data Modeling

A typical free market tries to establish an equilibrium between buying and selling forces. This dynamic mechanism of trading has inspired traders to predict the future trends of the market. Two kinds of analytical approaches are common (Murphy 1986):

1. *Fundamental analysis.* Forecasting is based on macroeconomic data, such as exports and imports, money supply, interest rates, foreign exchange rates, inflationary rates, and unemployment figures, and the basic financial status of companies.
2. *Technical analysis.* Forecasting is based on the rationale that history will repeat itself and that the correlation between price and volume reveals market behavior. Prediction is made by exploiting implications hidden in past trading activities, by analyzing patterns and trends shown in price and volume charts. This micro-level scrutiny does not consider any external factors, such as news about wars in the Middle East.

We have established the stock data model from the technical analyst's point of view. Two arguments support this technical approach. First, short-term trends of price movement, the subject of the predictive model, are considered to be dependent on the difference between the buying and selling forces of the market. Although the efficient market hypothesis holds that prices fluctuate randomly about their intrinsic value, and that the best market strategy to follow would be a simple "buy and hold" strategy as opposed to any attempt to "beat the market," the fact that the intrinsic value of a stock is not fixed forever encourages us to find a model that can indeed "beat the market." We believe that once the intrinsic value of a stock shifts due to changes in macroeconomic factors, the only way for the market to compensate for the discrepancy between the market price and the new intrinsic value of that stock is either to pull up or to push down the market price. This is reflected on the price and volume charts significantly enough for the neural networks to build a computational model that can correlate the short-term trends of price movement with the retrospective technical indicators.

Second, extensive computation and a long training period are needed for an expert to build the technical views of the market. Even well-trained investors cannot easily predict the impending market movement from several charts and figures for one stock, not to mention the hundreds or even thousands of different stocks listed on the stock market. Therefore, if a neural network model could be trained to simulate the experience-based intuition of a successful technician, it could substantially increase the number of stocks that could be analyzed in real time (Hawley et al. 1990).

The input vector of the DAS net consists of technical indices preferred by human experts. The output vector, on the other hand, models the predictive short-term trends of price movement of the chosen stock. These retrospective input vectors and predictive output vectors are discussed next.

Retrospective Feature Extraction

A technical view of a stock for the nth trading day can be defined as a 4-tuple $\mathbf{S}_n = <H_n, L_n, C_n, V_n>$, where n is the index of the nth trading day; H_n, L_n, and C_n are the highest, the lowest, and the closing prices, respectively, of the nth trading day; and V_n is the daily trading volume for the nth trading day. Let

$$\mathbf{T}_{n,\alpha} = \{\mathbf{S}_i \mid i = n, n-1, \ldots, n-\alpha+1\}$$

where α is the number of daily data $\mathbf{S}_n$ in time series $\mathbf{T}_{n,\alpha}$. Although $\mathbf{T}_{n,\alpha}$ is usually viewed as the retrospective characteristic of a stock for the latest α trading days, this simple model has two shortcomings that make it unsuitable for use directly as input data for a neural network. First, if $\mathbf{T}_{n,\alpha}$ is used as input data for a neural network, then there should be no less than 4α neurons in the input layer of the neural network. For a large α, the computation time for training the neural network is lengthened by the large numbers of neurons in the neural network. Second, $\mathbf{T}_{n,\alpha}$ represents retrospective trading information of the latest α trading days. It has no memory of trading information out of the scope of α trading days.

To extract temporal information from $\mathbf{T}_{n,\alpha}$, a transformation is used to figure rates of price changes for consecutive trading days. We have developed a transformation F: $\mathbf{T}_{n,\alpha} \rightarrow \mathbf{R}_n$ to extract retrospective features from the time series data set $\mathbf{T}_{n,\alpha}$ to form a simple 16-tuple vector $\mathbf{R}_n$. The 16-tuple vector $\mathbf{R}_n = (r_n^0, r_n^1, \ldots, r_n^{15})^T$ is composed of 16 technical indices chosen a priori by analysts. Keeping the retrospective characteristics of stock trading information in the 16 dimension vector $\mathbf{R}_n$, the number of neurons in the input layer of the neural network is reduced from 4α to 16 by the transformation F: $\mathbf{T}_{n,\alpha} \rightarrow \mathbf{R}_n$.

Four equations are used as atomic operations to generate 16 technical indicators of the 16-tuple vector $\mathbf{R}_n = (r_n^0, r_n^1, \ldots, r_n^{15})^T$. The general infinite impulse response tranformations are defined in Eqs. 5.1 to 5.3. Equation 5.4 denotes the relative rate of change for X_n between k consecutive trading days.

$$MA_k(X_n) = \frac{1}{k}(X_n) + \frac{k-1}{k}MA_k(X_{n-1}) \tag{5.1}$$

$$BIAS_k(X_n) = \frac{X_n - MA_k(X_n)}{MA_k(X_n)} \tag{5.2}$$

$$OSC_{j,k}(X_n) = \frac{MA_j(X_n) - MA_k(X_n)}{MA_k(X_n)} \tag{5.3}$$

$$ROC_k(X_n) = \frac{X_n - X_{n-k}}{X_n} \tag{5.4}$$

Some indicators used to generate the 16-tuple vector $\mathbf{R}_n$ are listed below:

$$K_n^k = \frac{C_n - MIN_{i=n-k-1}^{n}(L_i)}{MAX_{i=n-k-1}^{n}(H_i) - MIN_{i=n-k-1}^{n}(L_i)} \tag{5.5}$$

$$D_n^k = MA_3(K_n^k) \tag{5.6}$$

$$FR_n = \frac{H_n - L_n}{C_{n-1}} \tag{5.7}$$

$$RSI_n^k = \frac{\sum_{\substack{i=n-k-1,\\ C_i > C_{i-1}}}^{n} (|C_i - C_{i-1}|)}{\sum_{i=n-k-1}^{n} (|C_i - C_{i-1}|)} \tag{5.8}$$

$$VA_n = VA_{n-1} + \frac{(C_n - L_n) - (H_n - C_n)}{H_n - L_n} \times V_n \tag{5.9}$$

Definitions of 16 components of the 16-tuple vector $\mathbf{R}_n = (r_n^0, r_n^1, \ldots, r_n^{15})^T$ are shown in Eqs. 5.10 to 5.25:

$$r_n^0 = K_n^{12} \tag{5.10}$$

$$r_n^1 = K_n^{12} - D_n^{12} \tag{5.11}$$

$$r_n^2 = ROC_1\,(K_n^{12}) \tag{5.12}$$

$$r_n^3 = OSC_{3,10}\,(FR_n) \tag{5.13}$$

$$r_n^4 = ROC_1\,(OSC_{3,10}\,(FR_n)) \tag{5.14}$$

$$r_n^5 = RSI_n^6 \tag{5.15}$$

$$r_n^6 = RSI_n^6 - RSI_n^{12} \tag{5.16}$$

$$r_n^7 = ROC_1\,(RSI_n^6) \tag{5.17}$$

$$r_n^8 = OSC_{3,6}(C_n) \tag{5.18}$$

$$r_n^9 = ROC_1(OSC_{j,k}(C_n)) \tag{5.19}$$

$$r_n^{10} = \frac{\sum_{\substack{i=n-11,\\ VA_i > VA_{i-1}}}^{n} (|VA_i - VA_{i-1}|)}{\sum_{i=n-11}^{n} (|VA_i - VA_{i-1}|)} \tag{5.20}$$

$$r_n^{11} = \frac{\sum_{\substack{i=n-5,\\ VA_i > VA_{i-1}}}^{n} (|VA_i - VA_{i-1}|)}{\sum_{i=n-5}^{n} (|VA_i - VA_{i-1}|)} - \frac{\sum_{\substack{i=n-11,\\ VA_i > VA_{i-1}}}^{n} (|VA_i - VA_{i-1}|)}{\sum_{i=n-11}^{n} (|VA_i - VA_{i-1}|)} \tag{5.21}$$

$$r_n^{12} = ROC_1\left(\frac{\sum_{\substack{i=n-11,\\ VA_i > VA_{i-1}}}^{n} (|VA_i - VA_{i-1}|)}{\sum_{i=n-11}^{n} (|VA_i - VA_{i-1}|)}\right) \tag{5.22}$$

$$r_n^{13} = ROC_1(BIAS_{10}(P_n)) \tag{5.23}$$

$$r_n^{14} = BIAS_{10}(V_n) \tag{5.24}$$

$$r_n^{15} = ROC_1(BIAS_{10}(V_n)) \tag{5.25}$$

Predictive Trend Modeling

The desired output vector P_n of the DAS net is chosen to reveal the trend of the price movement during the next six trading days. The value of traditional $\%K_n^k$ of the stochastic process invented by Lane (1984) reveals where the closing price of the current trading day stands relative to the retrospective fluctuation range of prices in the last k trading days. To represent where the closing price of the current trading day will stand in relation to the fluctuation range of prices for the consecutive k trading days, the forward-calculated FK_n^k, is defined by Eq. 5.26:

$$FK_n^k = \frac{MAX_{i=n}^{n+k}(H_i) - C_n}{MAX_{i=n}^{n+k}(H_i) - MIN_{i=n}^{n+k}(L_i)} \quad (5.26)$$

The FK_n^k, ranging from 0 to 1, is the only element of the output vector P_n. Traders can take the value of the FK_n^k as the probability of a profitable trade during the next k trading days from the position of buying stocks at the closing price of the current trading day. As shown in Figure 5.1, a bullish trend can be identified by a very high value of the FK_n^k (>0.7), which puts the closing price of the nth trading day near the bottom of the price fluctuation range of the consecutive k trading days. Conversely, in down trends, the FK_n^k will have a low value (<0.3), and the closing price of the nth trading day will be located near the top of the price fluctuation range of the consecutive k trading days.

System Architecture

The problem contemplated here is to predict the trend of price movement, the FK_n^6, based on the retrospective feature vector R_n. Dual mod-

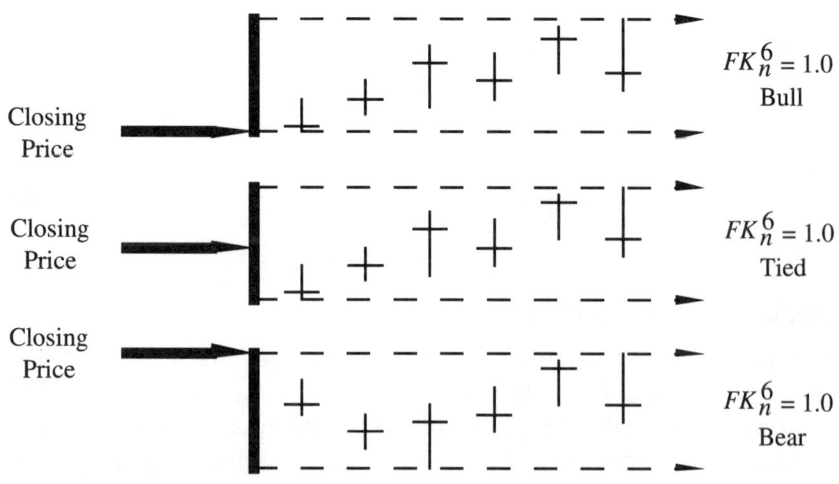

FIGURE 5.1
Predictive trend modeling

ules of structure-level adaptive feed-forward neural networks are used to learn the mapping function between the input vector R_n and the desired value of FK_n^6 of the desired output vector P_n. Two output vectors, O_{nS} and O_{nL}, are generated by the short-term module and the long-term module of neural networks, respectively. The architecture of the proposed system is shown in Figure 5.2.

Feature Extraction and Training Scheme

First of all, it is necessary to generate training patterns suitable for the *DAS net* architecture. Each training pattern is a 2-tuple $\langle \mathbf{R}_n, \mathbf{P}_n \rangle$ that consists of the input vector $\mathbf{R}_n$ and the desired output vector $\mathbf{P}_n$. The retrospective feature extraction unit, as shown in Figure 5.2, generates selected technical indices. This unit is used to establish the transformation $F: T_{n,\alpha} \rightarrow R_n$ that can extract retrospective features from the raw time series data set $T_{n,\alpha}$ to form the 16-tuple input vector R_n. The predictive feature extraction unit, designed according to Eq. 5.26, is used to generate the desired output vector P_n of the training pattern.

For the output vector $\hat{Y}_i$ generated by the *DAS net* model, the quantitative measure of the goodness-of-fit for n training patterns is given by *GF*, defined in Eq. 5.27:

$$GF = \frac{\sum_{i=1}^{n} \| \mathbf{Y}_i - \overline{\mathbf{Y}} \|^2 - \sum_{i=1}^{n} \| \mathbf{Y}_i - \hat{\mathbf{Y}}_i \|^2}{\sum_{i=1}^{n} \| \mathbf{Y}_i - \overline{\mathbf{Y}} \|^2} \tag{5.27}$$

where $\mathbf{Y}_i$ is the desired output vector and $\bar{Y}$ is the mean of $\bar{Y}_i$ for n training patterns. Thus, *GF* measures the goodness-of-fit between the *DAS net* model and actual short-term trends of the stock market in the sense that it gives the proportionate reduction in the sum of squares of deviations obtained using the *DAS net* relative to the naive predictor $\bar{Y}$. Assume n is the number of training patterns in the training window. Then, *GF* can be used to evaluate the goodness-of-fit between n pairs of output vectors $\hat{Y}_i$ generated by the DAS net model and the desired output vector Y_i.

$$MSE = \frac{\sum_{i=1}^{n} \| \mathbf{Y}_i - \hat{\mathbf{Y}}_i \|^2}{n} \tag{5.28}$$

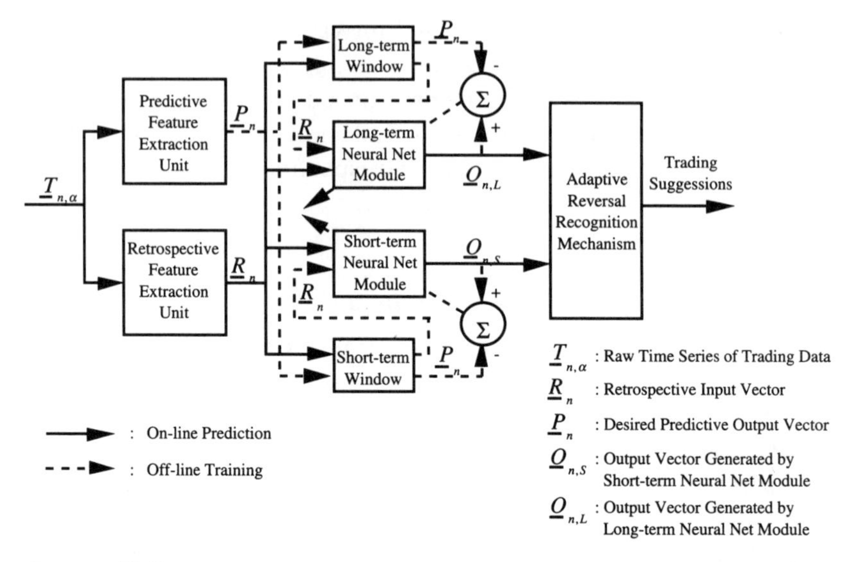

FIGURE 5.2
System architecture

The generalized mean squared error (*MSE*) between n pairs of output vectors $\hat{Y}_i$ of the network and the desired output vector Y_i is defined in Eq. 5.28:

$$MSE = \frac{\sum_{i=1}^{n} \| \mathbf{Y}_i - \hat{\mathbf{Y}}_i \|^2}{n} \tag{5.28}$$

The generalization ability of the *DAS net* model can be represented by the MSE between n pairs of output vector $\hat{\mathbf{Y}}_i$ generated by the *DAS net* and the desired output vector $\mathbf{Y}_i$ when the *DAS net* model is tested with n test patterns.

A fixed stock market model may fail when the market behavior changes. To adapt the *DAS net* according to current market trading dynamics, a moving-window training scheme is used to tune the weights of the *DAS net* according to training data filtered by two-fixed windows (Jang et al. 1991a, 1991b, 1993).

When the initial training process is completed, the *DAS net* is trained through a moving-window training scheme. A moving-window $MW_n^m = \{\langle \mathbf{R}_i, \mathbf{P}_i \rangle \mid i = n - m + 1, n - m + 2, \ldots, n\}$ is a collection of the latest m training patterns at the time index n. The training scheme is composed of three phases: training, prediction, and moving patterns from the test set to the training set. This three-phase, moving-window training scheme is shown in Figure 5.3. In the training phase, the desired output vector $\mathbf{P}_i$ for all 2-tuple $\langle \mathbf{R}_i, \mathbf{P}_i \rangle$ in the moving window of the training set is compared with the output vectors $\mathbf{O}_{i,S}$ and $\mathbf{O}_{i,L}$ generated by the dual net, and the differences are back-propagated to modify the interconnection weights of the dual net. To control the goodness-of-fit, the training process for a specific window of training patterns stops when the calculated value of *GF* is larger than a predefined threshold GF_{th}. When the training process stops, the prediction phase is started by presenting the input vector $\mathbf{R}_{n+1}$ of the first 2-tuple $\langle \mathbf{R}_{n+1}, \mathbf{P}_{n+1} \rangle$ in the test set to the input layer of the dual net. Then the output of the short-term and long-term modules of neural networks of the dual net, $\mathbf{O}_{n+1,S}$ and $\mathbf{O}_{n+1,L}$, are appended to a prediction history file. The output vectors $\mathbf{O}_{i,S}$ and $\mathbf{O}_{i,L}$ of the dual net collected in the prediction history file are used to analyze the generalization ability of the dual net. In the third phase, the 2-tuple $\langle \mathbf{R}_{n+1}, \mathbf{P}_{n+1} \rangle$ just used to generate the output vectors $\mathbf{O}_{n+1,S}$ and $\mathbf{O}_{n+1,L}$ is moved from the test set to the moving window of the training set and the time index n is increased by one. Thus, the moving window $MW_n^m = \{\langle \mathbf{R}_i, \mathbf{P}_i \rangle \mid i = n - m + 1, n - m + 2, \ldots, n\}$ becomes $MW_{n+1}^m = \{\langle \mathbf{R}_i, \mathbf{P}_i \rangle \mid i = n - m + 2, n - m + 3, \ldots, n + 1\}$. Finally, an iteration is started.

As shown in Figure 5.3, the short-term module of the neural network uses a 24-day moving window to keep the latest 24 2-tuple $\langle \mathbf{R}_n, \mathbf{P}_n \rangle$ training patterns in the training set. This 24-day moving window keeps the short-term module sensitive to the latest changes in market behavior. The long-term module of the neural network, on the other hand, concentrates on the training patterns collected from the latest 72 trading days. Thus, the *DAS net* keeps track of both short-term and long-term views for the mapping between the retrospective technical indices and the short-term trends of price movements of the Taiwan stock market.

Primitive Neural Network Modules

The *DAS net* used in the present system is composed of two modules of multilayer feed-forward neural networks. The function of this primitive neural network is to form a multidimensional mapping between the input vector $\mathbf{R}_n$ and the desired output vector $\mathbf{P}_n$. Each primitive

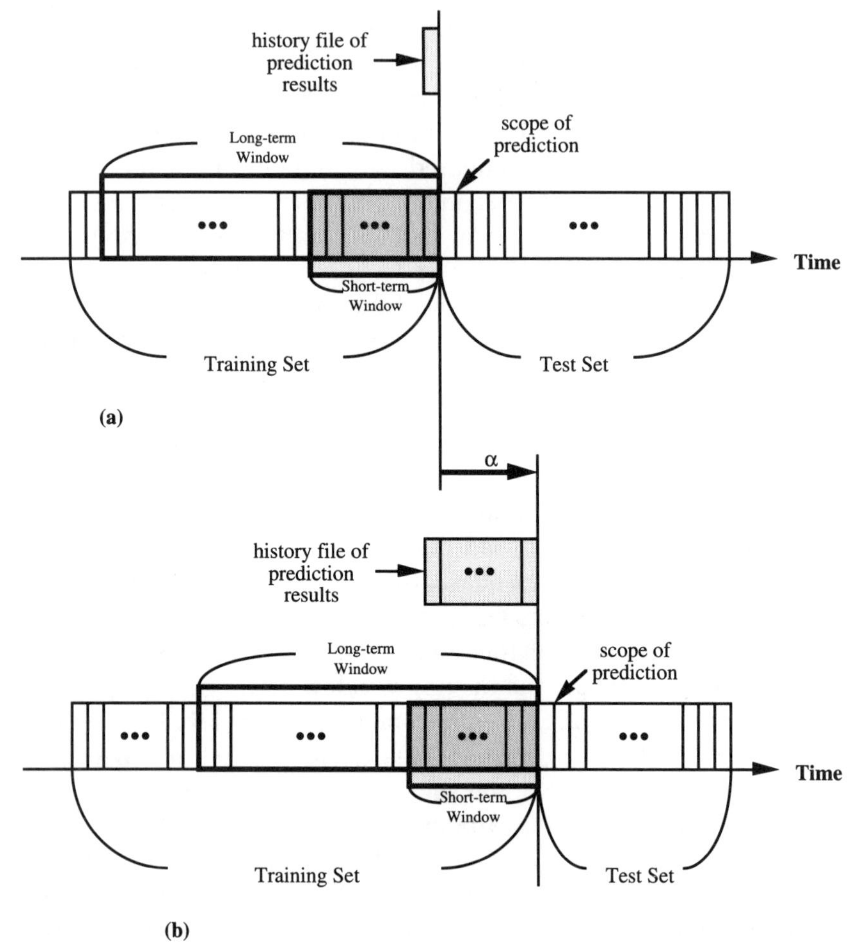

FIGURE 5.3
Moving-window simulation. (a) Initial status. (b) After moving α pattern from the test set into both the short-term and the long-term windows of training patterns.

network consists of three layers: the input layer, one hidden layer, and the output layer. The input layer is made up of 16 units, each related to a projection value of the technical indicator of a selected stock on one of the 16 axes spanning $\mathbf{R}_n$. The output layer contains only one neuron that generates the predicted value of the FK_n^6. The architecture of the primitive network is illustrated in Figure 5.4.

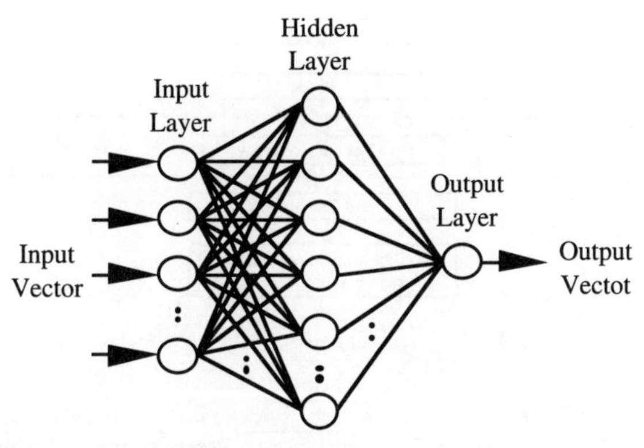

FIGURE 5.4
Primitive neural network module

Learning Algorithm for Weight Adaptation

After the output vector $\hat{\mathbf{Y}}_i$ of the network is formed, it is compared with a desired output vector $\mathbf{Y}_i$ and the error vector $\mathbf{E}_i = \mathbf{Y}_i - \hat{\mathbf{Y}}_i$ is calculated. The learning of the neural network is done by adjusting the interconnection weights toward the direction of minimizing the MSE $\epsilon = \langle \|\mathbf{E}_i\| \rangle$ using the back-propagation algorithm (Rumelhart et al. 1986).

Structure-Level Adaptation

The structure-level adaptation of the primitive neural networks is accomplished by adjusting the number of neurons in the hidden layer through two major procedures: neuron generation and neuron annihilation (Lee 1991). An auxiliary neuron randomization procedure is also included to address the problem of a local minimum. Figure 5.5 shows the control flow chart of the proposed structure-level adaptation techniques for multilayer feed-forward neural networks.

Neuron Generation

The basic criterion for generating a new neuron in the network is that the representation power of the network become insufficient. We use the stabilized goodness-of-fit as an indicator to determine whether the network needs to generate a new neuron. If after a certain period of weight adaptation, the GF has stabilized but is smaller than the

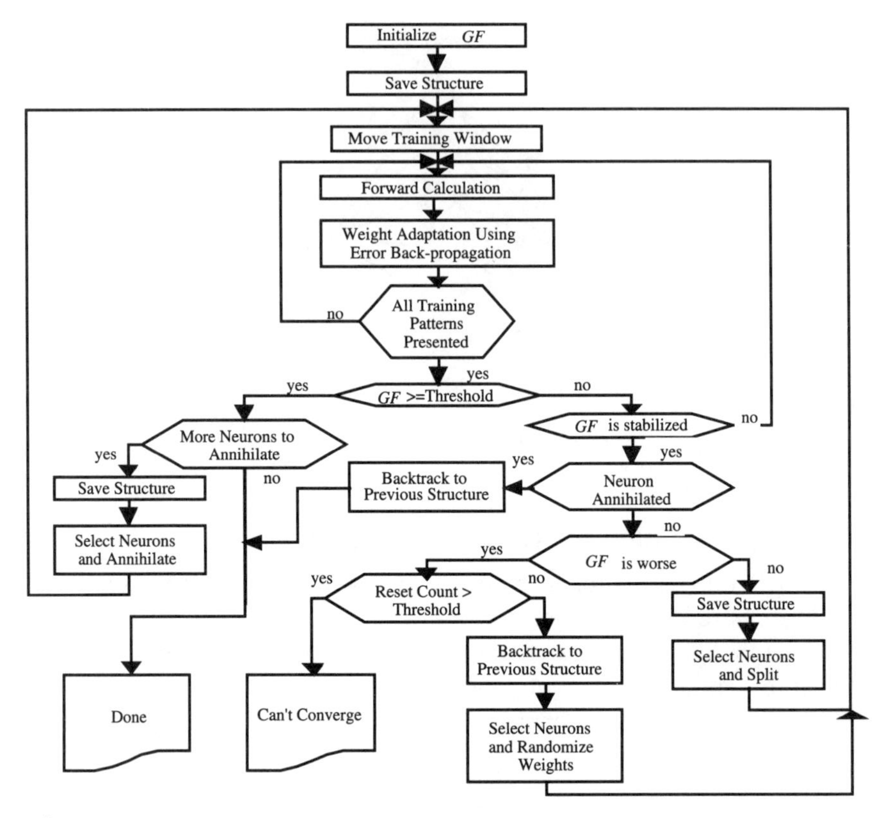

FIGURE 5.5
Flow chart of structure-level adaptive learning algorithm

desired value GF_{th}, then a new neuron is added to increase the discriminating power of the network. In other words, the network may have been trapped in a local minimum point. Therefore, adding a new neuron into the network provides the network with a chance to escape from the local minimum (Lee 1991).

When a new neuron is generated, it is created with random interconnection weights (Hirose et al. 1991; Murase et al. 1991), or it can carry the same interconnection weights as one of the original neurons (Lee 1991). We use a modified version of Lee's approach to determine the interconnection weights of the newly generated neuron. Through monitoring the behavior of each neuron during the learning phase, we identify the neuron with the least power and the largest influence on the final system error; this neuron can be used as a seed to generate the new neuron. The input interconnection weights of the newly generated neuron are duplicated from the neuron that contributes most to the output fluctuation of the network. To increase the probability of escaping from the local minimum, the output interconnection weights are generated randomly.

Neuron Annihilation

After a certain period of weight adaptation, if the approximation GF is stabilized and is larger than the desired value, then a neuron may be annihilated without affecting the performance of the network. For a set of training patterns, if the forward signal entropy of a particular neuron is very close to zero, then this neuron is not a functioning element in the network and may be annihilated without degrading the performance of the network.

Neuron Randomization

If the stabilized GF of a network is not only smaller than a predefined value GF_{th}, but also smaller than the GF value before training, then we conclude that the performance of the network cannot be improved through weight adaptation on the current interconnection weights. Therefore, we introduce a weight randomization procedure into the process of structure-level adaptation. When the performance of the network only worsens after the normal weight adaptations, the input and output interconnection weights of the neuron with the largest influence to the final system error are randomized. Then the training process is restarted with the modified interconnection weights.

The neuron randomization procedure alters the position of the standing point of the network's learning process in the hyperplane spanned by the interconnection weights of neurons. Therefore, the neuron randomization procedure may increase the probability for the learning process of the network to escape from the local minimum point.

Market Prediction and Stock Trading Decision Support

The objectives of a stock market prediction system are to extract useful knowledge from a large-scale database containing time-series price and volume data of stocks, and to provide real-time stock market prediction. The retrospective feature extraction unit shown in Figure 5.2 is used to extract useful features $\mathbf{R}_n$ representing current market status from a large database of time-series data $\mathbf{T}_{n,\alpha}$. The *DAS net* architecture is used to build a function relating the retrospective feature vector $\mathbf{R}_n$ representing market status, and the future trends of price movements $\mathbf{P}_n$.

The *DAS net* provides stock traders with the prediction of future trends of price movements. However, if we want to further automate the decision-making process, it is necessary to provide stock traders with timely and accurate stock market prediction, such as recommendations for buying, holding, or selling certain kinds of stocks at certain times. We have developed an adaptive mechanism that can recognize reversals of price trends and generate stock trading recommendations for buying, holding, or selling according to the trend prediction generated by the *DAS net* (Jang and Lai 1993).

Performance Evaluation

The performance of the intelligent stock market prediction system based on the *DAS net* is examined by simulated trading of the TSEWPI of the Taiwan stock market from 1990 to 1991 on a Sun SPARC Station 2 workstation. In this simulation, the initial training set is composed of the 2-tuple $\langle \mathbf{R}_n, \mathbf{P}_n \rangle$ patterns generated from daily TSEWPI data collected from 1987 to 1989, while the test set consists of the daily TSEWPI data collected from 1990 to 1991. The $MA_k(X_n)$ defined in Eq. 5.1 is used to define the k-day moving average of time-series variable X_n of the nth trading day.

The algorithm for generating buy-and-sell signals using the *DAS net* is as follows:

```
if (MA_3(FK^6_{n-1}) < lower threshold of BUY and MA_3(FK^6_n) > upper
threshold of BUY)
then {
  if (BUY existed)
  then
          HOLD;
  else
          BUY;
}
```

```
if (MA3(FK6n-1) > upper threshold of SELL and MA3(FK6n) < lower
threshold of SELL)
then {
  if (BUY existed)
  then
          SELL;
}
otherwise {
          HOLD;
}
```

A one-point buying and selling strategy is used in this simulation. One-point buying and selling means all available money is used to buy stocks at a particular time. It also means that all stocks held are sold all together.

The annual performance of buying and selling the TSEWPI from 1990 to 1991 is shown in Table 5.4. All the rates of return shown in the table were calculated after taking the transaction cost of 1% for each transaction into consideration. To evaluate the quality of the proposed stock market prediction system, two categories of criteria, namely, profitability and consistency, were applied.

Profitability criteria comprised the total number of trades over the test period, the number of profitable trades, the percentage of profitable trades, the average rate of return per profitable (unprofitable) trade, the average rate of return per trade, and the total rate of return over the test period.

Consistency criteria comprised the standard deviation of the rate of return per trade, the Sharpe ratio (the average rate of return divided by the standard deviation), the maximum rate of return per profitable (unprofitable) trade, and the maximum drawdown (the cumulative rate of return of successive unprofitable trades).

Rather than an actual object for trading listed on the Taiwan Stock Exchange (TSE), the TSEWPI is a weighted index of prices of stocks of significant companies listed on the TSE. Therefore, buying and selling of the TSEWPI is equivalent to buying and selling a portfolio consisting of stocks of significant companies listed on the TSE, of which the ratio of the investment amount in each firm's stocks is the same as the weighting factor of its contribution to the calculation of the TSEWPI.

The annual rates of return obtained by two other trading strategies are also listed in Table 5.4. The annual rates of return of the four

TABLE 5.4 *Performance Evaluation of Trading on the Taiwan Stock Exchange*

	1990		1991	
Type of neural network	Fixed-structure neural networks	DAS Net	Fixed-structure neural networks	DAS Net
Total number of trades	13	4	5	4
Number of profitable trades	3	2	3	3
Percentage of profitable trades, %	23.08	50.00	60.00	75.00
Average rate of return per profitable trade, %	+20.81	+22.23	+13.40	+14.77
Average rate of return per unprofitable trade, %	−8.78	−8.93	−6.20	−13.75
Average rate of return per trade, %	−1.95	+6.65	+5.56	+7.64
S.D. of rate of return per trade	0.1200	0.1608	0.1204	0.1035
Sharpe ratio	−0.1625	0.4138	0.4621	0.7380
Maximum rate of return per profitable trade, %	+27.95	+38.21	+32.22	+25.84
Maximum rate of return per unprofitable trade, %	−14.92	−12.67	−8.10	−13.75
Maximum drawdown, %	−80.33	−17.85	−8.10	−13.75
Total rate of return, %	−30.66	+21.61	+25.76	+29.20
Total rate of return of CSITC Growth Fund, %	−39.52		+23.42	
Total rate of return of Kwang Hua Growth Fund, %	−41.13		+3.75	
Total rate of return of NITC FuYuan Fund, %	−45.21		+28.30	
Total rate of return of Citizens Securities Investment Trust Fund, %	−42.71		+9.86	
Annual TSEWPI change, %	−52.93		+1.56	

closed-end funds issued in Taiwan are used to show the performance of trading decisions made by human experts. All of the four funds are domestic securities investment trust funds for domestic investors and invest mainly in stocks listed on the TSE. The principal objective of the funds is capital growth through investments in listed firms, subject to certain restrictions under the "Securities Investment Trust Contract" and related regulations. For example, the annual turnover of each fund is not allowed to exceed 60% of the annual average turnover of all stocks listed on the TSE. Under this restriction on the turnover of funds, fund managers cannot trade as frequently as using our one-point buying and selling strategy would require. However, they can increase the rate of return by adjusting the portfolio to include stocks with high expected returns. The annual rate of change of the TSEWPI is used to represent the raw annual rates of return of trading on the TSEWPI according to the buy-and-hold strategy.

The one-point buying and selling trading strategy used in our simulation allows high turnover. However, the portfolio is fixed, investing in stocks of all significant companies listed on the TSE. Fund managers of the closed-end funds have stricter limitations on the turnover of their funds than our trading strategy. However, they can adjust their portfolio to increase the rate of return through stock selection methods. Both strategies are more aggressive than the buy-and-hold strategy.

The performance of the proposed system is better than that of fixed-structure neural networks. Furthermore, the proposed system outperforms all of the closed-end funds in the testing period from 1990 to 1991. According to Table 5.4, it is obvious that the annual rates of return (after transaction costs) gained by using the trading decisions generated according to the weighted output of the *DAS net* are larger than those gained using the buy-and-hold strategy over the testing period. We can conclude that an intelligent stock trading decision-support system using dual-module neural networks can outperform the simple buy-and-hold trading strategy.

Summary

Dual adaptive-structure neural networks that can predict the short-term trends of price movement and recognize the effective reversals have been utilized to develop an innovative, intelligent, and profitable stock trading decision-support system for the Taiwan stock market. Transformations used to identify both retrospective and predictive features from raw data gathered from the market have been presented.

Reinforcing the temporary correlation between the neural weights and the training patterns, the structure of the DAS net is self-synthesized based on training patterns collected from windows of different sizes. The proposed adaptation techniques also allow the *DAS* net to continuously adapt its structure to follow statistical changes in the problem domain. According to our analyses of the generalized MSE, we have shown that a portfolio of dual-module neural networks generalizes better than a single-module neural network. In addition, we have justified the effectiveness of the proposed structure-level adaptive neural networks by comparing their generalization ability with that of fixed-structure neural networks. Finally, in the testing period from 1990 to 1991, the annual rates of return from using trading decision support generated by the proposed system are larger than those generated by buy-and-hold strategy, as well as superior to all the closed-end funds managed under different disciplines by human experts. The above-average rates of return shown indicate that an intelligent stock trading decision-support system with price trend prediction and reversal recognition, can be realized using structure-level adaptive neural networks.

References

Cybento, G. 1989. "Approximation by Superpositions of a Sigmoidal Function." In *Mathematics of Control, Signals and Systems.* New York: Springer-Verlag.

Elder, J. F., IV, and M. T. Finn. 1991. "Creating 'Optimally Complex' Models for Forecasting." *Financial Analysts Journal,* (Jan.-Feb.).

Felsen, J. 1975. "Learning Pattern Recognition Techniques Applied to Stock Market Forecasting." *IEEE Transactions on Systems, Man and Cybernetics* SMC-5 (6).

Hawley, D. D., J. D. Johnson, and D. Raina. 1990. "Artificial Neural Systems: A New Tool for Financial Decision-Making." *Financial Analysts Journal* (Nov.-Dec.).

Hirose, Y., K. Yamashita and S. Hijiya. 1991. "Back-Propagation Algorithm Which Varies the Number of Hidden Units." *Neural Networks* 4.

Jang, G. S., F. Lai, B. W. Jiang, C. C. Pan, and L. H. Chien. 1991a. "An Intelligent Stock Portfolio Management System Based on Short-Term Trend Prediction Using Dual-Module Neural Networks." In *Proceedings International Conference on Artificial Neural Networks* (Finland).

Jang G. S., F. Lai, B. W. Jiang, and L. H. Chien. 1991b. "An Intelligent Trend Prediction and Reversal Recognition System Using Dual-Module Neural Networks." In *Proceedings First International Conference on Artificial Intelligence Applications on Wall Street* (New York).

Jang, G. S., and F. Lai. 1993. "Intelligent Stock Market Prediction System Using Dual Adaptive-Structure Neural Networks." In *Proceedings Second International Conference on Artificial Intelligence Applications on Wall Street* (New York).

Lane, G. C. 1984. *Trading Strategies*. Future Symposium International.

Lee, T. C. 1991. *Structure Level Adaptation for Artificial Neural Networks*. Kluwer Academic Publishers.

Murase, K., Y. Matsunaga, and Y. Nakade. 1991. "A Back-Propagation Algorithm Which Automatically Determines the Number of Association Units." In *Proceedings IEEE International Joint Conference on Neural Networks*.

Murphy, J. J. 1986. *Technical Analysis of the Futures Markets: A Comprehensive Guide to Trading Methods and Applications*. New York: Institute of Finance.

Rumelhart, D. E., J. L. McClelland, and the PDP Research Group. 1986. *Parallel Distributed Processing. Explorations in the Microstructure of Cognition*. Vol. 1. *Foundations*. Cambridge: The MIT Press.

Notes

1. The authors thank the Taiwan Fuji Xerox Foundation for a research award. They are indebted to Jack J. Y. Yeh, president of National Investment Trust Company, Ltd., and to Thomas Huang, manager of the Planning Department of Jih Sun Securities Company, Ltd., for discussions and critiques of early versions of this system. Finally, they also thank Dr. Gerald L. Chan for his inspiration and kind help.
2. Background on the Taiwanese stock market can be found in "Taiwan," by *Sam Chang*, in *The World's Emerging Stock Markets*, Keith K. H. Park and Antoine W. Van Agtmael, ed. Chicago: Probus (1993), pp. 101–19.

CHAPTER 6

Trading U.S. Treasury Notes with a Portfolio of Neural Net Models

Guido J. Deboeck and Masud Cader

People who sleep better at night because they own bonds and not stocks are susceptible to rude awakenings.

Peter Lynch
Beating the Market, *1993*

Chapters 3 to 5 provided illustrations of neural networks applied to stock markets. Chapter 6 shows how to design neural network models for trading U.S. Treasury notes. The first part of this chapter demonstrates significant differences between securities of different maturities. Nonlinear dynamics and rescaled range analysis to detect these differences. Based on the discovery of different patterns in U.S. Treasury securities, the focus then shifts to the modeling of two-year and ten-year Treasury notes. The second part of the chapter discusses the design and performance of these neural net models, the advantages of combining models into a portfolio, and the robustness and stability of a portfolio of neural net models.

Daily Returns of U.S. Treasury Securities

Traditional assumptions regarding the behavior of U.S. Treasury markets no longer hold. Several authors have suggested that financial markets are nonlinear dynamic systems and that a fractal market hypothesis is more useful than other hypotheses for market analysis (Peters 1991a, 1991b, 1994). To test this on U.S. Treasury securities we applied several techniques, which are discussed in detail in Part Four.

The data used in this chapter are daily closing yields on U.S. Treasury securities from January 1975 to December 1992. Statistical and nonlinear dynamic analysis techniques were applied on closing yields through December 1991; yields from August 1988 to the end of December 1992 were used for modeling.

As a starting point we computed descriptive statistics, including the mean, variance, standard deviation, skewness, kurtosis of yields, and Pearson correlations among securities. Tables 6.1 to 6.4 provide these basic statistics. Computations on the first-order differences of the closing yields[1] showed that the Pearson correlation between Fed Funds and one-year T-bills is small; the correlation between T-bills

TABLE 6.1 *Descriptive Statistics on U.S. Treasury Yields (August 1988 to December 1991)*

Statistic	Fed Funds	1-year T-bill	2-year T-note	5-year T-note	10-year T-note	30-year T-bond
Minimum	4.11	3.97	4.77	6.01	6.75	7.45
Maximum	10.71	9.66	9.91	9.76	9.53	9.46
Range	6.60	5.69	5.14	3.75	2.78	2.01
Mean	7.78	7.44	7.85	8.17	8.39	8.48
Variance	2.17	1.59	1.13	0.52	0.30	0.19
Standard deviation	1.47	1.26	1.06	0.72	0.54	0.43
Standard error	0.05	0.04	0.04	0.03	0.02	0.02
Skewness	−0.61	−0.73	−0.62	−0.45	−0.10	0.18
Kurtosis	−0.69	−0.37	−0.02	0.25	−0.53	−1.03
Coefficient of variation	0.19	0.17	0.13	0.09	0.06	0.05

TABLE 6.2 *Pearson Correlation Matrix on U.S. Treasury Yields (August 1988 to December 1991)*

	Fed Funds	1-year T-bill	2-year T-note	5-year T-note	10-year T-note	30-year T-bond
Fed Funds	1.0000					
1-year T-bill	0.9605	1.0000				
2-year T-note	0.9050	0.9747	1.0000			
5-year T-note	0.7950	0.8919	0.9653	1.0000		
10-year T-note	0.6642	0.7719	0.8770	0.9636	1.0000	
30-year T-bond	0.5034	0.6152	0.7423	0.8698	0.9679	1.0000

TABLE 6.3 *Descriptive Statistics on First-Order Differences of U.S. Treasury Yields (August 1988 to December 1991)*

Statistic	Fed Funds	1-year T-bill	2-year T-note	5-year T-note	10-year T-note	30-year T-bond
Minimum	−0.31	−0.08	−0.04	−0.04	−0.03	−0.03
Maximum	0.35	0.04	0.05	0.04	0.03	0.03
Range	0.66	0.12	0.09	0.08	0.06	0.06
Mean	−0.0007	−0.0007	−0.0006	−0.0004	−0.0003	−0.0002
Variance	0.0019	0.0001	0.0001	0.0001	0.0001	0.0000
Standard deviation	0.0436	0.0096	0.0090	0.0080	0.0071	0.0068
Standard error	0.0015	0.0003	0.0003	0.0003	0.0002	0.0002
Skewness	0.9260	−0.6912	0.0027	−0.1357	−0.2489	−0.1550
Kurtosis	21.2718	7.7823	3.3770	1.9559	1.4632	2.2032
Coefficient of variation	−66.6091	−13.6517	−15.1342	−20.0824	−25.4351	−41.7844

TABLE 6.4 *Pearson Correlation Matrix on First-Order Differences of U.S. Treasury Yields (August 1988 to December 1991)*

	Fed Funds	1-year T-bill	2-year T-note	5-year T-note	10-year T-note	30-year T-bond
Fed Funds	1.0000					
1-year T-bill	0.0661	1.0000				
2-year T-note	0.0148	0.6952	1.0000			
5-year T-note	0.0023	0.5933	0.8251	1.0000		
10-year T-note	0.0129	0.5056	0.7168	0.8401	1.0000	
30-year T-bond	0.0142	0.4431	0.6487	0.7617	0.8380	1.0000

and two-year notes is 0.69; between two-year and ten-year notes, 0.71; and between ten-year and thirty-year bonds, 0.83. Thus, *risk management of a portfolio of U.S. Treasury securities cannot be based on risk equivalence or duration alone, but should incorporate the covariances between yield movements of securities with different maturities.* Chapter 19 shows a robust methodology for proper risk management of any portfolio, including one composed excusively of U.S. Treasury securities!

Frequency distributions over long periods of time show that the first-order differences of Treasury securities have a non-Gaussian distribution of yields and approximately Gaussian distributions of the first-order differences. Approximate Gaussian distributions for the first-order differences showing fat tails or small side lobes imply non-

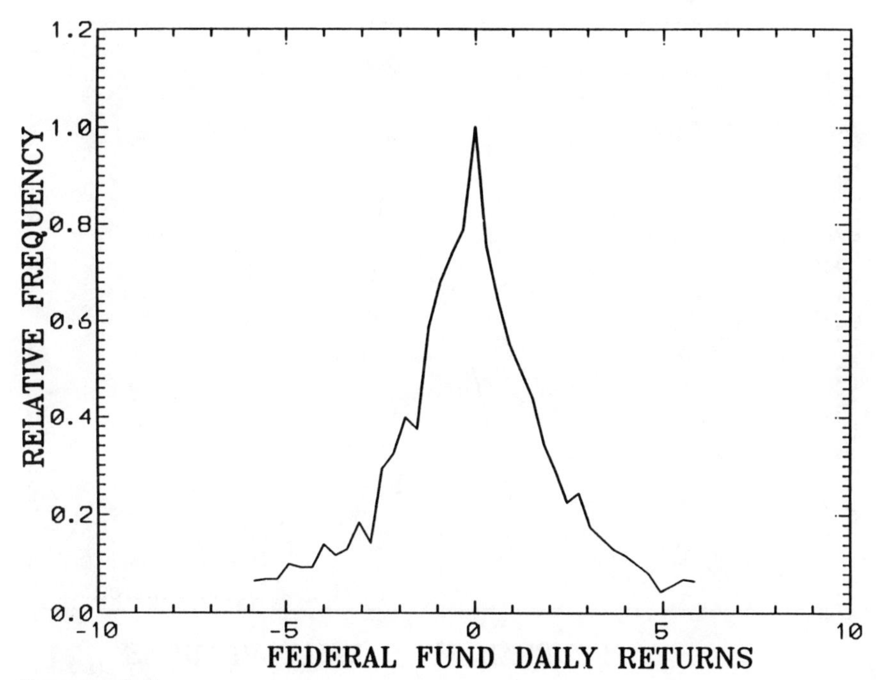

FIGURE 6.1
Distribution of Federal Fund daily returns (1975–1991)

random distributions. The approximate Gaussian distributions of the yield changes of U.S. Treasury securities with various maturities are shown in Figures 6.1 to 6.7. The irregularity of some of the frequency distributions provides a hint that U.S. Treasury markets may be chaotic.[2]

Trends in U.S. Treasuries

We also applied rescaled range analysis to estimate the Hurst coefficient, which measures the memory in a time series (see Chapter 15 for a detailed discussion). Coefficient variance from 0.5 is interpreted as a bias, or memory effect. A Hurst coefficient greater than 0.5 indicates a memory effect that is biased toward reinforcement of the trend, which is called *persistence*. A Hurst coefficient less than 0.5 indicates a negative bias or a tendency to reverse, which is called *antipersistence*. A Hurst coefficient equal to 0.5 indicates a *random*

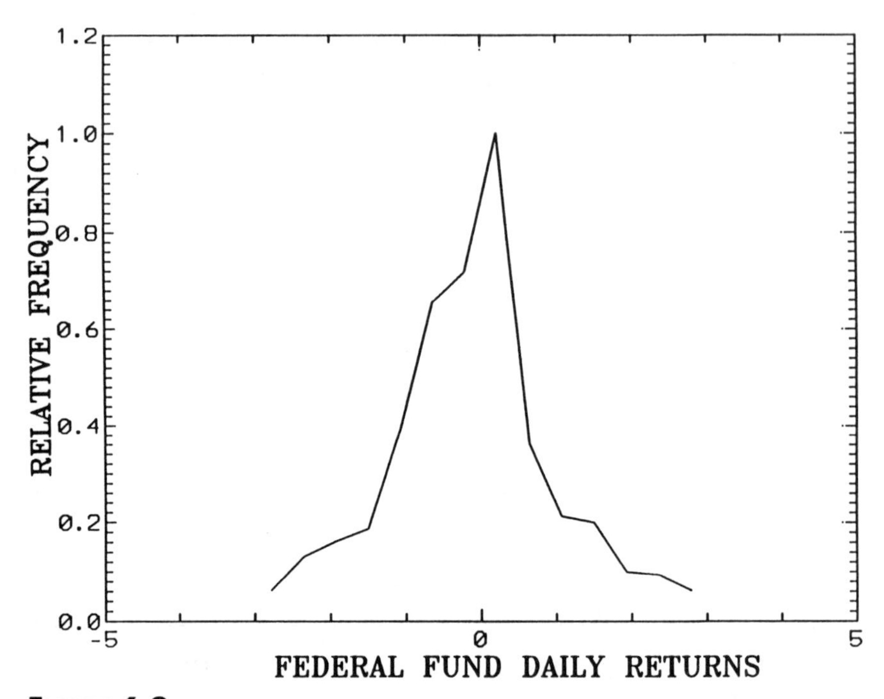

FIGURE 6.2
Distribution of Federal Fund daily returns (1988–1991)

walk. The benefit of rescaled range analysis is that for a finite number of observations the range depends on the number of observations.

Figures 6.8 to 6.11 show the results of applying rescaled range analysis to fifteen years of daily U.S. Treasury yield changes. The Hurst coefficient for Fed Funds is 0.41, which indicates antipersistence. The Hurst coefficient for Treasury bills and notes varies between 0.59 and 0.63, indicating persistent trends. The Hurst coefficient for the yield changes of thirty-year bonds is 0.54, which implies that bonds move according to a random walk, more so than any other security.

The different patterns of yield changes of U.S. Treasury securities across different maturities and over different time horizons are important to the design of trading models. Market signals that are random or are close to random may not be suitable for neural network modeling, which is generally more amenable to persistent or antipersistent trends. The remainder of this chapter focuses on securities that exhibit persistent trends.

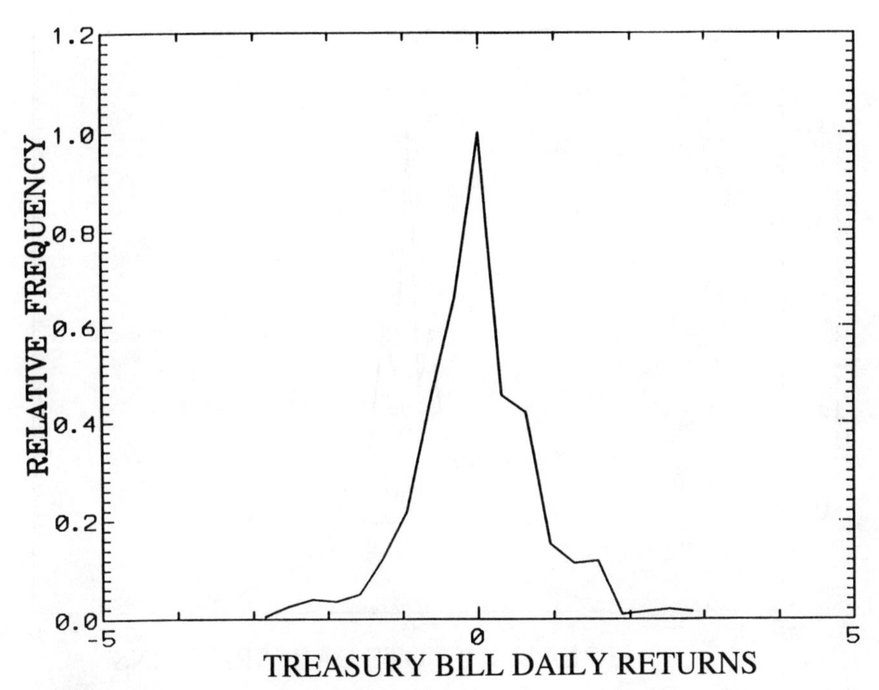

FIGURE 6.3
Distribution of one-year Treasury bill daily returns (1988–1991)

Design of Neural Net Models

The previous section demonstrated that yield changes of U.S. Treasury notes are not random, but instead show persistent trends. Some maturities are easier to predict than others. We will focus our discussion of the design of neural net models on two- and ten-year notes.

From Chapter 1 we recall that neural networks can be thought of as a method of parameterizing classes of nonlinear functions; for example, $f: X \rightarrow Y$, where, in the case of trading systems, X represents market indicators and their transformations, and Y is an output signal or measure of profitability that can be translated into a trading recommendation. Neural networks do nothing more than project the input space into a lower-dimensional space in which classification of patterns or nonlinear relations may be realized more effectively. The neural networks used here are classified as machine-learning tools that use a gradient descent algorithm to minimize the Euclidean distance (or error) between the desired output and the given input(s).

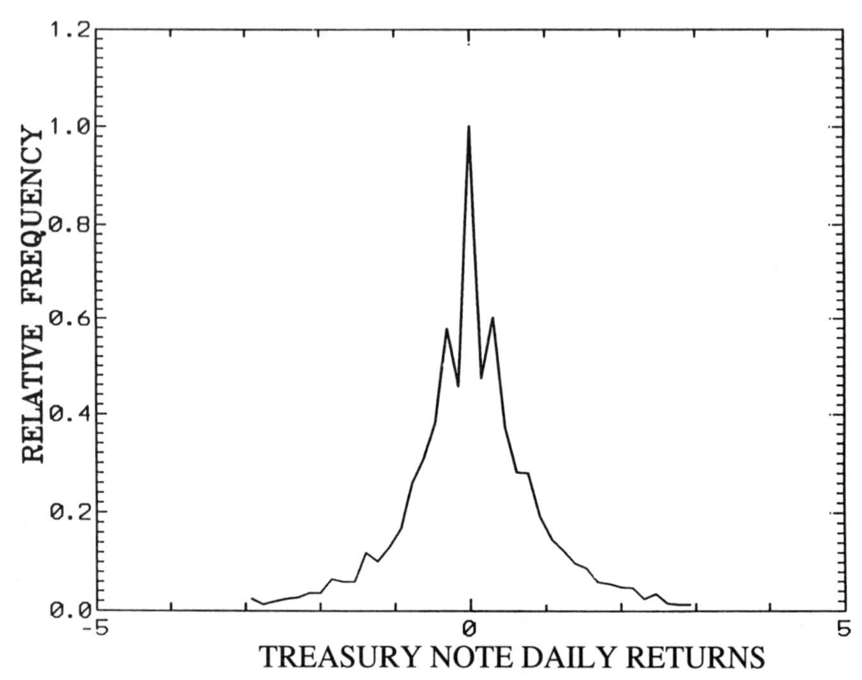

FIGURE 6.4
Distribution of two-year Treasury note daily returns (1975–1991)

The following sections describe neural net models for trading two- and ten-year notes.

Neural Net Inputs, Architecture, and Learning Algorithms

The first step in designing a neural net trading system is to define the objectives and inputs of the system. The objective is to design models that trade at a certain level of profitability without taking undue risks over time (e.g., six months to one year). To judge the profitability and risks we shall use the ratio of average profit divided by the maximum drawdown (or the maximum of the successive losses over a period of time).

The next step is to define the inputs and outputs. The basic inputs for these neural net models are daily closing yields, the spread between the instrument and the Fed Funds rate, and trend and volatility indicators. *Trend indicators* are the difference between the current yield and the yield ten, twenty, or thirty days earlier; *volatility in-*

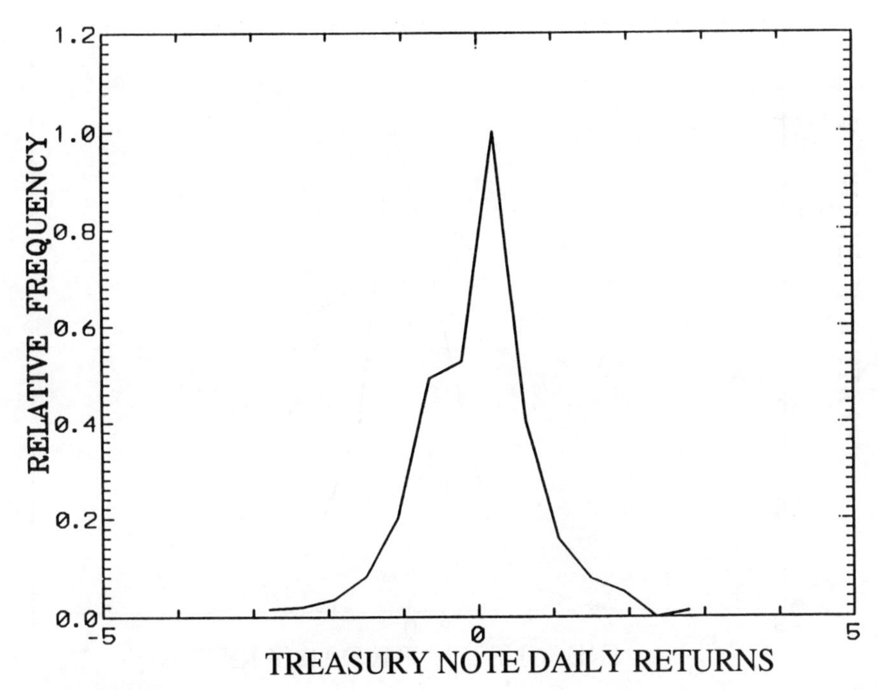

FIGURE 6.5
Distribution of two-year Treasury note daily returns (1988–1991)

dicators are the standard deviations of daily yield changes over the past ten, twenty, or thirty days.

The third step in the design is to select a neural net architecture and learning algorithm. The architecture of these neural net models is very simple. They consist, in general, of an input layer with three to six inputs, a hidden layer with three to five processing elements, and an output layer with a single output. All nodes are fully connected between adjacent layers. For a small input space (less than ten inputs), the maximum size of the hidden layer is determined by the number of inputs minus 1 (or 0). The learning algorithm adopted is a variant of back-propagation that uses a modified gradient descent learning algorithm to dynamically evolve a relationship between inputs and the output. The current neural network models utilize an extended-delta-bar-delta (EDBD) learning algorithm. In general, these models utilize a general learning schedule and do not utilize the weight memory capability of the EDBD algorithm. We have found that the EDBD algorithm is sensitive to the learning rate parameters and tends to

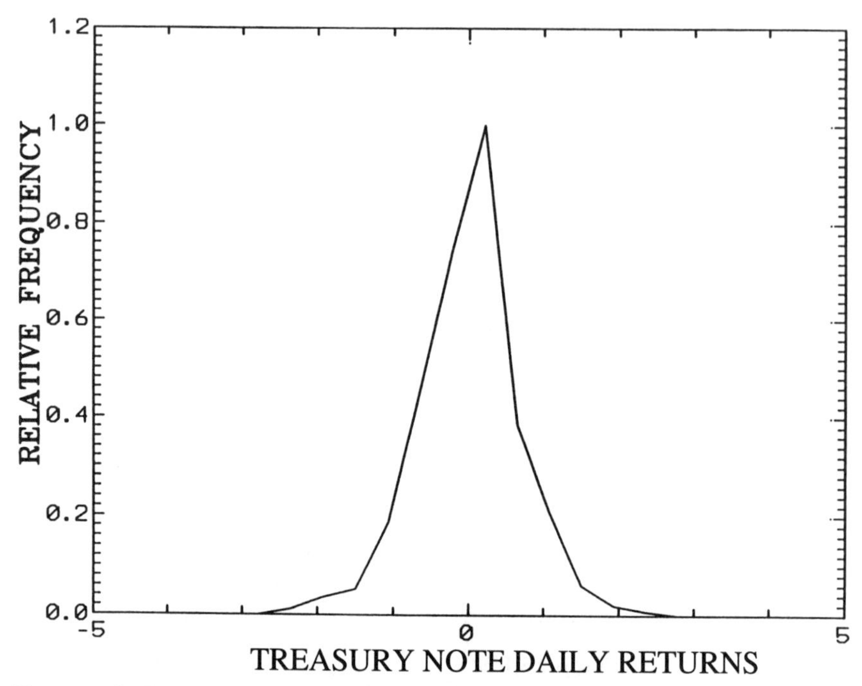

FIGURE 6.6
Distribution of ten-year Treasury note daily returns (1988–1991)

saturate into an unusable state if the number of hidden processing elements is too small and/or the learning rate is too high.

Neural Network Training

Designing numerous neural net models has taught us that the selection of a neural net architecture and learning algorithm is far less important than the selection and identification of appropriate samples for training a neural net. The sampling techniques for generating the training samples have a critical influence on the performance of the neural net trading system.

For the models described here, training files were constructed with about 100 records selected from a total of about 1,120 trading days in the period from FY89 to FY92 (FY stands for fiscal year, which starts July 1 and runs to June 30 of the next year). The data file on which the neural nets were tested included the full data history plus the period from July 1, 1992, to December 30, 1992 (which covers more

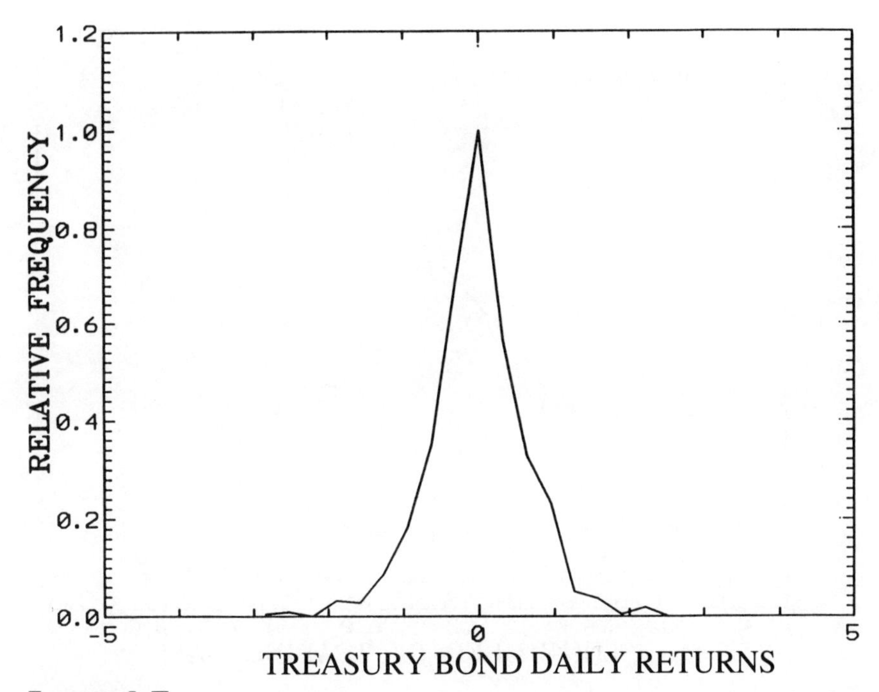

FIGURE 6.7
Distribution of thirty-year Treasury bond daily returns (1988–1991)

than 150 trading days). Generalization capabilities of each network were judged by the performance during the first six months of FY93. The training data files did not include records from FY93.

The criterion used to stop neural net training was based on "savebest." Savebest is a utility that checks the performance of neural nets based on a test data file after every N thousand training cycles. When the performance on the basis of the test data file starts to decrease, training of the neural net is automatically stopped.

One problem with this utility is that savebest is designed to halt training based on performance as measured by the reduction of the error or difference between actual and desired output. In the case of financial neural nets, a reduction in the root mean square (RMS) error of the actual versus the desired results may have little correlation with the financial performance of a trading system (as measured by the ratio of average profitability to maximum drawdowns). Thus, choosing models that exhibit lowest error does not necessarily produce models with robust financial performance statistics.

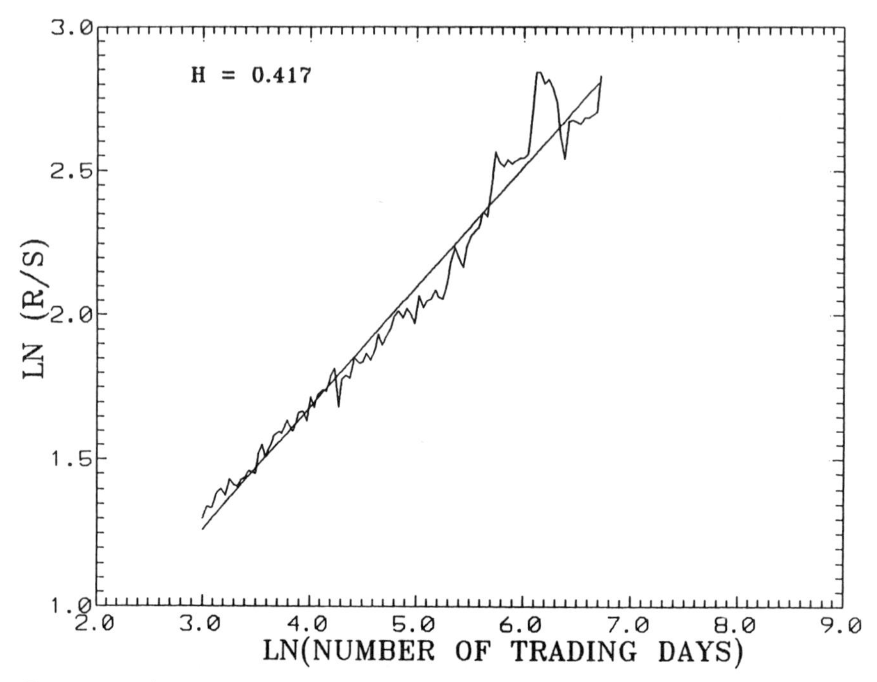

FIGURE 6.8
Hurst coefficient for Federal Funds

Refinements of the savebest process led us to halt training of neural models for ten-year notes after 10,000 to 15,000 learning cycles; for two-year notes the number of learning cycles was between 20,000 and 30,000 iterations. The RMS error of the neural nets at that point was about 0.3. To reduce the probability of the network not learning or saturating, the epoch size was varied between 10 and 32 for two-year models, whereas ten-year models were trained with an epoch of 16.

Neural Net Model for Two-Year Notes

This model uses as inputs the closing yields of the two-year notes, the ten-day trend of two-year yields (computed as the difference between the current yield and the yield ten days earlier), the spread between the two-year yields and the Fed Funds rate, and the ten- and thirty-day volatility of yield changes. The network structure includes a hidden layer with five processing elements and a single output.

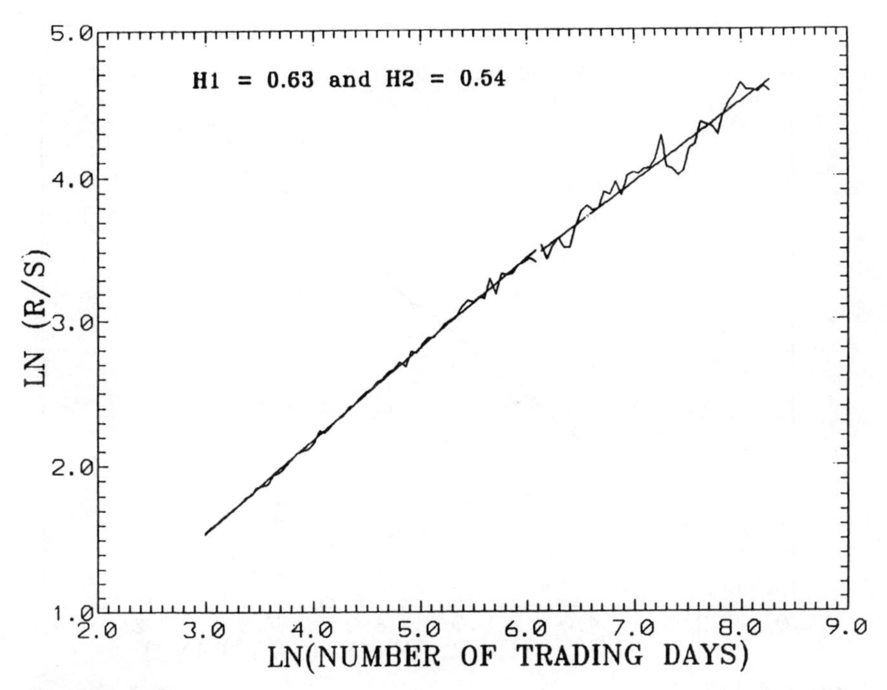

FIGURE 6.9
Hurst coefficient for two-year Treasury notes

The training file for this model contains about 100 records (selected through uniform sampling based on the output signal) from a historical database from July 1, 1988, to June 30, 1992. Training of this model while manipulating the learning rate provided a network after about 20,000 to 30,000 iterations.

Neural Net Model for Ten-Year Notes

This model uses as inputs the closing yields of the ten-year notes, the ten- and forty-day trends of ten-year yields (computed as the difference between the current yield and the yield ten/forty days earlier), the spread between the ten-year yields and the Fed Funds rate, and ten- and thirty-day volatility of yield changes. The network structure included a hidden layer with five processing elements and a single output.

The training file for this model contains about 100 records (i.e., all records with output signals 1.5 standard deviation from the mean)

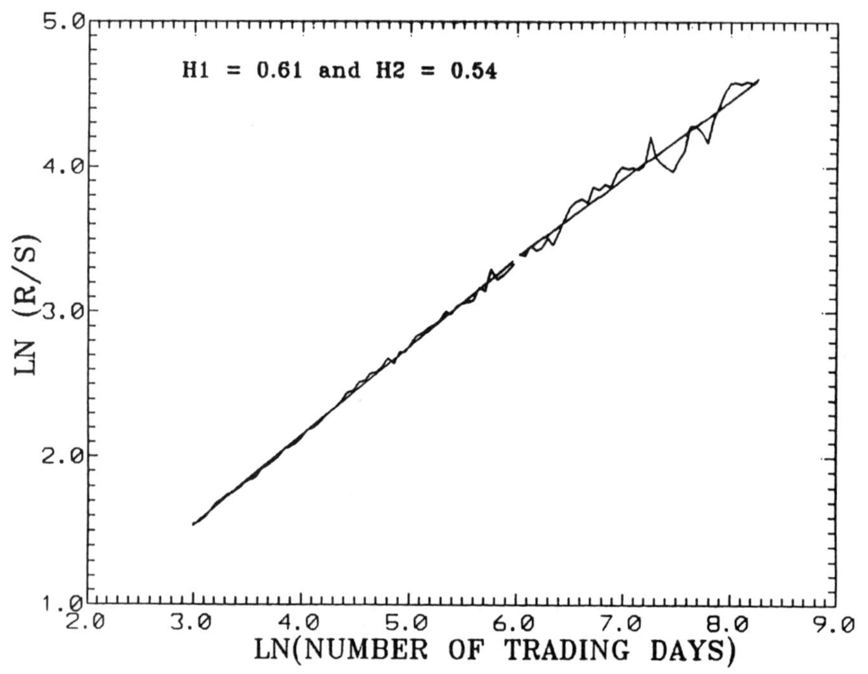

FIGURE 6.10
Hurst coefficient for ten-year Treasury notes

selected from a historical database from July 1988 to the end of June 1992. Training of this model took between 10,000 and 15,000 iterations.

Several other variations were tested before selecting the above model for trading ten-year notes. The two main variations were:

1. A model that uses as inputs the closing yields of the ten-year notes, the spread between the ten-year yields and the Fed Funds rate, and the ten- and thirty-day volatility of yield changes. The main difference is the absence of trend indicators. The network structure included a hidden layer with four processing elements and a single output. The training file for this model contained about 81 records (or 10% selected by uniform sampling from the distribution of the output signal). Training of this model took only 5,000 iterations.
2. A model that uses as inputs the closing yields of the ten-year notes, the spread between the ten-year yields and the Fed Funds rate, and the ten- and forty-day trends of yield changes.

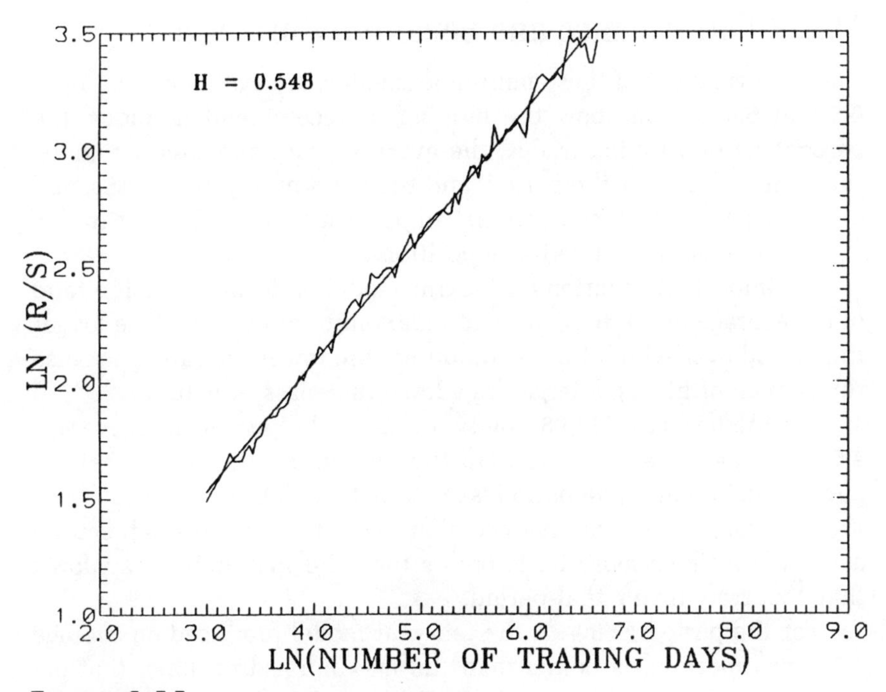

FIGURE 6.11
Hurst coefficient for thirty-year Treasury bonds

> The main difference is the absence of volatility indicators. The network structure included a hidden layer with three processing elements and a single output. The training file for this model contained about 80 records (or 15% selected by uniform sampling from the distribution of the output signal). Training of this model with tweaking of the epoch size was accomplished in about 5,000 iterations.

All models exhibit trade-offs between average profit, maximum drawdown, and frequency of trading. In addition, networks trained with both trend and volatility indicators generally had larger returns than those trained on volatility or trend indicators only; models trained only on trends produced better results than those trained only on volatility. Both types of models produce good results with conservative average returns and minimal drawdowns, but low frequency of trades (approximately ten trades per year).

Model Performance and Sensitivity of Results

The performances of these neural net models are evaluated in Tables 6.5 and 6.6, which show the number of recommended trades, the percentage of winning trades, the average profit and loss per trade, the total return over the period, and the maximum gains, losses, and drawdowns (or successive losses). All performance evaluations in this chapter are based on fixed-size positions.

An important criterion for judging model performance is the ratio of the average profit divided by the maximum drawdown. The longer the period over which this is computed, the lower the ratio—because the chance of hitting a large drawdown increases over time.

For the period FY92-93, the two-year model produced on average 4.6 times more basis points than the maximum drawdown that occurred during the same period (see Table 6.5). In the test period from July 1, 1992, to the end of December 1992, this model produced on average 1.9 times more basis points than the maximum drawdown that occurred during that period.

For the period FY89-93, the ten-year model produced on average 1.3 times more basis points than the maximum drawdown that occurred during the same period (see Table 6.6). In the test period from

TABLE 6.5 *Performance of Neural Net Model for Two-Year Notes*

Parameter	FY92-93	FY93
Period (years)	1.52	0.51
Total number of trades	48	18
Number of profitable trades	35	11
Profitable trades, %	72.9	61.1
Trades/year	32	35
Average gains/profitable trade	10	7
Average loss/losing trade	−9	−8
Average P&L/trade	5	1
S.D. of P&Ls	5	4
Total realized P&L (in basis points)	231	25
Total return (in basis points)	253	47
Max. gain (single trade)	36	21
Max. loss (single trade)	−21	−18
Max. unrealized drawdown	−33	−13
Sharpe ratio	1.05	0.38
Basis points/year	167	92
Avg. basis points/max. drawdown	4.64	1.96

TABLE 6.6 *Performance of Neural Net Model for Ten-Year Notes*

Parameter	Original model		Volatility variant model		Trend variant model	
	FY89-93	FY93	FY92-93	FY93	FY92-93	FY93
Period (years)	4.5	0.50	1.52	0.51	1.52	0.51
Total number of trades	125	13	20	8	13	5
Number of profitable trades	80	8	17	6	10	5
Profitable trades, %	64.0	61.5	85.0	75.0	76.9	100
Trades/year	28	26	13	16	9	10
Average gains/profitable trade	8	7	5	4	4	5
Average loss/losing trade	−9	−5	−5	−5	−4	0.0
Average P&L/trade	2	2	3	2	2	5
S.D. of P&Ls	4	2	1	1	1	1
Total realized P&L (in basis points)	264	29	66	16	31	24
Total return (in basis points)	271	37	66	16	31	25
Max. gain (single trade)	38	21	9	9	17	9
Max. loss (single trade)	−35	−7	−8	−8	−6	0.0
Max. unrealized drawdown	−24	−22	−14	−14	−15	−14
Sharpe ratio	0.53	0.91	2.47	1.35	1.93	4.03
Basis points/year	60	74	44	31	20	48
Avg. basis points/max. drawdown	1.32	2.83	5.80	2.13	2.01	N/A

July 1, 1992, to the end of December 1992, this model produced on average 2.8 times more basis points than the maximum drawdown that occurred during that period.

The volatility variant model produced on average 5.8 times more basis points than the maximum drawdown that occurred during the period FY92-93 (see Table 6.6). In the test period it produced on average 2.1 times more basis points than the maximum drawdown. As also shown in Table 6.6, the trend variant model averaged twice as many basis points as the maximum drawdown in the period FY92-93; in the test period it produced 25 basis points and had no drawdown. A similar network with one additional processing element at the hidden layer increased the return of the network in the test period to 36 basis points with no drawdown.

Models trained on FY91-92 data tended to produce better results than those trained on the data since FY89 (which can be explained by the relationship between input space and the number of samples; given more hidden processing elements and a longer experience period, the performance characteristics of the models could be similar). The sampling method for the training file was invariant to the input space for ten-year notes, whereas it was much more important for two-year notes. Uniform sampling over the entire period produced much better performance characteristics than models trained on the basis of standard deviation sampling.

A sampling method is intended to account for the population of data (e.g., similar means and variances). The simplest sampling methodology, the *simple sampler*, samples the distribution such that the probability of the occurrence of each record is equally probable. This is known as *uniform sampling*, as it selects samples uniformly over the distribution. It is implemented by dividing the distribution into N ordered bins and selecting one sample randomly from each bin. We instead used the first element of each bin, rather than a randomized record from within each. Given a uniformly distributed random variable X, it is possible to create another random variable that has any desired distribution. For example, with $X \sim$ uniform(0,1), one can generate a random variable $Y = \sqrt{-2\log(X)}\sin(2\pi X)$ that is normally or Gaussian distributed with mean 0 and variance 1, i.e., $Y \sim$ normal(0,1).

Other sampling methodologies can be used to elicit interesting facets of the data set. One such method is to sample all records outside or within a certain multiple of the standard deviation. This method assumes that the distribution of data is Gaussian or nearly symmetric. *Standard deviation sampling*, based on the output signal, produces more appropriate samples because the objective of the model is learning properties about outliers.

With regard to sensitivity to training time, all models exhibited similar features: the equity curve of the two- and ten-year models showed rapid increases in equity in FY89-90, slower increases in FY91-92, and again large equity gains during FY92-93. Undertrained or overtrained models tended to perform less well. When the model is undertrained there tends to be no recovery (e.g., in the period FY92-93), whereas models that are overtrained generally have larger drawdowns (a double-humped equity curve).

Performance of a Portfolio of Neural Net Models

The neural net models described above can be combined in a portfolio. Various approaches can be taken to the actual weighting of each model. We have adopted a weighting based on duration so as to give equal risk to each model in the portfolio.

Table 6.7 shows the combined results of both models. The percentage of winning trades for this portfolio is about 60 to 65%; average returns are 150 basis points per year over 4.5 years. In FY93, this portfolio produced (without retraining) 89 basis points after six

TABLE 6.7 *Portfolio Performance—July 1988 to December 1992*

Parameter	FY89-93	FY92-93	FY93 (6 months)
Period (years)	4.5	1.5	0.5
Total number of trades	133	43	15
Number of profitable trades	86	28	9
Profitable trades, %	65	66.3	61.3
Trades/year	29	28	30
Total basis points	677.2	251.0	88.8
Max. gain	52.4	36.7	36.7
Max. loss	−47.3	−23.9	−16.7
Max. drawdown	−64.0	−26.2	−26.2
Avg. basis points/year	150	165	89
Basis points/max. drawdown	2.3	6.3	3.4

months and 74 basis points after 12 months. The maximum drawdown of the combined results was 64 basis points for the past 4.5 years and 26 basis points in FY93. The performance ratios are 2.3 over 4.5 years, 6.3 over the past 1.5 years, and 3.4 in the first six months of FY93.

To test the robustness and stability of the neural net portfolio results, the average return and the maximum drawdowns were examined in moving windows (see Table 6.8). The average returns over approximately 1,000 windows varied from 34 basis points in three months, to 69 basis points in six months, to 127 basis points for one year. The standard deviation of these average returns was 36, 51, and 62 basis points for three-, six-, and twelve-month windows, respectively. This portfolio of neural net models had a 66% chance of producing a minimum of 65 basis points (i.e., 127 basis points minus 62) per year.

The average of the drawdowns in three-, six-, and twelve-month overlapping windows varied from 30 to 48 basis points. Thus, based on 912 overlapping windows of one-year periods, *this portfolio produced on average 2.6 times more basis points than the maximum drawdowns.*

Figure 6.12 shows the cumulative equity curve of this portfolio (right axis) together with the evolution of the two- and ten-year yields from August 1988 to December 1992. A blow-up of this plot for the period July 1992 to December 1992 (Figure 6.13) demonstrates the performance of this portfolio based on 150 trading days, during which no records were used for training the neural net models.

TABLE 6.8 *Robustness and Stability of Portfolio Results (Measured by 3-, 6-, and 12-Month Windows)—July 1988 to December 1992*

Parameter	3-month windows	6-month windows	12-month windows
Number of windows	1,098	1,033	912
Average P&L (1)	34.8	69.3	127.4
S.D. of P&L	35.9	51.3	62.0
Avg. max. drawdown (2)	29.8	37.2	48.4
Avg. ratio (1)/(2)	1.1	1.8	2.6

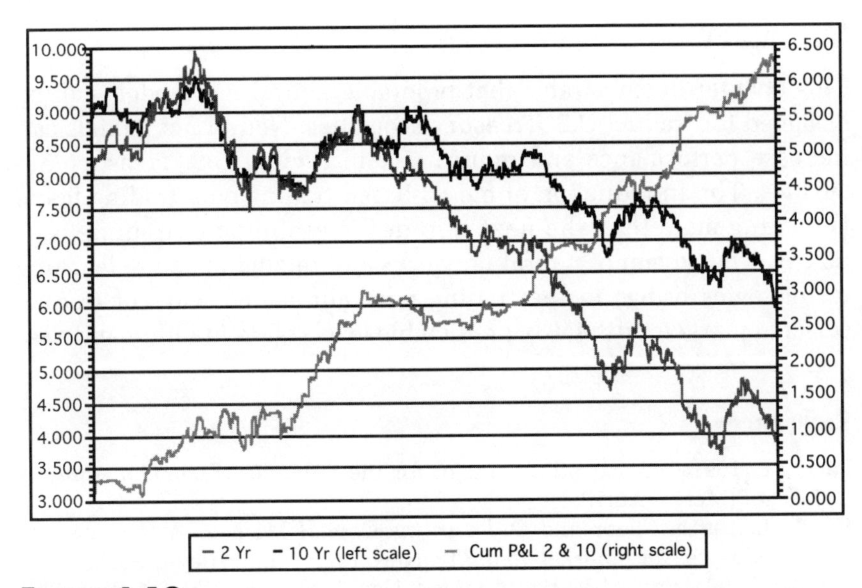

FIGURE 6.12
Equity curve of portfolio of neural net models for two- and ten-year notes (July 1988 to December 1992)

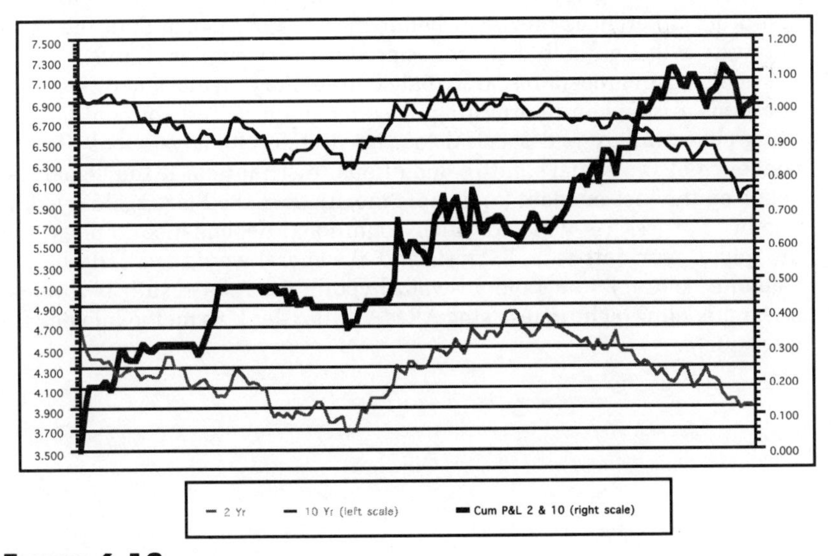

FIGURE 6.13
Blow-up of equity curve of portfolio of neural models for two- and ten-year notes (July 1992 to December 1992)

Summary

This chapter demonstrates that profitable neural net models can be designed for trading U.S. Treasury securities. Neural net models can increase performance and reduce risk in trading U.S. Treasury Securities. The introduction of model-based trading onto trading floors is nothing more than the next step in the evolution of trading technology. The advent of neural networks and related machine learning technologies brings to the trading floor automated ways of trading selected markets with fairly predictable levels of profitability and risk.

References

Peters, E. 1991a. "A Chaotic Attractor for the S&P 500." *Financial Analysts Journal* (March–April).

Peters, E. 1991b. *Chaos and Order in the Capital Markets: A New View of Cycles, Prices and Market Volatility.* New York: John Wiley & Sons.

Peters, E. 1994. *Fractal Market Analysis.* New York: John Wiley & Sons.

Notes

1. The first-order differences were computed by taking the log of the ratio of the current closing yield divided by the previous-day closing yields. The first-order differences provide a better measure of change than the rates of return, although both approaches produce very similar results.
2. It is evident from recent research that contradictory results and unresolved questions arise when interpreting these techniques. Tests of nonlinearity using chaos theory and spectral (or signal energy) analysis look for dependence of the signal's entire conditional distribution. If the signal depends on the first moment (or sample mean), then the Martingale hypothesis of "fairness" is violated. Such conditional dependence implies that the higher moments (e.g., kurtosis and skewness) would similarly be dependent, contrary to a random-walk hypothesis. As a result, traditional linear modeling techniques using ARMA or Box-Jenkins methodologies are inadequate.

CHAPTER 7

Neural Nets for Foreign Exchange Trading

Henry Green and Michael Pearson

Fundamentally, currencies are not securities, although they are actively traded. The largest participants, the central banks, are not return maximizers. Therefore, their objectives are not necessarily those of rational investors.

Edgar Peters
Fractal Market Analysis, 1994

In this chapter Henry Green, formerly with Hill Samuel Bank and currently with Financial Technology Systems Development Ltd. in London, and Michael Pearson of the Royal Bank of Scotland provide an overview of the capabilities and limitations of neural networks for forecasting foreign exchange rates. As an example, Green and Pearson use trading of the "cable rate," or the U.S. dollar to U.K. pound rate. The overall findings suggest that an 18.6% return on capital is possible.

Foreign Exchange Markets

Foreign exchanges are traded on a nearly continuous basis. This type of dealing generates information about worldwide financial markets and takes place in one of the least regulated of marketplaces. In London more than 11,000 dealers work the spreads between the buy and sell of foreign exchange (FX) rates. The City of London represents the single largest global market for trading foreign exchange. On average, over 300 billion dollars (U.S.) are traded daily, which represents nearly the total combined value of the New York and Tokyo markets. In essence, many trades made on the foreign exchange markets represent a game or a preoccupation with beating the market for rewards that can be substantial, without employing physical capital.

Many argue that economic fundamentals play a large role in the establishment of trends and rates in foreign exchange markets. Fundamentals do play a role—but their significance is questionable. With the exception of those who are required to manage specific currency risks, economic fundamentals do not play a role in determining ultimate buy/sell decisions.

Design Strategy

One of the key motivations for development of a neural network is to improve performance and decrease risk. Risk management is not only about control, it is also about optimization of resources. Neural networks can make contributions to maximizing return, reducing costs, and limiting risks (i.e., by minimizing capital at risk, and thus reducing the amount of economic capital consumed). In the remainder of this section, we will discuss the inputs and results obtained by applying neural nets to the U.S. dollar and U.K. pound rate.

Input Data

There is no best input data set for neural network modeling. The input data selected in Table 7.1 is for forecasting a one-day forward U.S. dollar to U.K. pound rate (i.e., the "cable rate"), with emphasis on direction as opposed to magnitude. The cable rate that actually

TABLE 7.1 *Typical Set of Input Data for a Specific FX Rate Forecast Problem*

Currency	Input data for each
French franc	FX rate (to U.S. dollar)
German Deutschmark	Volatility of FX rate
Japanese yen	Interest rates
Swiss franc	Oscillators (various)
U.K. pound	Output feedback (U.S. dollar)
	Forecast error (=err)[a]

(a) err $= \Sigma$ (target $-$ prediction)^2/Σ (target $-$ mean)^2
$=$ arg. relative variance

results in the market is used to derive the average relative variance error, which is used as feedback to the neural network input.

Table 7.2 shows the general data domain that is appropriate for FX rate forecasting using neural networks. The different data types can be rationalized to be relevant in combinations that are "meaningful" to a target forecast problem. It is clear from the number of combinations that much trial and error is involved just in the selection of the input data. Many so-called key indicators can be constructed from the basic economic and price data series, including timing indicators, moving averages, point and figures, turnover ratios, and/or others.

As there is no central market exchange for trading FX, daily exchange volumes are impossible to measure accurately. In addition, the information that is typically broadcast by the electronic information vendors is not always correct. Considerable care must be taken to clean (or scrub) data before actual use.

Four principal rules have been established for using market data in a neural network (these apply in general and not just to FX forecasting). These rules are summarized below.

1. Select key data that are representative of the market

- Market rates
- Liquidity
- Volatility
- Continuous trading frequency
- Cross-rates and spreads

2. Select or measure the most exact data from the market.

Data type	Market representation
Foreign exchange	High
Bonds	Medium
Equity markets	Low
Economic	High uncertainty as to reported estimates

3. Verify the liquidity of the market

It is fundamental that the assumption of the trading models be realizable in practice. If a model assumes that any required transactions are possible and that the market is not liquid, then the net outcome can be anticipated to be poor.

TABLE 7.2 *Initial Input Data Selection for an FX Neural Network Trading System*

Data type	Examples	Maturity	Frequency	Detail	History
Foreign exchange rates	Majors, minors, EMS	Spot	Tick-by-tick, hourly, weekly, monthly	Open, close, high, low	As far back as possible

4. *Give preference to markets and instruments that trade continuously*

It is best to select a market and instruments that closely approximate true continuous trading. Markets that have less than a true type of continuous trading are challenging, and the cost and effort to use data corresponding to such markets is very high. Although many smoothing and rollover date strategies exist to create an apparently continuous activity, the underlying processes of the market cannot be expected to be adequately resolved by synthetic data series.

Development Experience

During the design and development of a neural network application, input data identification, smoothing, filtering, and scaling require considerable effort. Objectives and utility of the data must be defined, interrelationships established, and an input data vector found that is minimal in size yet preserves the maximum amount of information.

The structure of the neural network can be optimized only through a trial-and-error approach. Keeping a fixed network design and limiting the selection of neural network algorithms to tried and tested ones provides a manageable initial approach. After the performance of the neural network has been stabilized by using a finite learning data set, the data set must be tuned by perturbing selected variables (a subjective process) and then by systematically eliminating redundant input variables. The process of identifying the best neural network algorithm(s) requires experimentation and patience. Because of regime changes in market data, the optimal design and resulting output achieved can be expected to remain viable only for a finite time.

The experience we have gained can be summarized as follows: If a given neural network implementation forecasts price movements precisely, then the neural network design and input data are subject to a finite life span. Maintaining the neural network's optimal performance requires developing and maintaining human expertise at the input, design, and output levels.

Results

Forecasting the U.S. dollar against the U.K. pound exchange rate (the "cable rate") will be demonstrated using a neural network with a hybrid error back-propagation and Cauchy algorithm. The network was run on a SUN SPARC UNIX workstation and initially used 26 input variables. After elimination of redundant input variables, the final network was run using only five input variables. Network learning employed historical data ranging back to January 1988. The aim of the demonstrated forecasting was to establish sensitivity to historical data range and learning iterations to provide a basis for qualifying different input data vectors. The principal results are compared with forecasts using single-variable time-series modeling.

Figure 7.1 presents the results of a neural network trained with approximately four and a half years of historical data. The results indicate a generally acceptable fit of the historical time-series, as evidenced by the trading performance. For comparison, forecasts have also been made based on one-, three-, six-, and nine-period autocorrelation ARMA models. The forecasts made using the ARMA models are less accurate and, although not explained in detail here, can be shown to require optimization for new data whenever the forecast range exceeds two or three days. This short-term performance decay is not as strong with the neural network results.

The results of Figure 7.1 show the combined measures of historical volatility of the cable rate averaged over 30 days and the cumulative profit or loss from trades carried out using turning points for the trading strategy. The annualized return on capital achieved for the trading period from April 1992 to the end of March 1993 was 18.6%. The performance attribution was 14% return on outright foreign exchange rate positions and 4.6% due to gains from interest rate income. The average Sharpe ratio for the trading period was 3.5. The percentage of winning trades was 63%. The maximum drawdown over the simulated trading was 2.2% of initial capital. A total of 157 trades were carried out over a 12-month period. Profit and loss calculations

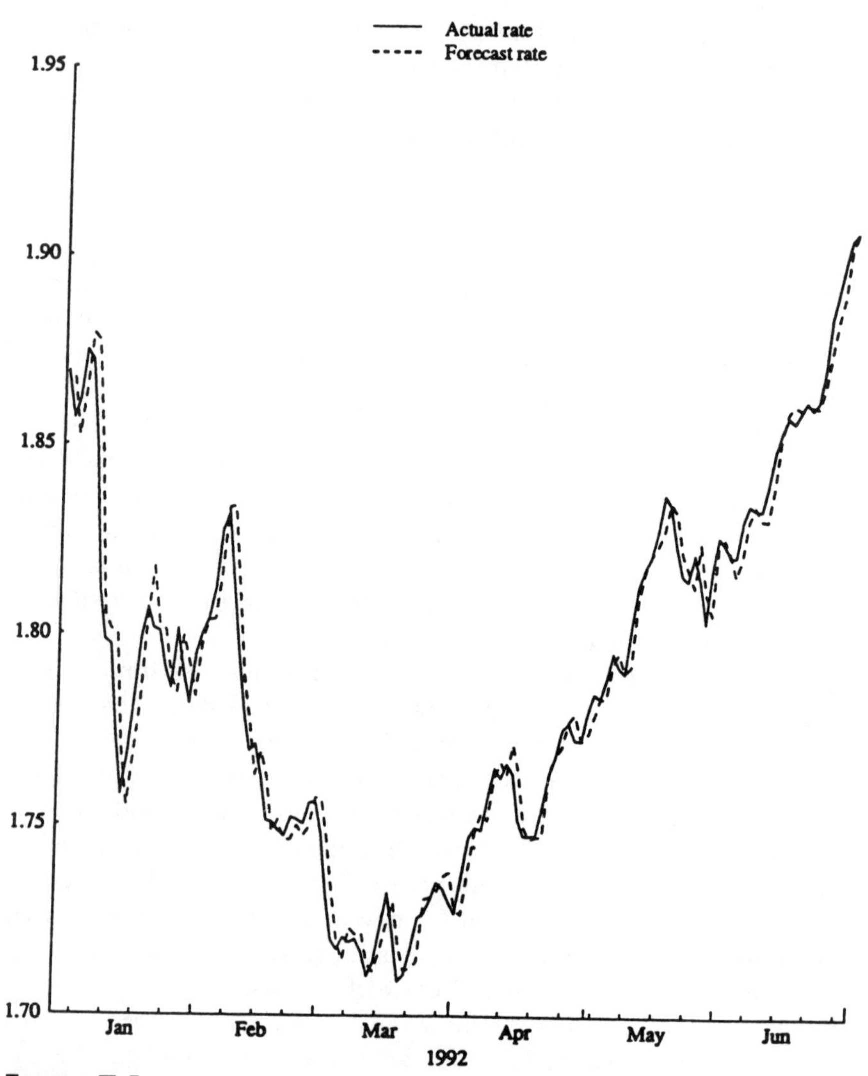

FIGURE 7.1

Cable rate forecast for 1992 based on monthly retraining

incorporated real market transaction costs and used a foreign exchange rate slippage factor on both the bid and offer rates. Overall, the cable rate forecasting model behaved well and remained stable, and the trading decision model (turning point strategy) provided for an overall improvement of 2.2% return on capital in contrast to outright trading every forecast signal.

In principle, a neural network can be used to forecast a complex process such as foreign exchange rate movements. Establishing the proper number and mix of input parameters is a complex process, and no single input data vector can be guaranteed to result in a unique and optimal forecast. The testing phases of the network used included variations in the choice of algorithms, length of historical data, number of learning iterations, and input data vector. The decision to start with a small number of input variables was based on a number of factors. The most practical was that a small-size input vector facilitated changes to the vector components and also with identification of dominant or key input variables. The final input data vector is, of course, expected to be larger and to be a balance between obvious (in the sense of an expert's opinion) and not so obvious input parameters.

In summary, the developments leading to a trial demonstration model help to harness experience and to facilitate the more rigorous steps that must be taken in the development of a working or production system. There are no shortcuts. Once started, a systematic development program must be followed.

PART TWO

Strategy Optimization with Genetic Algorithms

zi qiáng bù xi:

A person should never stop striving for self-improvement.

CHAPTER 8

Genetic Algorithms and Financial Applications

Laurence Davis

Genetic programming may be more powerful than neural networks or other machine learning techniques.

John R. Koza
Genetic Programming, 1992

Genetic algorithms (GAs) are problem-solving techniques with an astonishing property: they solve problems by evolving solutions as nature does, rather than by looking for solutions in a more principled way. Genetic algorithms, sometimes hybridized with other optimization algorithms, can be applied to a wide range of problems—primarily because of several brilliant design decisions made by John Holland when he invented GAs in the early 1970s. Laurence Davis, president of Tica Associates, has written several books on genetic algorithms and has applied GAs to real problems for 11 years. In this chapter Davis discusses ways in which GAs are used in the financial area.[1]

The Origin of Genetic Algorithms

Why Simulate Evolution?

In the early 1970s John Holland, a founder and innovator in the field of computer science, was inspired by the theory of evolution to create a computer algorithm (Holland 1975). Holland was obsessed by the fact that, despite early pronouncements and expectations, computer scientists in the early 1970s had not come close to duplicating with computers various natural abilities that seemed rather simple and mundane: the ability to see, to understand speech, to navigate in the world, and so forth. Twenty years after the first widely publicized predictions that those capabilities were just around the corner, scientists were still learning just how hard it would be to endow computers with such skills. Yet some of these capabilities exist in "unintelligent" creatures in the natural world, and biologists argue that they have arisen naturally through evolutionary processes.

According to the theory of evolution, such capabilities were produced by chemical operations on chromosomes made up of DNA. Holland wondered whether he could make evolution work in a different way, by applying computerized processes to computerized chromosomes. It is not at all obvious how to do this, and other theoreticians have constructed different computer algorithms while looking to evolution for inspiration. (Their approaches and Holland's genetic algorithm approach are currently merging, in a hybridization process reminiscent of those that occur in nature.) Here, I shall concentrate on Holland's approach, the one with which I am most familiar.

How to Simulate Evolution

There are two critical features of Holland's version of computerized evolution that have practical and theoretical power: *selection by fitness* and *crossover*. The combination of these two is unique to the genetic algorithm. Later on, I will explain what these features are and how they work together in the context of simulated evolution to make the development of a genetic algorithm population strikingly similar to that of a natural species.

First, let us note that the theory of evolution includes these features, which Holland built into his genetic algorithms:

Evolution operates on encodings of biological entities, rather than on the entities themselves. Evolution takes place at the level of the chromosome. To understand what a chromosome encodes, it is necessary to build the animal whose cells contain it. Similarly, Holland's genetic algorithms operate on encodings of solutions to problems, and those encodings must be decoded in order for them to be understood and evaluated.

Nature tends to make more descendants of chromosomes that are more fit. This principle, sometimes called "survival of the fittest," says that the chromosomes in a population associated with the fittest individuals will, on average, be reproduced more often than chromosomes associated with unfit individuals. The notion of fitness is a relative one. The fittest individual in a species early on may be greatly inferior to the fittest individual much later in evolution. Fitness is determined by comparing an individual with the other members of its current population. In nature this comparison takes place by seeing which individuals eat well enough and live long enough in their environment to reproduce. In a genetic algorithm, the user who has a problem to solve provides a function that evaluates chromosomes. This *evaluation function* plays the role of the environment for a genetic algorithm.

Variation is introduced when reproduction occurs. Mutation and recombination occur in nature during reproduction. Holland designed his genetic algorithms so that analogues of these two types of operations take place in a genetic algorithm when parent chromosomes reproduce. Mutation operations introduce random diversity into the population. Recombination operations allow chromosomal material from different parents to be combined in a single child.

Nature has no memory. Whatever nature "knows" about its search for good chromosomes is contained in the genetic makeup of the DNA that currently exists. Similarly, whatever a genetic algorithm "knows" about its search for a good solution to the problems posed is contained in the genetic material currently in its population of chromosomes.

How the Algorithm Works

These principles underlie Holland's basic genetic algorithm. Now we can consider how the algorithm works, with reference to Figure 8.1. To begin with, an initial population of chromosomes is created, and each chromosome is evaluated with the user-supplied evaluation function. The subsequent operation of the algorithm is like the operation of an engine that cycles through a series of four steps again and again until it is shut off. A genetic algorithm cycles again and again through the following four steps:

1. *Reproduction* occurs. Parents are selected from the population using a random selection procedure. Each parent's selection

chances are biased so that those with the highest evaluations are most likely to reproduce. Children are made by copying the parents, and the parents are returned to the population.

2. Possible *modification* of children occurs. Mutation and recombination operators are applied to the children. These operators are described below. Their chances of being applied and the points on the children at which they are applied are randomly determined.
3. The children undergo *evaluation* by the user-supplied evaluation function.
4. Room is made for the children by *discarding* members of the population of chromosomes. (The weakest population members should be the most likely to be discarded.) The children are then inserted into the population.

Steps 1 to 4 are repeated until the algorithm is halted. The composition of the final population of chromosomes—the best chromosome or chromosomes—is the genetic algorithm's solution to the problem.

If this is your first exposure to genetic algorithms, there are some important things that you should know. First, note that new chromosomes are constructed and old ones discarded on each pass through the cycle. Since the reproduction process tends to choose the "fittest"

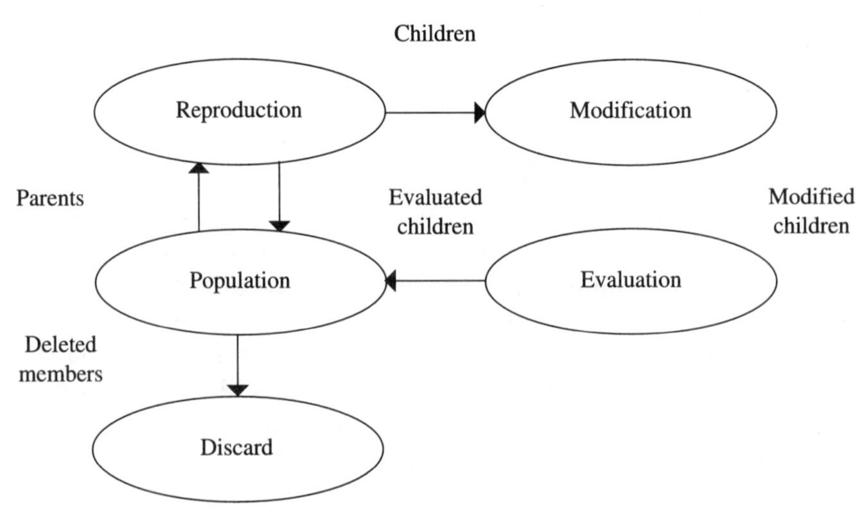

FIGURE 8.1
GA cycle of reproduction

members of the population as parents, the population tends to evolve. If all goes well, an initial population of relatively undistinguished solutions may evolve to yield the algorithmic equivalent of an Einstein or a Michelangelo. Second, note that nothing has been said about how chromosomes encode solutions. Holland's encoding technique will be explained below, and some others will be described later. A surprisingly diverse array of encoding techniques has been developed and employed by engineers in the field. Finally, note that little has been mentioned about how parents are chosen or about how mutation and recombination occur. We turn to those topics next.

Specifics of Holland's Genetic Algorithm

Now that we have considered how the algorithm works at a high level, let us consider the components of Holland's first genetic algorithm. There are at least as many ways to implement the various components of a genetic algorithm as there are genetic algorithm researchers, but the techniques described here were the first ones, and they still are used in some genetic algorithm applications today.

First, Holland's representation was remarkably simple and general. The topic of what representation to use is a complicated one, and may be the most critical decision to make when designing genetic algorithms. A chromosome in Holland's genetic algorithms was a list of zeroes and ones—a *bit string*. Computers represent everything they do with bit strings, a simple testament to the generalizability of this *binary* representation strategy. A further strength of this representation technique is that theoretical results may be easiest to prove when applied to this representation. A two-character alphabet is the simplest possible one, and this fact leads to theoretical and practical advantages when binary notation is employed.

Second, Holland's selection procedures were simple. In Holland's genetic algorithms, all evaluation functions returned positive numbers; the higher the number, the better the chromosome. Holland used a technique now called *roulette wheel* parent selection to determine which population members are chosen for a reproduction event. Using this technique, each chromosome's evaluation is proportional to the size of its slice on a roulette wheel, as shown in Figure 8.2. Selection of parents to reproduce is carried out through successive spins of the wheel. After a spin, the chromosome chosen is the one whose slice the arrow points to. Thus, in Figure 8.2, chromosome 2, with an evalu-

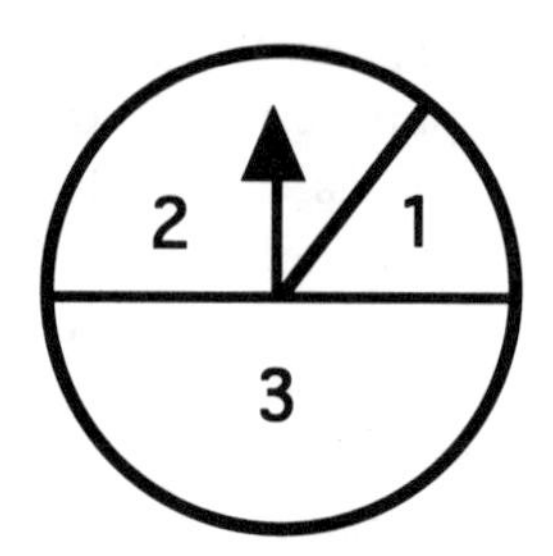

FIGURE 8.2

Parent selection by the roulette wheel technique. If chromosome 1 has an evaluation of 5, then chromosome 2 has an evaluation of 10 and chromosome 3 an evaluation of 15. The chances of a chromosome being chosen as a parent are proportional to its evaluation.

ation twice that of chromosome 1, is twice as likely as chromosome 1 to be chosen as a parent in each reproduction event. Chromosome 3 is one and a half times more likely to be chosen than chromosome 2, and three times more likely to be chosen than chromosome 1.

Third, Holland replaced his entire population during each reproduction event. This technique, now called *generational replacement*, causes each chromosome to have only one set of chances to reproduce—at the point when the children of its generation are being created. Good chromosomes will tend to have more than one child, while bad ones will tend to have none. An important benefit of this technique is that it is relatively immune to noise in the evaluation function. The genetic algorithm may be slowed down by variation in the evaluation function, but the use of generational replacement causes the population to evolve, on average, to include chromosomes that repeatedly tend to evaluate well.

Fourth, Holland used two operators, *binary mutation* and *one-point crossover*, to introduce diversity during reproduction. Both of these operators are applied with probabilities that are parameters of the algorithm. At a very low level of probability, binary mutation replaces bits on a chromosome with randomly generated bits. When one-point crossover is applied, it takes two child chromosomes, selects at random a point on those children, and swaps the genetic material on the children at that point. Figure 8.3 shows the result of applying one-point crossover to two binary chromosomes. The bottom portions of the two chromosomes have been swapped by the one-point crossover operator. The use of crossover is a very important feature of the genetic algorithm, and may be critical to its success.

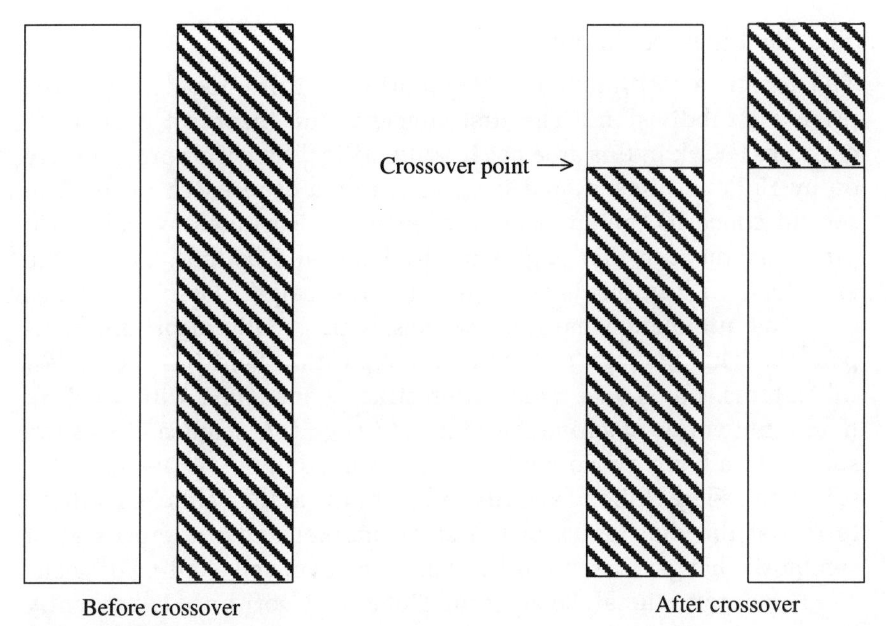

FIGURE 8.3
One-point crossover

To summarize, Holland's first genetic algorithm used generational replacement, roulette wheel parent selection based on evaluation, binary representation, binary mutation, and one-point crossover. It evolved its initial populations by randomly generating bit strings of the desired length. Specifications of population size, probability that a bit will be mutated, and probability that crossover will occur would be sufficient to allow a competent genetic algorithm developer to recreate Holland's genetic algorithm, without further knowledge of the original system.

This genetic algorithm and slight variants on it are sometimes called *traditional genetic algorithms*. It is important to note that all genetic algorithms in use today descend directly or through a series of evolutionary changes from this initial genetic algorithm. Other genetic algorithms may be better adapted to the environmental niche of the problem they are solving, but they would fail miserably on problems of other types. Holland's algorithm is remarkably robust, in that it is applicable to a wide variety of problems and performs impressively across a variety of domains.

The Sociology of the Field

The genetic algorithm field, now entering its third decade, has three principal subdivisions. The first concerns the theory of genetic algorithms. Work in this area began with critical theorems proved early on by Holland himself, and is an active area of research today. The second concerns the application of genetic algorithms to real-world problems, an area that is growing by leaps and bounds. This is the area dealt with in this book. The third area, called *classifier systems*, concerns automated learning systems with genetic algorithm components. This division of the genetic algorithm field has the fewest inhabitants, but it has many interesting connections with work in neural networks, machine learning, and cognitive science. It was the subject of a book by Holland and others (Holland et al. 1986), and it is being used by Holland and his collaborators at the Santa Fe Institute to model the mechanics of the stock market and the processes of symbiosis that gave rise to cells. These are novel and exciting projects.

The International Society for Genetic Algorithms is currently being formed. *Evolutionary Computation*, a journal reporting research on genetic algorithms and other algorithms inspired by evolutionary phenomena, is published by The MIT Press. The International Conference on Genetic Algorithms is held during the summer of odd-numbered years, and the Parallel Problem Solving in Nature conference in Europe is held during the summer of even-numbered years. Genetic algorithm topics often are discussed at neural network, operations research, artificial intelligence, artificial life, and machine learning conferences.

Finally, an electronic mail message sent to ga-list-request-@aic.nrl.navy.mil on the Internet will add you to the readers of the Ga-List, an electronic bulletin board on which GA issues, products, papers, and conferences are announced and discussed.

For Further Reading

For an excellent overview of the history and theory of the genetic algorithm field, see David E. Goldberg's book, *Genetic Algorithms in Search, Optimization, and Machine Learning* (Goldberg 1989). For more material on applications of genetic algorithms, see *Handbook of Genetic Algorithms* (Davis 1991). *Advanced Technology for Developers* (High-Tech Publications, Sewickley, PA) also contains articles on genetic algorithm applications.

Genetic Algorithms and Financial Applications

Genetic algorithms have been applied in several ways in the financial field, and there are a number of other potential avenues of application. Genetic algorithm researchers and consultants are becoming increasingly involved with financial institutions, and the number of fielded financial applications of genetic algorithms can be expected to grow dramatically over the next few years. A selected subset of real and potential applications is surveyed below to acquaint you with some of the techniques that can be used to apply genetic algorithms to financial problems, and to set the stage for the more detailed genetic algorithm applications in chapters to follow.

Genetic Algorithms for Determining Model Applicability

Dr. Norman Packard, a founder of the Prediction Company in Santa Fe, observed that the techniques he and his associates use for predicting the behavior of nonlinear time series are better in some kinds of situations than others. The Prediction Company buys and sells stocks, and the ability to predict confidently the behavior of time series containing market data would be quite valuable. Packard investigated genetic algorithms for determining when their prediction tools would work well.

To illustrate Packard's approach, let us assume that his company is predicting market behavior based on 18 indicators. These indicators might be moving averages, measures of volume, algebraic combinations of more basic indicators, and so forth. Packard would use a genetic algorithm in such a case to evolve descriptions of indicator combinations that are correlated with good predictions from his prediction system.

The methodology is as follows. Packard's chromosome is a list of 36 real-valued numbers. Two of these numbers are associated with each of his 18 indicators, one denoting a low value of the indicator and one denoting a high value. Each chromosome, or list of 36 numbers, is evaluated by determining how well, on average, Packard's prediction system has done historically when each indicator value has fallen between the low and high values associated with it on the chromosome.

The genetic algorithm can evolve populations of chromosomes that home in on good combinations of low and high parameter values.

The result of successive runs of the system is a set of chromosomes that describes combinations of parameter values correlated with high performance of the prediction system. To use such chromosomes, Packard's group need only wait until current indicator values fall into the ranges specified by one of these chromosomes, and then trade based on the predictions of the system. That the predictions will be accurate can be assumed with some confidence.

Genetic Algorithms for Evolving Market Indicators

One of the most interesting and fastest-growing areas of the genetic algorithm field at present is the area called *genetic programming* by John Koza, one of its pioneers. In Koza's seminal book (1993), techniques are described for using genetic algorithms to evolve computer programs to satisfy a wide range of goals. A financial application using this type of technique, but developed independently of Koza's work, is described in Chapter 9 of this book. Andrew Colin's use of genetic programming techniques to create algebraic combinations of market indicators that produce higher predictive performance when input to a neural network is innovative and suggestive. (Colin has adapted the traditional mutation and crossover techniques so that they can be applied to his tree-structured chromosomes.) Many more applications of these techniques can be anticipated as the successes of pioneers like Colin and Koza become better known.

Genetic Algorithms for Training Neural Networks

I am aware of at least two commercially available neural network systems for financial applications that include genetic algorithm training techniques.

One neural network/genetic algorithm system was built by Casey Klimasauskas of NeuralWare™, who has written a number of articles on genetic algorithm applications for the financial industry, generally in conjunction with neural networks (see also Chapter 1 by Klimasauskas in this book). Thanks to Klimasauskas, NeuralWare is one of the first neural network systems to include a genetic algorithm training module. NeuroForecaster™, a neural network/genetic algorithm system developed by NIBS Pte Ltd. in Singapore, was also designed with financial applications in mind and includes a genetic algorithm training option.

There is presently much interaction between researchers in the genetic algorithm and neural network communities. Readers interested in pursuing this topic are directed to Schaffer et al. (1993), which provides a good survey.

Spreadsheet Genetic Algorithms and Financial Applications

Most publicly available genetic algorithm software systems require some programming ability, if only to link the genetic algorithm with the software that evaluates chromosomes. An ingenious exception is the Evolver™ system, created by Matt Jensen of Axcelis in Seattle. Evolver interacts with the Excel™ and WingZ™ spreadsheet programs so that users who are familiar with these programs can define their evaluation functions through the spreadsheet interface, thus making the genetic algorithm accessible to nonprogramming users.

More than 800 copies of this genetic algorithm have been sold, and I have spoken with users who have used it to predict prepayment rates for mortgage-based securities, to refine models of market movement, and to select securities for portfolios. Evolver™ is becoming known as a genetic algorithm tool for those who do not want to work with the genetic algorithm software directly.

Credit Application Software

Genetic algorithms have been used to build criteria for approving or rejecting credit applications. KiQ, a London company, has built and is marketing such a software system, and reports dramatic savings for some of its customers.

Database Mining

Many financial institutions maintain databases of information on customers, transactions, and securities behavior. There are a number of ways that the information in such databases may be used to good effect: to determine whether to approve or reject loans or credit increases; to decide who should receive marketing materials or who should be contacted by telephone; or to decide which securities to buy or sell. Some of the techniques for extracting information from such databases fall into a field called *database mining*, which is the enterprise of searching for undiscovered (and valuable) correlations or generalizations that are buried in the gigabytes of information the databases contain.

Genetic algorithms can be used to improve the performance of these database mining techniques. Approaches include improved training of neural networks, using techniques referred to above; improved classification techniques, including improvement of case-based and nearest-neighbor classification routines; improvement of the parameters in rules of expert systems that accomplish database mining; and generation of such rules using genetic programming techniques.

Application Principles

Suppose you have an optimization problem, and you wonder whether a genetic algorithm should be used to solve it. Below are a number of considerations that are often helpful in reaching a decision.

Is a Small Improvement in Optimization Sufficient?

Most real-world optimization problems of any importance already have optimization approaches that have been handcrafted especially for them. The amount of additional optimization that might be produced by a genetic algorithm, perhaps in conjunction with specialized algorithms, may not be great. Accordingly, it is important to look for problems where a small improvement (1 to 10%, for instance) is worth the time and trouble of building a genetic algorithm.

Examples of domains for which this is the case include currency or securities trading, where small performance improvements can yield significant payoffs; database mining, where decreased rates of loan failure or improved hit rates in marketing are also very significant; and portfolio management, where portfolios that better satisfy the competing criteria with which they are assembled may attract significantly higher levels of investment.

Can the Problem Be Formulated So That the Chromosome Is a List of Numbers?

Most genetic algorithm applications use chromosomes that are lists of numbers. Every publicly available genetic algorithm software system I am aware of (except those for genetic programming) is built to handle this sort of representation as the default case. It can be a quick learning experience to experiment with genetic algorithms of this type, since user requirements are minimal; in general, the user need only specify the number and types of the chromosome fields, and an evaluation function. Examples of this type of application are widespread and can be found in Davis (1991), Goldberg (1989), and genetic algorithm conference proceedings.

Can the Genetic Algorithm Be Hybridized with Other Algorithms?

The traditional genetic algorithm described earlier has been tested successfully on a wide range of problems, but is rarely the best known solution. However, *tailoring* a genetic algorithm to a particular problem is often a very effective way to produce an algorithm better than any currently available. One approach to tailoring is to hybridize the genetic algorithm with algorithms currently used in the domain. The resulting program, incorporating the global search capability of the genetic algorithm and the domain-specific capabilities of the existing algorithms, generally performs better than its parents.

Other ways of tailoring a genetic algorithm to a domain include incorporating heuristics used to solve problems in the domain as mutation operators in the genetic algorithm; using fast, imprecise algorithms from the domain to create initial population members; and using hill-climbing techniques from the domain to optimize the result of a genetic algorithm run. Combining such techniques with a genetic algorithm is at present as much an art as a science.

Tailoring a genetic algorithm to a domain requires more knowledge than using a generic genetic algorithm, but it can be a high-payoff strategy when the benefits to be gained from increased optimization merit the additional effort.

Is There an Evaluation Function?

A genetic algorithm requires an evaluation function, in order to assign each chromosome a rating that will determine its reproductive fitness. If an evaluation function does not currently exist, producing it can be a very time-consuming process. However, production of an evaluation function is required for nearly any principled optimization system, so this step is mandatory.

Examples of problems for which the user probably has or can easily create evaluation functions include learning to trade from historical data; derivation of principles for finding productive entries in a database of consumers; and learning to approve or reject credit card transactions. Examples of problems for which it would be difficult to derive evaluation functions include learning to design corporate logos; making strategic decisions for Fortune 500 companies; and discovering the proof of a mathematical conjecture.

In these latter three cases, the evaluation function is visual and difficult to state in computer terms; intuitive and subject to a great many factors that computers cannot at present parse; or all-or-nothing (we do not know how to rate partially correct proof of theorems; such proofs are good or bad). The status of these problems may change as the field of computer science evolves, but as things currently stand, I personally would shy away from a fixed-price contract to deliver a genetic algorithm to accomplish any of them.

Will CPU Time Be Available?

Genetic algorithms can consume great amounts of CPU time, because they require the evaluation of a great many chromosomes. For this reason, there are very few real-time genetic algorithms. In deciding whether to use a genetic algorithm, it is important to estimate how much time the algorithm will require, and whether the potential improvement in solution quality justifies an increase in CPU time. Many of the existing applications of genetic algorithms concern design problems (machine parts, networks, or network inputs are designed by the algorithm) or optimization of other algorithms. Very few include genetic algorithms in their day-to-day operation. As computing resources become cheaper and more parallel, this trend will probably change.

Conclusions

Genetic algorithm is an approach to optimization and learning invented by John Holland and inspired by the biological theory of evolution. Genetic algorithms (sometimes in conjunction with other techniques) have proved able to solve some problems better than any competing approach. Financial institutions are paying a good deal of attention to genetic algorithms, because various problems faced by such institutions would handsomely repay even small increments of improvement. The number of financial applications of genetic algorithms is expected to increase greatly over the next few years.

References

Davis, L. 1991. *Handbook of Genetic Algorithms.* New York: Van Nostrand Reinhold.

Goldberg, D.E. 1989. *Genetic Algorithms in Search, Optimization, and Machine Learning.* Reading, MA: Addison-Wesley.

Holland, J. 1975. *Adaptation in Natural and Artificial Systems.* Originally published by the University of Michigan Press in 1975. Reissued by The MIT Press, Cambridge, 1992.

Holland, J., K.J. Holyoak, R.E. Nisbett, and P.R. Thagard. 1986. *Induction: Processes of Inference, Learning, and Discovery.* Cambridge: The MIT Press.

Koza, J. 1993. *Genetic Programming.* Cambridge: The MIT Press.

Schaffer, J.D., W. Darrell, and L.J. Eshelman. 1993. "Combinations of Genetic Algorithms and Neural Networks: A Survey of the State of the Art," *IIEE Workshop Proceedings.*

Note

1. Some of the material in this chapter originally appeared in Davis's column on genetic algorithm applications in *Advanced Technology for Developers.*

CHAPTER 9

Genetic Algorithms for Financial Modeling

Andrew M. Colin

It is hard to predict,
especially the future.

Frederick Pohl

This chapter introduces applications of genetic algorithms (GAs) in trading. Andrew Colin, who at the time of this writing was a manager with the Treasury of Citibank in London, wrote this chapter for fund managers, traders, and senior executives who want to acquire knowledge about GAs, and for computer-literate traders and applications programmers who want to start building GA applications. Those who merely desire basic know-how about GA applications may skip the more technical sections of this chapter. For those who would like to build their own applications, Colin supplies source codes along with tips for getting applications up and running. The structure of this chapter is as follows. First, a typical trading optimization problem that is simple enough to be easily implemented is described. Colin shows how a genetic algorithm finds a near-optimal solution in a fraction of the time required by less sophisticated means. The ideas can be easily extended to more complex trading problems. Second, an example of strategy acquisition or rule induction is outlined. Given a target (e.g., maximizing profit while minimizing drawdowns), Colin shows how to construct simple, logical rules that can be used for trading. The third example builds on the results of the second: given a set of plausible models, Colin shows how to combine forecasts.

At the end of the chapter, Colin reviews a list of useful GA-based computer packages for constructing computer trading models. Various exercises are also provided. The degree of difficulty of these exercises, as suggested by the number of stars, rises logarithmically rather than linearly; thus, one-star problems require only a little work, while five-star problems involve more effort. Completion of these exercises, however, is not necessary to gain an understanding of the usefulness of GAs for financial optimizations.

Background

The unifying feature behind the advanced technologies described in this book is that they can all be used to gain a better understanding of complex nonlinear systems. Why are these technologies being applied to trading? There are two reasons. The first is that a market represents a cross section of a complex dynamic system that shows some deterministic behavior—and applying some of these ideas may help us to understand how that behavior comes about. However, our understanding does not have to be complete to produce useful results. In an active market, prices are usually as likely to move up as down, so being able to correctly forecast the market direction 55% of the time will very probably make money. Thus, even a slight improvement in forecasting accuracy can result in a very highly leveraged return.

The second reason is that every market trader, whether aware of it or not, has some form of model underpinning his or her views of market behavior. Whether an individual is a chartist, a fundamentalist, a technical trader, or a trend follower, all believe that at certain times there are rules that govern the behavior of price action. Using appropriate rule induction technology—such as our strategy acquisition example—understanding of the processes that lead to trading decisions can be enhanced.

We should not expect such rules to be simple in form. If a rule is easy to find, it will probably appear as a glaring inefficiency in the market and will promptly be arbitraged away, thereby invalidating it. For example, this may be a reason for the poor returns of simple moving average models in the foreign exchange markets over the last 15 years.

Some recent work in the field of neural networks has led to increased interest in connectionist ideas for time-series forecasting. In an important paper, Farmer and Sidorowich (1987) showed that a back-propagation neural network could be used to forecast a low-dimensional chaotic time series. Such time series have many properties in common with time series from nature, such as sunspot numbers and population densities over time. Unfortunately, this approach usually fails when applied to real financial time series, perhaps due to the high-dimensionality of the systems involved and the noise that always exists in markets.

The author's view—which has been reflected in the approach to forecasting models built at Citibank over the last four years—is that, while neural networks have their place and can form an important component of a forecasting system, the most important step in building such a system is the preprocessing of data. Before we can perform any useful forecasting, we must first extract leading indicators from our time series that have some demonstrable forecasting ability. The determination of working leading indicators from price data and other exogenous sources is by far the most critical part of constructing a profitable model; unfortunately, it is also the most difficult and time consuming.

The approach shown in this chapter is to use technical indicators for forecasting. A technical indicator is a mathematical function of price that compresses useful information about recent behavior of the market into a single real number. This number may then, when used with the outputs of other indicators, produce useful information about future market behavior. This approach has the advantage of being completely objective, unlike human identification of chart patterns. For more information on technical indicators, see Kaufman (1987) or any of a multitude of publications on market forecasting. For simplicity, the only technical indicator we will use from here on is a moving average, or a combination of moving averages, of price.

Using a Genetic Algorithm

Genetic algorithm is one particular technique that can assist in the identification of leading indicator functions. In its most basic form, the GA is developed and implemented as follows:

1. Construct an encoding of a problem into a concise form, such as an array of ones and zeros, that supports the crossover and

mutation operators in a natural way. For the purposes of this chapter, we call the encoded structure a *chromosome*. The individual entries on the chromosome are referred to as *alleles*.

2. Set up a population of hypotheses about the problem. This is usually done by, for instance, writing ones and zeros at random into a set of chromosomes. In what follows, the terms "hypothesis" and "individual" are used interchangeably.
3. Next, construct an objective function that enables the comparison of one individual against another, so that the two may be ranked in *fitness*—a measure of how close the realized solution comes to the ideal solution. Then, evaluate the initial set of hypotheses and assign a fitness measure to each, according to how well each lies near the desired solution.
4. Now, allow the hypotheses to mutate and exchange material between each other via crossover. The likelihood of an individual being able to pass on genetic material is proportional to the previously evaluated fitness measure. Then, reevaluate the fitness measures over the population. This step encapsulates the three vital mechanisms of a genetic algorithm: *mutation*, *crossover*, and *selection*. These operators must be implemented for every problem treated.
5. Repeat step 3 for several hundred iterations, or generations. At the end of this period, the overall fitness of the population is likely to have drastically increased, so that there will be many hypotheses in the population that lie near the global optimum for the system.

This chapter is intended to be complementary to Chapter 8, so we will not discuss the basic mechanics of a GA in detail. Instead, we supply three examples from the domain of currency trading. The first solves the problem set out above for a variety of different fitness functions. The second illustrates how a GA may be used not just for optimization, but for the broader area of strategy acquisition, so that the computer designs its own trading rules as well as optimizing them. The third shows how to use a GA to combine forecasts from different models.

The language in which the examples have been implemented is ANSI C. The majority of GA applications seem to be coded in C or LISP, and most users have access to a C compiler in some form. Portability considerations suggest C as the language of choice, but GAs have been implemented in most modern computer languages.

Example 1: A Trend-Following Trading Model

Perhaps the simplest way to process price data is to take a moving average of the measurements, and to take long or short positions in the market according to whether this average is moving up or down. Figure 9.1 illustrates the technique working in a trending market.

A moving average is simply calculated by taking an average of the prices seen over the last N days, where N is called the period of the moving average. A moving average with period 1 is identical to the underlying price. This quantity is straightforward to calculate, but it is even simpler to calculate a related quantity called the *exponential moving average*, which is defined as follows:

$$MA(t) = \alpha \times p(t) + (1 - \alpha) \times MA(t - 1)$$

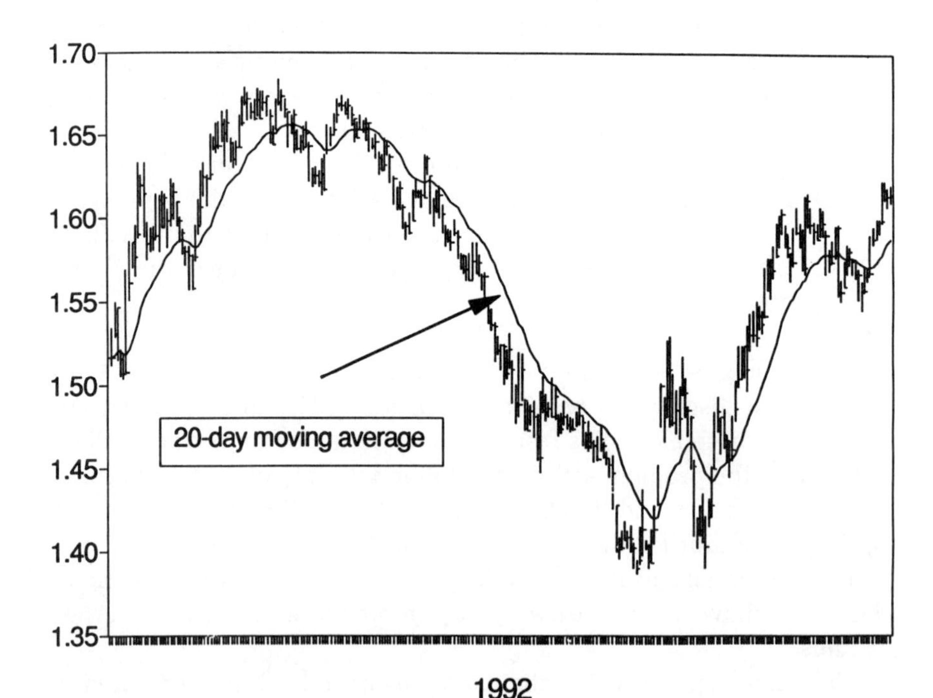

FIGURE 9.1
U.S. dollar/German Deutschmark exchange rate

where α, the smoothing coefficient, is calculated from N, the moving average period, by

$$\alpha = \frac{2}{1 + N}$$

Exponential moving averages have a number of advantages over ordinary moving averages, which are detailed by Makridakis et al. (1983).

Moving averages can work extremely well when the market is in a trending phase. When a large move is occurring, this type of price filter allows the trader to ignore short-term fluctuations and follow the large moves, rather than be distracted by small fluctuations. Unfortunately, for much of the time, exchange rates "range trade," or move up and down by small amounts, and in this case a moving average is likely to lose money.

To avoid this effect, we might use a combination of long- and short-term moving averages in a form called a *price oscillator*. To use this technical indicator, we buy when the shorter-term moving average has crossed the longer-term moving average, and sell when the reverse occurs (Figure 9.2). It is not clear whether up-moves are governed by the same dynamics as down-moves, and so we might wish to have two price oscillators governing our position-taking—one for long positions and one for short positions. Note that, in this case, neither or both of the price oscillators may be suggesting that a certain position should be taken. In the first case, the model will have no position, and in the second again no position will be entered into, as the long and short forecasts will cancel each other out.

Frequently, it is also a good idea to implement a stop-loss and/or a profit stop. A stop-loss attempts to cut down on the number of losing positions by setting a loss level such that, if the model loses this amount on a particular trade, then the position is cut out. A profit stop works in reverse, by attempting to catch profits seen in fleeting market moves. For this model, we shall incorporate four different stop-loss/profit stop options: (1) no stop, (2) 1% stop, (3) 2% stop, and (4) 3% stop.

The essential parameters that define our model, then, are the four integer periods of the moving averages that make up the two price oscillators, and two more lying between 1 and 4 that give a suitable stop-loss. For the purposes of this example we can restrict the moving average periods to lie between 1 and 50 days; longer periods are un-

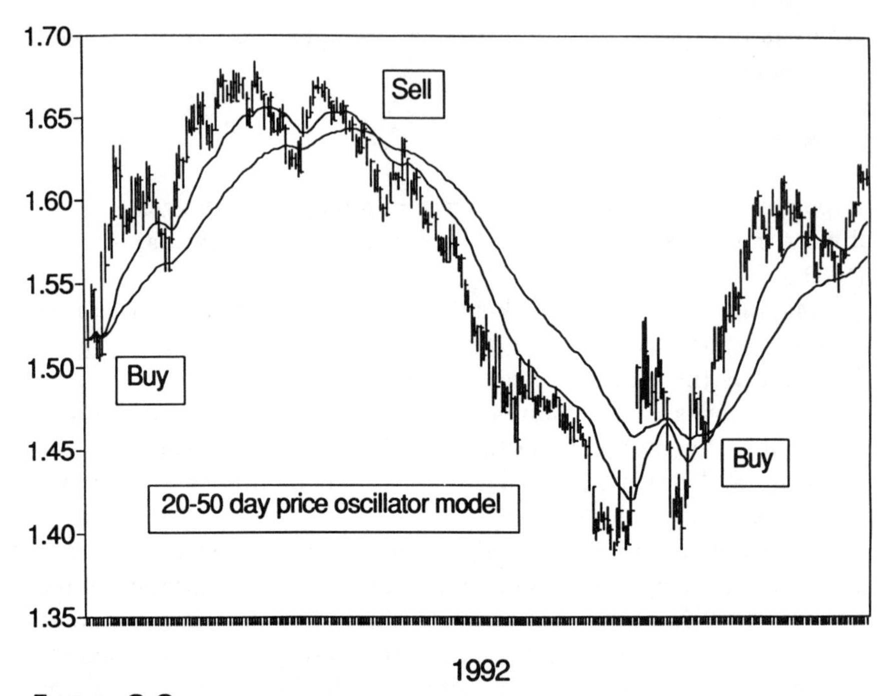

FIGURE 9.2
U.S. dollar/German Deutschmark exchange rate

likely to add much extra value to the model, but are easily implemented if the user wishes.

Is it feasible to try every different case in order to select the one that gives the highest profit over our historical database? For this simple model there are

$$50 \times 50 \times 50 \times 50 \times 4 \times 4 = 100{,}000{,}000$$

different cases. Suppose, further, that we have a fast computer and can evaluate the profit from 100 models per second. Then to run through every single case will take about 11.5 days of computer time!

One doesn't necessarily have to work through every single case to find a good answer for this sort of problem. There are many effective "hill-climbing" techniques that can be used to optimize problems of this type; Press et al. (1988) present a good overview. However, trading problems are characterized by an abundance of "false optima," or solutions that appear to be the best possible while in fact being suboptimal, and with a hill-climbing technique it is more than likely that the eventual solution will not be the best.

This example is fairly typical of the sorts of optimization problems encountered in setting up trading models. For a one-off experiment, we might be willing to tie up a computer for this length of time; but usually many runs are required, and the situation is worsened when we start to consider more realistic models, with many more degrees of freedom. An alternative approach is necessary.

A genetic algorithm is a powerful and efficient way to solve problems of the type we have just sketched. Shorn of its biological origins, a GA provides an ingenious technique for optimizing a supplied objective function over a data set. Although more complex to implement than the standard hill-climbing algorithms, a GA has the important property of being able to discard local optima within the search space.

The first step in implementing a GA is to choose an appropriate data structure with which to represent the problem. A good data structure will map the salient features of the problem into a readily manipulable form in a concise and efficient manner, and it will allow straightforward implementation of crossover and mutation. Suitable data structures include arrays of ones and zeros, arrays of real numbers, binary trees, linked lists, or combinations of all of these.

For this problem a chromosome consists of six integers

(MA_period1) . . . (MA_period4) (Stop-loss) (Profit stop)

and these can be efficiently mapped onto an array of bits, using a decimal-to-binary conversion routine.[1] Opinions differ as to whether the canonical conversion or gray-scale conversion (Koza 1992) is more effective. For simplicity, we use the former (see Goldberg 1989, p. 110). A suitable chromosome structure would be

```
typedef struct {
    int *chromosome;
    float score;
} CHROMOSOME;
```

where "chromosome" gives the address of an array of integers that will contain the chromosome's bits, and "score" will contain the fitness of this individual.

The next step is to consider the resolution required. The moving average periods will be allowed to vary down to 1 and up to around 60. In this case we can use eight bits per period, which gives 64 different values. For the profit stops and stop-losses, we have restricted ourselves to four different values in increments of 0.5%. To represent four different values we only need two bits. Clearly, if you feel finer resolution will help, then another bit will double the number of potential values. However, an excessive number of bits will only slow down the convergence of the program, and will not necessarily lead to higher resultant profits in reality, since it can often be difficult to cut out a position at exact stop-loss levels in a turbulent market.

Thus, our initial population has storage set aside for it as follows:

```
CHROMOSOME parent[POPULATION_SIZE];
int i;
for (i=0; i<POPULATION_SIZE;i++) {
    parent[i].chromosome = (int*) malloc (CHROMOSOME_LENGTH * sizeof
        (int));
    if (parent[i].chromosome == NULL) {
        printf("\nMalloc error at line %i", __LINE__);
        exit(0);
    }
}
```

where POPULATION_SIZE is the number of individuals per generation, and CHROMOSOME_LENGTH is the number of bits per chromosome. This latter quantity equals $8 + 8 + 8 + 8 + 2 + 2 = 36$, from the analysis above.

What sort of data structure might not suit a particular problem? Consider a ratio optimization problem, where we have a set of non-negative quantities that are constrained to add up to 1. A sample individual from a population of such entities might be

[0.1 0.4 0.3 0.2]

This sort of problem might occur in, for instance, determining the optimal ratios of chemicals to be used in some industrial process.

If the resolution required is not likely to be less than 0.01, we could map each entry onto an array of seven bits and decode that number by dividing by $27 = 128$. This implementation is likely to be efficient and fast. However, the design of crossover and mutation operators that maintain the normalization condition is not straightforward, and in fact this consideration means that a different data structure is indicated: an array of real numbers. Our third example below illustrates the solution to a problem of this type.

Similarly, suppose we would like to solve a traveling salesman-type problem, in which the optimal order in which to visit a number of towns has to be determined, traveling along the shortest possible route and visiting each town only once. In this case, a trial solution for seven towns might be

(1 4 7 6 5 2 3)

A mutation operator would exchange the contents of two locations. In this case the best data structure is an array of integers, rather than bits. The alphabet in which the solution is expressed has to be constrained to disallow the same site being visited twice, which violates the constraints of the problem.

We have stressed the validity of alternative types of data structures. Many first attempts of using GAs considered that the only permissible data structure is a bit string. As you can see, this is by no means true.

Let's return to the sample problem. The next step is to decide on the form of mutation and crossover operators, which, fortunately, are straightforward for bit strings. For instance, mutating the first bit in

(00011000)

gives

(10011000)

In order to cross over two bit strings, we first have to choose a crossover site. If this is taken as the string midpoint, we have

(0111) (1100)

which give, when crossed over at their midpoint,

(0100) (1111)

```
/*
* Crossover function. Requires storage to have been previously put aside for
    both
* parent and child genes
*/
void crossover (CHROMOSOME *parent1, CHROMOSOME *parent2,
        CHROMOSOME *child1, CHROMOSOME *child2)
{
    int i, xover_position;
    xover_position = random (CHROMOSOME_LENGTH);
    for (i=0; i<xover_position; i++) {
        child1->chromosome[i] = parent1->chromosome[i];
        child2->chromosome[i] = parent2->chromosome[i];
```

```
    }
    for (i=xover_position; i<CHROMOSOME_LENGTH; i++) {
        child1->chromosome[i] = parent2->chromosome[i];
        child2->chromosome[i] = parent1->chromosome[i];
    }
}
void mutate (CHROMOSOME *chrom)
{
    int mutate_position;
    mutate_position = random (CHROMOSOME_LENGTH);
    if (chrom->chromosome[mutate_position]==1)
        chrom->chromosome[mutate_position] = 0;
    else
        chrom->chromosome[mutate_position] = 1;
}
```

In the next section we'll consider some of the points that arise when our crossover operators are more constrained.

We can now set up our initial population, with random values, and assign a fitness measure according to the strength of each chromosome (hypothesis). It is a salutary experience to set up a GA with the initial population being the best-known solution to a problem, and then watch the system evolve a better, completely different solution!

Initial conditions are a critical feature for a GA, influencing the subsequent path of the population through the search space. It is important to ensure that initial sampling is not restricted to one region of the search space, particularly in the case where crossover operators are constrained, or there is a danger that some regions may never be examined.

Regarding population size, the general rule is: the larger, the better. A large population can help to compensate, to some degree, for an inadequate initial sampling. Usually, a population of between 30 and 100 individuals suffices for most problems.

```
/* Set up initial random population */
int i;
for (i=0; i<POPULATION_SIZE; i++) {
    for (j=0; j<CHROMOSOME_LENGTH; j++)
        if (rand < RAND_MAX/2)
            parent[i].chromosome[j] = 1;
        else
            parent[i].chromosome[j] = 0;
}
/* Evaluate parent population */
```

```
for (i=0; i<POPULATION_SIZE; i++)
    parent[i].score = eval_chrom ( &parent[i], price );
/*
* Rank population in order and assign fitness measures between 0.1 (worst) and 0.9
* (best)
*/
qsort (&chromosome[0], POPULATION_SIZE, sizeof(chromosome[0]), chromcmp
    );
max_fitness = chromosome[POPULATION_SIZE-1].score;
min_fitness = chromosome[0].score;
for (i=0; i<POPULATION_SIZE; i++)
    parent[i].score = 0.1 + 0.8 * (parent[i].score − min_fitness)/(max-fitness -
min_fitness)
```

where "chromcmp" is a user-supplied function that gives relative rankings of chromosomes, and "price" is an array of prices (open, high, low) over which the model is to be optimized.

We now have all the components necessary to run our GA, except one—the objective function. This is where the abstractions of the GA mechanism meet the outside world, and defines to the system the difference between a good solution and a bad one. Note that we don't always have to know about the absolute fitness of a chromosome; to make a GA work, being able to rank solutions in order of fitness usually suffices.

What defines a good solution for our problem? Since the intention is (presumably) to make as much money as possible, the fitness measure can be taken as the amount of money made; this is implemented in routine FITNESS1.C. However, reality often places other constraints on models, and minimizing drawdowns often is just as important. We have implemented an alternative objective function in FITNESS2.C.[2]

The last stage in using an objective function is to convert the fitness of an individual into a probability—the likelihood that an individual will pass on its genetic material to subsequent generations. The way we have done this is, simply, to rank the solutions in ascending order and to assign the highest probability to the best individual, the lowest to the worst, and then to linearly interpolate between these two measures for all the rest. A system with this propagation mechanism is called a generational GA and is well suited for noisy problem domains. It is also a good choice for our objective functions, where profitability of a model can be negative. However, note that if a "cluster" of solutions arises that dramatically exceeds the fitness of the other solutions, it will take longer for these individuals to spread their characteristics into the surrounding population.[3] For more on population niches and elitism, see Goldberg (1989).

Example 2: Strategy Acquisition

A difficulty with the moving average model described above is that we are constrained to one fixed type of model. It would be convenient to be able to let the computer develop its own strategies, based on a number of different inputs, that could best model the market dynamics.

For instance, three potential models might be

if (10-day moving average > 0) then buy $ else sell $ (model 1)

if (15-day moving average > 0) then buy $ else sell $ (model 2)

if (25-day moving average > 0) then buy $ else sell $ (model 3)

In this example, we show how to write a GA-based strategy acquisition system that produces statements such as:

if (model 1 says BUY) then BUY else SELL

if (model 1 and model 2 say BUY) then BUY else SELL

if (model 1 and model 2 say BUY) OR (model 3 says BUY)
then BUY else SELL

It is not clear how to represent such statements in terms of a binary string, still less so as to how to perform crossover and mutation. An added difficulty is that the individuals in our population can all have different lengths and complexities, which suggests that the simple crossover mechanism will be at a disadvantage.

Fortunately, there is a simple and direct way to model such statements within a computer. This is done by utilizing a natural mapping, or isomorphism, between algebraic statements of the form shown above and a binary tree structure. Figure 9.3 suggests an extremely elegant way to implement genetic crossover. Many effective algorithms for the manipulation of binary trees are known (see Knuth 1973).

The first step is to define a data structure for the nodes that make up the binary trees. In C, we can write:

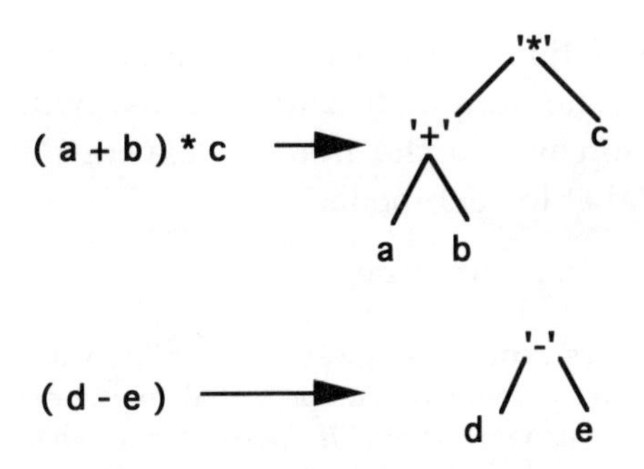

Crossing the expressions over could give:

FIGURE 9.3
Genetic algorithms for financial forecasting

```
struct node {
    int info;
    int label;
    struct node *left;
    struct node *right;
};
typedef struct node NODE;
```

The four fields allow each node to contain some data, a unique label used to help perform indexing operations, and pointers to the left and right subnodes. If a node lies at the end of a branch, then the information contained within must refer to a model forecast; if a node lies within a tree, then the information must refer to a logical operator. A simple way to distinguish between the two types of nodes is to set the left and right pointers of a terminal node (at the end of a branch) to NULL. Ameraal (1987) gives a fuller discussion of how to implement binary trees in C.

The main difference between this problem and its predecessor involves the crossover and mutation operators. In this example we will concentrate on implementing these operators. The full program is available on CD-ROM.

We first look at mutation: if it is decided to mutate an individual, we have to randomly choose a node within a tree, with likelihood independent of position within the tree. To expedite this, routine LabelTree assigns a label to each node:

```
void LabelTree (NODE *node, int *treesize)
{
  /* Assigns a unique integer, counting upward from initial value,
     to each node. The binary tree may then be sorted, and nodes
     retrieved, on the key "node->label." If *treesize = 0 when this
     routine is called, the number of nodes is returned in *treesize.
  */

  if (node->right) LabelTree(node->right, treesize);
  if (node) node->label = ++*treesize; /* check for null tree */
  if (node->left) LabelTree(node->left, treesize);
}
```

Now that the tree has been labeled and the total number of nodes counted, we may choose one at random. Given an index number, routine SearchTree returns the address of the associated node within a tree:

```
NODE *SearchTree (NODE *node, int key)
{
  /* Returns the node with label == key. Uses the previously
     sorted tree for fastest searching */
  if (node->label > key)
    return SearchTree (node->right, key);
  else
    if (node->label < key)
      return SearchTree (node->left, key);
    else
      return node;
}
```

Routines RandomNode and Mutate pull these functions together. RandomNode returns the address of a randomly chosen node within the tree, and Mutate uses the returned address to perform mutation:

```
NODE *RandomNode (NODE *root)
{
  /* Given a binary tree with root node "*root," returns a randomly
     chosen node from the tree. All nodes have equal likelihood of
     being selected.
  */
  NODE *temp;
```

```
  int treesize = 0, rnum;
  LabelTree (root, &treesize);
  rnum = random (treesize) + 1;
  temp = SearchTree (root, rnum);
  return temp;
}
void Mutate (NODE *root)
/*
* procedure: Mutate
*
* Alters the contents of a randomly chosen node from the
* specified tree.
*
*/
{
  NODE *temp;
  temp = RandomNode (root);
  temp->info = random (MAX_INT);
}
```

Next, we implement crossover between two binary trees. To do this, we first select a site within each at random and copy part of one tree into another:

```
NODE *xover (NODE *node, NODE *site1, NODE *site2)
/*
  Performs genetic crossover on binary trees.
  "site 1" is the location within the first (upper) tree at which
  crossover will occur;
  "site 2" is the location within the second (lower) tree which
  overwrites "site1" and acts as root node to the rest of the
  tree.
*/
{
  NODE *temp;
  temp = (NODE*) malloc (sizeof(NODE));
  if (!temp)
    err_exit (__FILE__, __LINE__);
  if (node==site 1)
  {
    temp->info = site2->info;
    if ((site2->left) && (site 2->right))
    {
      temp->left = xover (site2->left, site1, site2);
      temp->right = xover (site2->right, site1, site2);
    }
    else
```

```
      {
        temp->left = NULL;
        temp->right = NULL;
      }
    }
  else
    {
      temp->info = node->info;
      if ((node->left) && (node->right))
      {
        temp->left = xover (node->left, site1, site2);
        temp->right = xover (node->right, site1, site2);
      }
      else
      {
        temp->left = NULL;
        temp->right = NULL;
      }
    }
    return temp;
}
```

Population initialization is a more complex question in this case.

At Citibank we have written a sophisticated package for rule induction based on these ideas. The program is designed to manipulate databases of potential leading indicators, using arithmetic as well as logical operators. This enables the production and assessment of statements such as:

> if (15-day stochastic > 0.43) AND
> (12, 15-day price oscillator −0.01) then buy $

One of the most valuable uses of this package has not been the explicit form of the rules produced, but rather the variables that the system highlights by occurring in many "fit" rules, even when starting from a variety of different initial conditions. When these variables are isolated and their values used as inputs into a back-propagation-type neural network, the network can then model the interrelationships between them and build a powerful forecasting system (see Colin 1992).

For a further illustration on how to implement this type of system, see Koza (1992), who gives many examples of strategy acquisition, and also BEAGLE, a software package that is described below.

Example 3: Optimizing Combinations of Rules

Consider a moving average with an n-day period. At a given time t, the moving average forecast will be given by the sign of the difference between the value of the MA at time $t-1$ and the value at time t:

$$\text{MA}(n)\_\text{forecast} = \text{sign}[\text{MA}(n, t) - \text{MA}(n, t-1)]$$

It is instructive to use a genetic algorithm to combine forecasts from different models. To construct a joint forecast from n different moving averages, we can look for a set of real coefficients w such that:

$$\text{MA}\_f = \text{sign}\ [w(1)\text{MA}(\text{period}\_1)\_f + w(2)\text{MA}(\text{period}\_2)\_f + \ldots + w(n)\text{MA}(\text{period}\_n)\_f)] = \Sigma w(i)\ \text{MA}(\text{period}\ i)$$

where $\Sigma\ w(i) = 1$. Each individual forecast is weighted according to its contribution to the overall forecast. The GA's task is to decide how to optimize the coefficients $w(i)$. This constitutes a ratio allocation problem, with a solution that is a vector of real numbers:

$$(a[1], a[2], a[3], \ldots a[n])$$

such that

1. The individual components $a[i]$ lie in the closed interval $[0, 1]$.
2. The sum of all the components must equal 1.
3. The solution vector will, when substituted into some evaluation function, give a global maximum of that function.

An appropriate data structure is usually a trade-off between simplicity of representation and the complexity of the evolution operators. In this case, translating a vector of real-valued numbers to a sequence of binary digits is fairly easy. Unfortunately, as we mentioned above, constructing crossover and mutation operators that satisfy the "summation to 1" constraint is not at all straightforward. This is a case where a different data structure, one that does not involve bit strings, turns out to be much more suitable for the task at hand.

Suppose, instead, that we keep each individual as an array of real numbers. The selection mechanism is independent of representation. We shall show below how to build crossover and mutation operators for these individuals.

We can mutate a vector by changing the value of the entry at the *i*th position. In the accompanying routine we do this by adding a random number δ between 0 and 1 to the contents of the *i*th position, and by subtracting 1 if the result exceeds 1. For instance,

$$(0.1\ 0.2\ 0.4\ 0.3\ 0.0) \rightarrow (0.55\ 0.2\ 0.4\ 0.3\ 0.0)$$

on adding a δ of 0.45 to the first entry. The resulting vector violates the "summation to 1" constraint. To compensate, we can either choose a second site in the vector and subtract the same random number δ, or rescale all the other numbers in the vector. Although the first approach would probably work, it violates nonlocality of mutation, which is probably undesirable. We have therefore implemented the second option in the routine "mutate." This mutation mechanism also preserves the ratio of the nonmutated genes.

To perform the rescaling, we define a variable S which is set to the sum of the nonmutated entries. Here, $S = 0.2 + 0.4 + 0.3 + 0.0 = 0.9$. Then, all nonmutated entries are rescaled by the following factor:

$$1 - \frac{\delta}{S} = 1 - \frac{0.45}{0.9} = 0.5$$

In this case, the vector above becomes

$$(0.55\ 0.1\ 0.2\ 0.15\ 0.0)$$

and the sum of the entries is 1, as required.

```
void mutate (CHROM *chrom, GA_PARAMS *ga_params)
/*
* Performs "single-site" mutation, preserving ratios between
* unmutated gene values
*/
{
   int site, i;
   float delta, S;
   /* Choose mutation site */
   site = random (ga_params->n_genes);
   /*Choose amount by which to scale mutated gene (between 0 and 1) */
   delta = (float) rand / RAND_MAX;
   /* Perform mutation, ensuring that new value of gene <= 1.0 */
   chrom->genes[site] += delta;
   if (chrom->genes[site] > 1.0)
   {
     chrom->genes[site] -= 1.0;
```

```
      delta -= 1.0;
    }
    /* Rescale all other genes in chromosome */
    S = 0.0;
    for (i=0; i<ga_params->n_genes; i++)
      if (i != site)
        S += chrom->genes[i];
    for (i=0; i<ga_params->n_genes; i++)
      if (i != site)
        chrom->genes[i] *= (1.0 - delta/S);
}
```

To perform crossover, we select two individuals, A and B, and two crossover sites within them. The contents of the genes between the crossover sites are then exchanged. Both vectors will now violate the normalization constraint. For instance, let

A = (0.4 0.1 0.4 0.1 0.0)
B = (0.6 0.1 0.0 0.2 0.1)

and let the crossover sites be at the second and fourth locations within the vectors. After crossover, the vectors become

A = (0.4 0.1 0.0 0.2 0.0)
B = (0.6 0.1 0.4 0.1 0.1)

In order to adjust these individuals, we define a scaling factor R, given by the sum of the crossed-over sites within A divided by the sum of the crossed-over sites within B. Here,

$$R = \frac{0.1 + 0.0 + 0.2}{0.1 + 0.4 + 0.1} = 0.5$$

If we now divide the crossed-over components of A by 0.5, and multiply the corresponding components of B by the same amount, the vectors become

A = (0.4 0.2 0.0 0.4 0.0)
B = (0.6 0.05 0.2 0.05 0.1)

and the final result is that, no matter how much crossover is performed, the "summation to 1" constraint is always satisfied.

```
void crossover (CHROM *parent1, CHROM *parent2,
    CHROM *child1, CHROM *child2,
    GA_PARAMS *ga_params)
/*
```

```
* Performs ratio crossover between two parent chromosomes
*/
{
  int lower_site, upper_site, i;
  float upper = 0.0, lower = 0.0, temp, R;
  /* Choose mutation sites */
  lower_site = random (ga_params->n_genes);
  upper_site = random (ga_params->n_genes);
  if (lower_site > upper_site)
  {
    temp = lower_site;
    lower_site = upper_site;
    upper_site = temp;
  }
  /* Ensure that crossover sites do not coincide */
  if (lower_site == upper_site)
  {
    if (upper_site == 0)
      upper_site++;
    else
      lower_site--;
  }
  /* Calculate scaling ratio */
  for (i=lower_site; i<=upper_site; i++)
  {
    upper += parent1->genes[i];
    lower += parent2->genes[i];
  }
  R = upper / lower;
  /* Perform crossover, rescaling to conserve overall sum of ratios */
  for (i=0; i<ga_params->n_genes; i++)
  {
    child1->genes[i] = parent1->genes[i];
    child2->genes[i] = parent2->genes[i];
  }
  for (i=lower_site; i<=upper_site; i++)
  {
    child1->genes[i] = parent2->genes[i] * R;
    child2->genes[i] = parent1->genes[i] / R;
  }
}
```

Commercially Available GA Packages

Implementation of a genetic algorithm depends on user-supplied data structures and evaluation functions, as well as on more mundane topics such as file formats and computer resources. This problem-specificity suggests that GA-based software packages are unlikely to find the same horizontal markets as word processors or spreadsheets. In this section we survey several off-the-shelf packages that can be used to ease the task of getting a GA up and running. The list is not intended to be exhaustive.

BEAGLE

One of the earliest commercially available GA-based programs was BEAGLE (Biological Evolutionary Algorithm Generating Logical Expressions). Originally written in 1980 by Richard Forsyth, then at the University of Nottingham, and released as a PC-based program in 1987, this rule-finder system examines sets of attributes from a user-supplied data set and forms hypotheses from them to identify pre-defined objectives. BEAGLE takes as its underlying data structure a binary tree, which can be mapped isomorphically onto an algebraic expression. The program allows arithmetic and algebraic expressions to be mixed, and it gives several rules at the completion of the learning process, whose forecasts are combined via a signature table to give a single answer (see Forsyth's contribution in Davis 1991). BEAGLE remains a remarkably powerful and versatile package. A demonstration version is provided on CD-ROM, together with ordering instructions. For more information see the *BEAGLE User's Guide*, included with the BEAGLE computer rule-induction package.

Evolver™

A difficulty with off-the-shelf GA systems is that it is often difficult to allow the user to express complex objective functions. Evolver from Axcelis Inc. in Seattle gets around this problem by implementing a general-purpose GA tool as an add-in DLL (dynamic link library) that can be used in conjunction with an Excel™ or WingZ™ spreadsheet. The spreadsheet's macro command language then allows the user to specify an objective function. This simple and elegant approach also allows the user to work directly with spreadsheet data.

This product resembles the back-solving capabilities supplied by many spreadsheets, but with global optimization abilities rather than the far less capable hill-climbing optimizers that are often supplied with most modern spreadsheets. Evolver looks after the various parameter settings, such as crossover rates and population size. The package has been used on problems varying from load-demand forecasting for power stations to grading of rare coins. More information on this software is available from Axcelis (4668 Eastern Avenue North, Seattle, WA 98203, USA; 206/632-0885).

Genitor

Genitor is a public-domain GA tool, written in C for a UNIX/PC environment. To use the package on a particular platform, the user must build three libraries for the creation of GA programs. Each library manipulates a different type of data, which can be integer, floating point, or binary. These libraries contain functions that perform standard GA operations, such as selection (get_parents) as well as useful inspection and initialization routines (init_pool, sort_pool, show_progress, dump_status).

To use the Genitor routines, the user must first decide on a suitable encoding for the problem at hand and then write an objective function that accepts a chromosome and returns a real-valued fitness measure. The main program is then written, setting up a population and making appropriate library calls (several example programs are supplied with the package). Lastly, the program is linked together with the chosen library and run. The various system parameters, such as population size, can be specified at run-time to ease experimentation.

Genitor comes from a research environment and is not particularly user-friendly. The user should be capable of C programming. The advantage of using such a package is, of course, that the user has access to the system's source code and can port it to a variety of machines. A copy of Genitor is supplied on CD-ROM, together with documentation in TeX format.

GA-Tool

GA-Tool is a package similar to Genitor, using C++ and a user interface written for Unix "curses." The range of options is much wider than for Genitor, with binary or gray-coded chromosomes selectable, and several alternatives provided for chromosome representation, replacement, and crossover. This software has a broader range of options for experimentation than Genitor, but is even more difficult to use.

A manual for this package is available as a technical report from the University of Missouri—Kansas City (Computer Science, 5100 Rockhill Road, Kansas City, MO 64110, USA). GA-Tool is also supplied on CD-ROM.

Genesis and OOGA

Genesis is similar to Genitor, in that it is a C-based system that requires a user-specified objective function. Genesis is available together with OOGA, which is a similar system written in Common LISP and CLOS (Common LISP Object System). They are available from The Software Partnership (P.O. Box 991, Melrose, MA 01276, USA) for a small sum. OOGA is intended to be used in conjunction with Laurence Davis's book (Davis 1991) and illustrates his tutorial.

Other Sources

Other useful sources of code and technical know-how are available from Internet. To obtain information on GAs, send mail to the GA address provided on page 140.

Exercises

1. Try running BITSTRING.C on another financial time series, such as the Dow-Jones Industrial Average. (Hint: You will have to make sure that your data set is in the same format as the supplied data, or modify the READOHLC routine.) [*]
2. Modify the program BITSTRING.C so that it can optimize a price-oscillator model over (say) 100 days and then test itself over the next 100 days, retrain itself over the test period and test over the subsequent period, and so on. In this way, you will be simulating real trading conditions by testing each model on new data. Can you generate consistent profits? If not, why not? [**]
3. Modify the program you developed in Exercise 2 to optimize the test and training periods via the GA mechanism. (Hint: Extend the chromosome structure to include these two integers, and modify eval_chrom.c accordingly.) [***]

4. Find a source code profiler and examine the performance of BITSTRING.C. In which section of the code is the computer spending most time? Isolate this section and rewrite it to make the code run faster. (Hint: See Chapter 10 on optimizing GA programs.) [*]
5. Extend the rule-induction system to find rules based on inequalities between moving averages, where the periods of the moving averages are variable. For instance, you should expect to produce statements such as:

   if (15-day MA>0) OR (4-day moving average>0) then buy

   where the periods are given as a variable within the NODE structure. [***]
6. Extend the rule-induction system to process real values, as in Example 5. [****]
7. Extend the rule-induction system in Exercise 6 to find fuzzy rules (see Chapter 8). [*****]

References

Ameraal, L. 1987. *Programs and Data Structures in C*. New York: John Wiley & Sons.

Colin, A. 1992. "Neural Networks and Genetic Algorithms for Exchange Rate Forecasting." *Proceedings IJCNN* (Beijing).

Davis, L., ed. 1991. *Handbook of Genetic Algorithms*. New York: Van Nostrand Reinhold.

Farmer, D., and J. Sidorowich. 1987. *Physical Review Letter* 59(8), p. 845.

Goldberg, D., 1989. *Genetic Algorithms in Search, Optimization and Machine Learning*. Reading, MA: Addison-Wesley.

Kaufman, P.J. 1987. *The New Commodity Trading Systems and Methods*. New York: John Wiley & Sons.

Knuth, D. 1973. *The Art of Computer Programming*. Vols. 1–3. Reading, MA: Addison-Wesley.

Koza, J. 1992. *Genetic Programming*. Cambridge, MA: The MIT Press.

Makridakis, S., S.C. Wheelwright, and V.E. McGee. 1983. *Forecasting: Methods and Applications*. New York: John Wiley & Sons.

Press, W.H., B.P. Flannery, S.A. Teukolsky, and W.T. Vetterling. 1988. *Numerical Recipes in C*. Cambridge, U.K.: Cambridge University Press.

Notes

1. More than 90% of the computer time consumed by your genetic algorithm will be used in the chromosome evaluation function. Therefore, when you

design your program, put a commensurate amount of effort into making this part of the code efficient! A little effort here can work wonders for the eventual speed of your program. Koza (1992) suggests that "for both the conventional genetic algorithm operating on fixed-length strings and genetic programming, the vast majority of the computer time is, for any non-trivial problem, consumed by the calculation of fitness. In fact, this concentration is usually so great that it will rarely pay to give any other consideration at all to any other aspect of the run."

2. Both FITNESS1.C and FITNESS2.C are programs containing the above code and are supplied on CD-ROM.
3. These ideas are brought together in the program BITSTRING.C. This is supplied in both source code form and compiled form for a PC on CD-ROM. The supplied data comprises five years' worth of exchange rate data between the U.S. dollar and the U.K. pound.

CHAPTER 10

Using GAs to Optimize a Trading System

Guido J. Deboeck

Technology periodically steals a leaf from nature's book.

Steven Harp and Tariq Samad
"Genetic Synthesis of Neural Network Architecture," in
Handbook of GA's, *(L. Davies, ed.), 1991*

Part One of this book showed that neural networks are excellent for estimating nonlinear relationships and detecting complex patterns. However, training neural nets is something of an art. Specialized tools such as outlined in Chapter 9 can reduce the need for programming, but knowledge and skill are required to set up neural nets to perform transformations and to determine which training strategies to adopt. Several papers have been written on how the training of neural nets can be facilitated using genetic algorithms (GAs). This chapter will focus on how to use GAs to design and optimize a trading system in a spreadsheet without programming, without writing scripts, and without using expensive tools. The method presented can easily be extended to many other applications.[1]

Building a Hybrid Trading System

Spreadsheets are an easy frame of reference for learning about neural networks and genetic algorithms. This chapter shows how to implement a hybrid trading system in a spreadsheet without programming, without writing scripts, and without using expensive tools.[2] The example given here uses the WingZ™ spreadsheet on a Macintosh but can easily be transposed to Excel™ on a PC.

The five main steps in building a trading system are:

1. Use market data to calculate data transformations
2. Set up a feed-forward neural net
3. Establish an evaluation scheme for a trading system
4. Formulate a genetic optimization problem
5. Summarize, evaluate, and plot key results.

An overview of this methodology is shown in Figure 10.1.

Step 1: Use Market Data to Calculate Data Transformations

Market data are easy to obtain: several information services[3] provide access to daily and/or intraday data. However, different sources provide different selections. Some are exclusively focused on futures data, others on stock or fixed-income data. Furthermore, accessibility accuracy, quality, and cost of the data vary significantly. It is thus important to select appropriate sources, to put in place data-capturing utilities (e.g., daily downloads of closing prices and/or a utility for

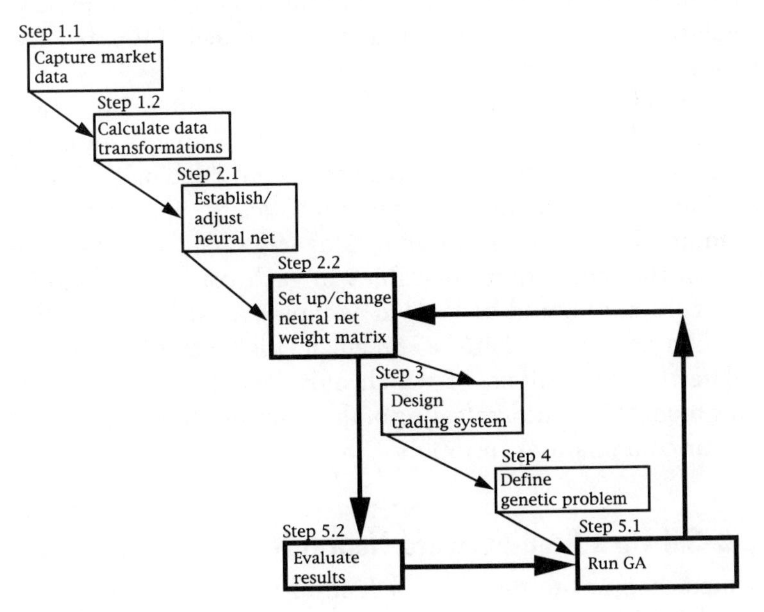

FIGURE 10.1
Methodology for developing a hybrid trading system in a spreadsheet

intraday prices capturing), and to organize the market data into a database for easy access by trading systems.

The importance of data transformations has been discussed in Chapter 2. Model-based trading allows a system to take into account many more market indicators and transformations than traders can.

Most traders look at the evolution of market prices (trends), volumes, the volatility of prices or yields, overbought or oversold conditions, momentum, liquidity, market participation, market risks, and the like. Market data transformations of trends, volatility, percentage differences, and technical analysis indicators are easy to implement. Fast Fourier and wavelet transforms require programming; however, they can be reduced to functions in a spreadsheet (see Chapter 2).

For our example we created a spreadsheet with yield data on U.S. Treasury securities. Column A contained dates; Columns B to F, closing yields on Fed Funds, 2-, 5-, and 10-year Treasury notes, and 30-year bonds, respectively. Next, we entered data transformations on closing yields. For example, data transformations on 10-year Treasury note yields were entered starting in column E. We computed 10- and 20-day trends (measured as the difference between the current yield and the yield 10 to 20 days earlier); a relative strength indicator (RSI); the spread between 10-year yields and the overnight Fed Funds rate; and volatility indicators, measured as the average of the daily rate of change in yields over 10 and 40 days.

All these data transformations can be computed by applying a simple formula and copying down the formula for the entire series. The RSI can be computed based on a script provided on the CD-ROM. In addition, the following summary statistics on each series should be computed: maximum, minimum, average, and the standard deviation of the entire range of values in each column. Data transformation columns should be labeled as inputs X_1 to X_n to facilitate the design of a neural net. Thus, a spreadsheet is created with yield data and data transformations, of which only a subset will be used in a trading model. This allows testing of the appropriateness of alternative inputs or combinations of inputs.

Step 2: Set Up a Feed-Forward Neural Net

The next step is the design of a feed-forward neural net within a spreadsheet. Figure 10.2 shows a simple structure of a neural net with an input layer for $X_1 \ldots X_4$ inputs, a hidden layer with $PE_1 \ldots PE_4$

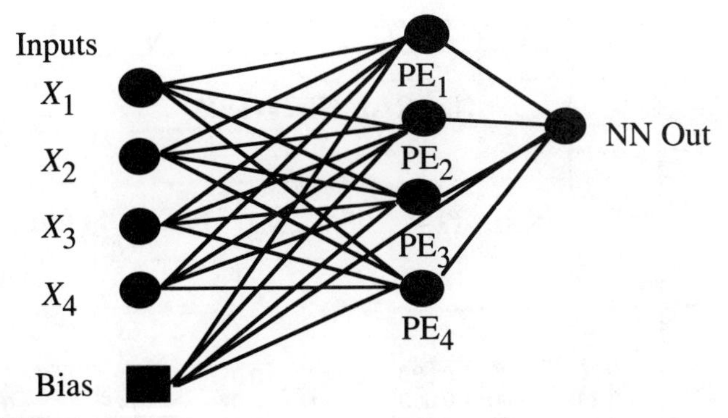

FIGURE 10.2
Simple feed-forward neural net with four inputs, four processing elements, and one output, fully connected.

processing elements, and an output layer with a single output. A feed-forward neural net can be set up in a spreadsheet by using separate columns for each PE on the hidden layer and a column for each output (NN Out). Thus, a neural net structure that has four PEs at the hidden layer and a single output requires five columns.

The value of each PE at the hidden layer is the hyperbolic tangent of the sum of connection weights (w_{ij}) with the data inputs (X_i). A constant or bias may be included. The hyperbolic tangent is used as a transformation function because it provides a broader range (-1 to $+1$) than a sigmoidal function (which is constrained between 0 and 1) in which to make the input/output transformations. This model can also be adapted to incorporate a radial basis transformation function. The formula for computing the PE values (columns Q to T in Figure 10.3) are

PE value = tanh ($AK$3+($AK$4*H12) + . . .
+ ($AK$7*K12))

The value of NN Out is the hyperbolic tangent of the sum of PE weights times the PE values (this assumes a bias of 1):

NN Out = tanh ($AK$8*Q12+$AL$8*R12+
$AM$8*S12+$AN$8*T12)

The connection weights and the bias are variables that need to be estimated during the "learning" process. All weights and biases

P	Q	R	S	T	U	V	W
			NEURAL NETWORK				
	PE1	PE2	PE3	PE4	OUT	DESIRED	ERROR
					0.56	0.63	1.0
					-0.68	-0.63	0
					-0.08	-0.02	0
					0.21	0.23	0
	-0.20	1.00	0.000	-0.627	0.02	0.32	0
	-0.23	1.00	0.000	-0.681	0.02	0.25	0
	-0.22	1.00	0.000	-0.667	0.02	0.21	0
	-0.28	1.00	0.000	-0.760	0.00	0.11	0
	-0.27	1.00	0.000	-0.741	0.01	0.13	0
	-0.29	1.00	0.000	-0.776	0.00	0.07	0
	-0.25	1.00	0.000	-0.716	0.01	-0.05	0
	-0.25	1.00	0.000	-0.712	0.01	-0.05	0
	-0.22	1.00	0.000	-0.673	0.02	0.09	0
	-0.24	1.00	0.000	-0.556	-0.05	-0.14	0
	-0.25	1.00	0.000	-0.627	-0.03	-0.10	0
	-0.26	1.00	0.000	-0.662	-0.02	-0.12	0

FIGURE 10.3
Neural network with configuration 4-4-1

are stored in the upper right part of the spreadsheet (e.g., cells AL3 to A08). The initial values for the weights and biases are random within user-defined ranges (e.g., −1 to +1). The weight matrix (see Figure 10.4) shows initial weights for a network model with four inputs, four PEs, and one output. This model is fully connected: all four inputs are connected with all four PEs, and all PEs are connected to the single output.

This feed-forward neural net model can be made more flexible in a couple of ways. One is to keep the architecture of the neural net open by *allowing a variable number of processing elements at the hidden layer*. To accomplish this, the weight matrix is extended with a set of "connect PE" values that are part of the parameters to be set. These connect PE values generate a floating point value within a narrow range, which is then translated via a simple if-condition into an integer value of 1 or 0. This integer value applied to a PE either includes it in or excludes it from the hidden layer (Figure 10.4). Since at least one PE is necessary, the PE value in column AK is set to 1 (otherwise, the GA will spend time investigating alternatives that exclude all PEs, which would be equivalent to estimating a linear regression between inputs and outputs).

AJ	AK	AL	AM	AN
1				
2	**Weights**	**PE1**	**PE2**	**PE3**
3	**Bias**	12.629	88.870	-59.076
4	**X1**	-93.453	9.048	-27.565
5	**X2**	-93.031	-36.393	-77.073
6	**X3**	-15.400	-70.163	4.098
7	**X4**	-6.081	-86.[illegible]46	93.919
8	**PE's**	-9.878	29[illegible]	5.545
9	Connect X1	0.06	-0.02	-0.06
10	Connect X2	-0.02	0.08	0.06
11	Connect X3	0.02	-0.07	-0.08
12	Connect X4	0.05	-0.07	-0.04
13	Connect PE	-0.09	-0.03	-0.08
14	X1	1	1	1
15	X2	1	1	1
16	X3	1	1	1

FIGURE 10.4
PE weights and biases of a neural network

A second way of making the structure of the neural network more flexible is by *allowing the connections between inputs and PEs to vary.* By extending the weight/bias matrix with parameters to determine the connectivity between inputs and PEs, a neural net architecture is created with a variable number of PEs and a variable number of connections. The values in rows AL9 to AO12 in Figure 10.5 are additional parameters that generate an integer value of 0 or 1 (in AL14 to AO17) in a manner similar to that described above. These integer values, multiplied with the product of weights and inputs, determine whether a connection is in or out. In sum, the fully connected structure in Figure 10.2 could be transformed into any combination of inputs, processing elements, and connections.

To account for a variable number of PEs and connections, the formula for each PE should be expanded as follows. A PE value is the hyperbolic tangent of: a bias (AK3), plus the sum of an input connect factor (AK14 for X_1), times a connection weight (AK4), times an input (H29). This PE value will be passed on to the output only if the integer in AK18 is 1. The formula for all PE columns is thus:

PE value = tanh ($AK$3+($$AK$14*$AK$4*H12) + . . . +
($AK$17*$AK$7*K12))*$AK$18

AK	AL	AM	AN	AO	AP	AQ	AR	
	Weights	**PE1**	**PE2**	**PE3**	**PE4**			
	Bias	1.502	-4.707	-8.812	-8.067	<< Wsonly		
	X1	5.952	2.458	2.137	-0.575			
	X2	7.637	5.748	-0.216	-2.434			
	X3	-9.261	5.445	2.207	1.697			
	X4	-5.183	-1.262	-9.471	1.539			
	PE's	-0.157	7.021	-3.581	-7.629			
	Connect X1	0.00	0.00	-0.02	0.03	<< wANDoon		
	Connect X2	-0.04	0.01	-0.09	-0.03			
	Connect X3	0.07	-0.09	0.00	0.01			
	Connect X4	-0.06	0.07	-0.07	-0.07			
	Connect PE	-0.07	-0.08	-0.08	-0.07			
	X1	1	1	0	1			
	X2	0	1	0	0			
	X3	1	0	1	1			
	X4	0	1	0	0		TRUE=1 FALSE=0	#PEs
	PE Opti	**1**	**0**	**0**	**0**	<< fixed >>>>	0	4
	B/S Signal	**-0.100**	**0.100**			<<Bssig		
	OBJECTIVE	0.249				<< fitt		
	CONSTRAINT	-1.794						

Genetic Optimized Neural Net Model for Trading 10 Year Notes FY90-93

Period	**FY90-93**	**FY90**	**FY91**	**FY92**	**FY93 till 12/31/92**
Basis Points Profit/year	-0.385	-1.337	-0.672	0.343	0.158
Maximum DrawDown or Risk	-1.794	-1.794	-0.890	-1.179	-0.653
Profit/Risk Ratio	-0.2	-0.7	-0.8	0.3	0.2
% Profitable Trades	66.7%	43.8%	58.3%	78.9%	70.0%
No Profitable Trades	36	7	7	15	7
No Lossing Trades	18	9	5	4	3
Average Trades/yr	15	16	12	19	10

Constraints		
Transactions	LONG & SHORT & OUT-OF-MARKET	
Transaction cost	0.005	per transaction
Slippage Cost	1%	
POSITION SIZE	FIXED	
Trading Signal for LONG	-0.10	
Trading Signal for SHORT	0.100	
RMS	0.249	

Design © 1993 Guido Deboeck : 8-Feb-93
Advanced Analytics Lab, World Bank 03:32:34PM

FIGURE 10.5
Summary of financial results

This representation of a feed-forward neural net in a spreadsheet may require two more columns: in the case of supervised learning, a column is required for a desired output and another for the computation of the error. Creating a desired output is not very difficult, requiring only that the patterns which the neural network should learn be defined.

Given a desired output, the error between the NN output and the desired output can be computed. In this application we defined the error as the square of the difference between the actual NN output and the desired output if the NN output completely "misses the boat" in terms of market direction. If the prediction result is contrary to the desired one, then an error is recorded; in other cases the absolute difference between the actual and desired output is unimportant. Thus a "penalty" is assigned only if the direction of the market is misjudged, not if the magnitude of the prediction is off. The root of the sum of all errors over a particular time period (the period on which the data are trained) is computed and stored as a potential objective to be minimized (cell AK20). The formula for the error column is thus:

Error = if((V12>=$AL$19 and U12>=$AL$19) or (V12<=$AK$19 and U12<=$AK$19) or (V12<$AL$19 and V12>$AK$19 and U12<$AL$19 and U12>$AK$19),0,(V12−U12)∧2)

Step 3: Establish an Evaluation Scheme for a Trading System

The previous step established a feed-forward neural network in a spreadsheet. This structure is essential for estimating nonlinear relationships among market indicators, their transformations, and the desired output. Given a learning algorithm this feed-forward neural network can find weights, which when applied to inputs will provide an output. This neural net output can then be translated into a trading recommendation.

An analog NN output is translated into a square wave or buy/sell function by applying the following simple rules:

1. If the output is less than a fixed lower limit, the trading recommendation will be: "go long" or "stay long" (+1).
2. If the output is greater than a fixed upper limit, the trading recommendation will be: "go short" or "stay short" (−1).
3. If the output falls between the lower and upper limits, then the recommendation is "stay out of the market" (0).

These rules are formulated as follows in the buy/sell (B/S) column (column Z):

=if(U13<0 and U13<=$AK$19,1, if (U13>0 and U13>= $AL$19,−1,0))

These rules provide a specific trading recommendation for each day. This B/S signal can be obtained by applying either a fixed or variable set of limits to the signal produced by the neural network. If variable B/S limits are preferred, the upper and lower limits become additional parameters to be optimized. Values for variable limits are set in AK19 to AL19.[4]

To evaluate the financial performance of the trading recommendations in this spreadsheet we adopted the framework for evaluation of neural nets provided in Chapter 2.

The remainder of this step is devoted to discussion of a methodology for computing the return and the risk of this trading model. Based on financial performance and risk measurements, a set of performance criteria can be specified for model evaluation.

To compute overall *financial performance* we start by computing the daily profit and loss (P&L). This is the difference between the current yield and the next day's yield if the position is long (or between the previous day's yield and the current yield if the position is short) multiplied by the position size and a factor that accounts for slippage costs, minus the transaction costs. Slippage costs is a small percentage deducted from the daily P&L to account for the fact that execution of the neural net recommendation may be at a price different from the close.

The daily P&L can be adjusted for "costs of carry" by adding the duration-weighted daily income minus the daily "repo rate" (i.e., the Fed Funds rate minus 10 to 20 basis points) for all long positions, and/or subtracting the duration-weighted daily income plus the daily repo rate for all short positions. The formula for daily P&L in column AB is

```
=if(Z13=0,0,if(Z13=-1,(E14-E13)*AA13*$AB$11-$AM$37,
(E13-E14)*AA13*$AB$11-$AM$37
```

The adjustment of daily P&L for cost of carry in column AC is

```
=if(Z13=1,(B12-B12-0.1)/(360*7),
if(Z13=-1,(B2-0.1-E12)/(360*7))
```

The profit and loss per trade (P&L/TR) is the sum of daily P&L for the length of each trade, that is, the length that a position remains long or short. To compute average gains for profitable trades or average losses for losing trades, columns AD and AE are used.

The cumulative P&L (column AF) for a year is the sum of the daily P&Ls for that period. The total return for a period is the cumulative P&L at the end of a period divided by the cumulative P&L from the start of the period, raised to the power of 1 over the period length. The total return takes into account both the realized and unrealized P&L from the start to the end of the period. The cumulative P&L can be plotted as the equity line of the trading system.

Financial risk is often measured in terms of volatility of daily returns, whereby volatility is the standard deviation of the daily returns (often annualized in order to make volatility measures comparable). Thus, to measure the risk of a trading system one could measure the standard deviation of the daily returns. Another approach is to measure the fluctuations of the daily returns against a benchmark. In this case it is important to compare "relative performance" versus "relative risk," that is, incremental performance versus incremental risk in comparison with the risk of the benchmark. Comparing relative performance with absolute risk is like comparing apples and oranges.

The risk of a trading system can also be measured in terms of the largest losses over a period of time, or the "maximum drawdown." The maximum drawdown is the cumulation of successive losses (worst-case scenario) or the degree of smoothness of the equity curve. The maximum drawdown is computed as follows (column AG):

=if(AG13<AG12, max($AG$13. .AG13)−AG13, AH12)

In sum, the above steps result in the formulation of a system for trading securities based on a feed-forward neural network with variable weights, processing elements, and connections. This trading system can be designed for different performance objectives and/or risk levels. For example:

- To maximize returns
- To minimize volatility of returns
- To minimize trades or transaction costs
- To maximize the ratio of the average profitability to the maximum drawdown
- Any combination of the above

Adoption of one or more of these evaluation criteria results in different trading strategies. All of the above prepares the way for a genetic optimization problem to meet user-defined performance criteria and risk limits.

Step 4: Formulate a Genetic Optimization Problem

A genetic algorithm is a procedure that generates a problem solution for a multiparameter optimization problem by considering in parallel

30 to 50 potential solutions and measuring the goodness-of-fit of each against the user-defined objective or fitness function (see Davis 1991; Goldberg 1989).

In this case, a problem solution is a neural net with a particular set of weights, PEs, and connections. A GA selects the best solutions from a collection of problem solutions and applies reproduction, crossover, and mutation to generate a new population of problem solutions. *Reproduction* takes the best solutions and considers them as "parents" to generate "children;" *crossover* mixes the attributes of the parents, and *mutation* introduces attributes that may not yet be represented. The new collection of problem solutions is cycled through the selection or "survival of the fittest" process. Poorer solutions are gradually weeded out until a solution meets particular criteria (e.g., N number of trials completed, time limit reached, or insignificant improvement between cycles). Thus, a GA is used to find a neural network that optimizes the user-defined performance objectives and meets user-defined constraints or risk limits.

An easy way to implement a GA in a spreadsheet, without programming macros or scripts (such as those described in Klimasauskas 1992), is to use Evolver™ from Axcelis Inc. Evolver is an inexpensive extension to both Microsoft Excel (under Windows) and WingZ (on the Macintosh). In this section we will show how to formulate a GA optimization problem using Evolver under WingZ (see also Deboeck and Deboeck 1992).

To set up the genetic optimization problem we need:

- An objective or fitness function
- Potential constraints to be met
- The parameters that are subject to change
- The ranges for the adjustable parameters

We should also specify:

- The number of problem solutions in each population
- The criteria for stopping the process, such as number of generations, time available to optimize, and/or reduction of improvement in successive generations (e.g., stop the process when the best solution no longer changes by more than x points)

The Evolver menu shows an item called Evolve. Clicking on it produces a window that allows the objective function, the constraints, and the range of adjustable parameters to be specified. The objective function in this example is the average annual return achieved over three years. It is computed from the figures in row 28 in Figure 10.5 (the sum of the number of basis points achieved in fiscal years 1990 to 1992 divided by 3). The constraints are the risks as measured by the maximum drawdown over three years. This is the minimum value of the maximum drawdowns shown in Figure 10.5. The adjustable parameters are the neural network weights and the biases, plus potentially the upper and lower limits for determining the B/S signal. The adjustable parameters are those stored in AN3 to AO8 in Figure 10.5 (or AO13 if the neural network structure can be fine-tuned, as described earlier in footnote 4).

Once the Evolver window is completed and the process is started by clicking on Evolve, the program will present a window that allows the user to set the ranges for the adjustable parameters (Figure 10.6). Setting these ranges too narrowly will accelerate the GA process but may constrain the GA to finding a solution in a small part of the hyperdimensional solution space. Thus, the ranges of the adjustable parameters must be carefully chosen. It should be noted that the ranges for the B/S signal upper and lower limits (in AL 19 and AK19) are constrained to ±0.20.

To monitor the progress of the GA, the "Evolver Progress" window shows the evolution toward the best solution and the increase (or decrease) in the average quality of solutions in each set of problem

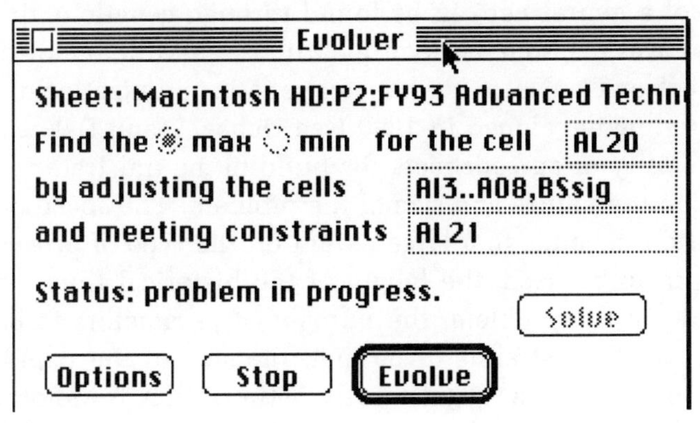

FIGURE 10.6
Setting up a genetic algorithm

solutions (population). The GA process can be paused or stopped at any point. The best solution or the ten best solutions obtained to that point can be retrieved and stored in a separate file.

Step 5: Summarize, Evaluate, and Plot Key Results

The final step in the design of this neural-based trading system is to build a summary evaluation report and to plot key results. The scope of this summary depends on user needs. An example is provided in Figure 10.7, which shows summary financial performance and risk measures by fiscal year, as well as for the entire period FY90–93. All of these are derived from the trading system section of the spreadsheet and are either simple additions or the drawdowns over a period specified in the column report. The main variables of the model are grouped to facilitate sensitivity analysis. The same type of summary could be compiled using quarterly or monthly results.

Any of the key results can be plotted (see Figure 10.7) A useful chart of a trading system is a graph of the market signal (in this case, the 10-year Treasury note) and the equity curve or cumulative P&L obtained by implementing the neural net trading recommendations.

Conclusions

This chapter demonstrates that a simple feed-forward neural net can be implemented in a spreadsheet and that the weights, PEs, and connections of a neural net can be found through genetic optimization. This work was inspired by Geoffrey Miller (see Miller et al. 1989) and by our work on GenNet, a program for genetic optimization of a neural net for trading developed in 1990 (see Deboeck and Deboeck 1992). The suitability of spreadsheets for building neural trading systems depends on the nature and size of the problem. The speed of genetic optimization depends to a large extent on the type of processor and math coprocessor used, the length of the historical time series, the complexity of the problem, the number of parameters to be found, the number of constraints to be met, the size of the population of problem solutions, and the criteria for optimization. A Motorola 68040 or an Intel 486, with math coprocessor, high clock speed, and lots of RAM are clearly minimum configurations.

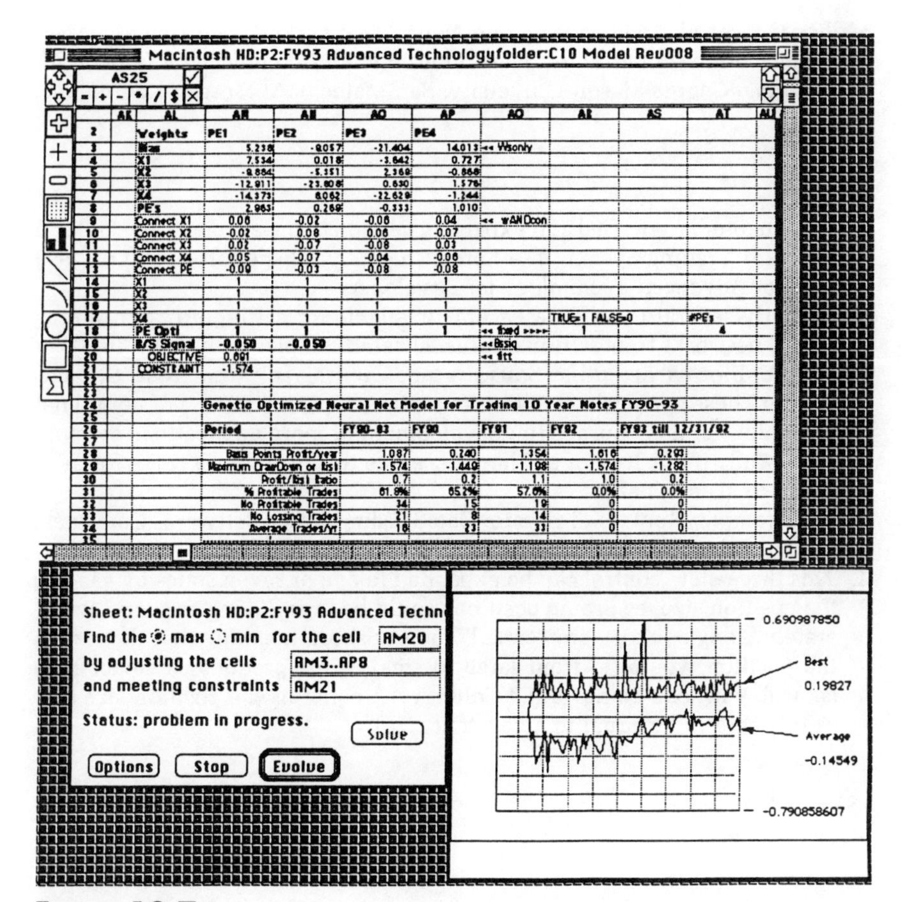

FIGURE 10.7
Summary screen on genetic optimization of a neural net in a spreadsheet

References

Davis, L. 1991. *Handbook of Genetic Algorithms.* New York: Van Nostrand Reinhold.

Deboeck, T., and G. Deboeck, 1992. "GenNet: Genetic Optimization of a Neural Net for Trading." *Advanced Technology for Developers* (Oct.).

Goldberg, D.E. 1989. *Genetic Algorithms in Search, Optimization and Machine Learning.* Addison-Wesley.

Klimasauskas, C.C. 1992. "Genetic Function Optimization for Time Series Prediction." *Advanced Technology for Developers* (July); "Hybrid Neuro-Genetic Approach to Trading Algorithms." *Advanced Technology for Developers* (Nov.) "An Excel Macro for Genetic Optimization of a Portfolio. "*Advanced Technology for Developers* (Dec.).

Miller, G.E. et al. 1989. "Designing Neural Networks Using Genetic Algorithms." *Proceedings of the Third International Conference on Genetic Algorithms* (George Mason University. San Mateo, CA: Morgan Kaufmann.

Notes

1. This chapter is a revised and expanded version of the article "How to Build a Hybrid Trading System in a Spreadsheet," published in the April 1993 issue of *Advanced Technology for Developers.*
2. Recently several tools have become available for setting up and training neural networks from within a spreadsheet. Examples include Brainmaker from California Scientific Software and Neuralyst™ from Epic Systems Group. These spreadsheet extensions stimulate the diffusion of neural net technology and encourage the development of more applications by more users. Only if the "rest of us" have access to this technology can we expect to see its true benefits.
3. For example, Reuters, Telerate, DataStream, Street Software, Comstock, Compuserve, and Dow Jones.
4. This three-state control can be expanded to five or seven states by varying the position size; eg., some positions could be "very" long or short, "moderately" long or short, and "small" long or short, meaning that the size of the position would go from large to small. This produces a seven-state control, with one being "out." Column AA contains the position size (left in this case to be equal to a constant size).

PART THREE

Portfolio Management Using Fuzzy Logic

zhòng Yòng:

change has an absolute limit, which produces two modes: yin and yang. These two modes produce four forms, which produce eight trigrams; the eight trigrams determine fortune and misfortune. Yin-yang is the emblem of fuzziness. It stands for the world of opposite forces.

CHAPTER 11

Why Use Fuzzy Modeling?[1]

Guido J. Deboeck

Machine wealth will help take duress out of our lives.

Bart Kosko
Fuzzy Thinking, *1993*

Part Three of this book is about fuzzy logic. We begin by explaining the difference between neural network and fuzzy logic approaches to system design and discuss the advantages and disadvantages of each. Next, we review methods for translating financial data into fuzzy memberships and for extracting fuzzy trading rules, and demonstrate how fuzzy expert systems actually produce crisp trading recommendations. The case studies presented apply these concepts and methods to the design of trading systems for the Shanghai stock market and for U.S. Treasury securities. Chapter 14 discusses hybrid systems that combine neural networks, genetic algorithms, and/or fuzzy logic to achieve better performance than is feasible with single-technology systems.

Neural Versus Fuzzy Models

A trading system is a disciplined method for trading financial markets that is usually based on one or more trading rules. Classical examples of trading rules derived from technical analysis were provided in the Introduction to this book.

A trading system based on a single trading rule seldom outperforms the market. A trading system with two or more rules has a better chance, depending on whether the rules fit the prevailing market regime. For example, a system based on trend-following rules will be effective only when the market is in a trending mode. Designing a trading system for both trending markets and nontrending markets (i.e., those with low volatility) requires two set of rules, as well as rules for switching from one set to the other.

Appropriate sets of trading rules for different markets, along with switching parameters, can be obtained using traditional techniques and tested using a simple spreadsheet program. A spreadsheet can compute weights for each market to maximize profitability over a given time horizon. Usually, mechanical trading systems are designed using elaborate simulations to find the parameters needed to achieve a desirable level of performance. However, the odds of finding stable parameters are small. A system with three variables—two inputs and a single output ($n = 3$)—where each variable can have only three values ($m = 3$), has m^n input/output combinations and 2^{m^n} possible relationships between them. Thus if $n = 3$ and $m = 3$, there are 27 (3^3) input/output combinations and 134 million (2^{27}) possible relationships. In reality, market prices tend to be influenced by many variables, each of which may fluctuate over a large range of values. In consequence, traditional techniques can explore only a very small part of the input/output product space.

Finding trading rules and parameters for a mechanical trading system through simulations should not be confused with training of a neural network system. A neural network uses a learning algorithm (e.g., back-propagation based on gradient descent) to map input variables ($X_1 \ldots X_n$) into the range of desired output(s) ($Y_1 \ldots Y_n$). In other words, a neural network establishes a mathematical relationship between input variables and one or more desired outputs. *Neural networks are model-free, nonlinear, dynamic systems that learn from examples* the relationships between inputs and outputs. Several point estimates of X_i, Y_i pairs are used to estimate the function $X \rightarrow Y$. Figure 11.1 shows the neural net mapping of X_i, Y_i pairs.

Neural nets are good at finding relationships in a huge input/output product space even if the patterns between inputs and outputs are ill defined (Deboeck et al. 1992). The performance of a neural trading system depends on:

- The information that flows into the system
- How the input information is preprocessed

- The desired output
- The quality of the neural net design

Neural net design depends on:

- The appropriateness of the neural net architecture
- The learning algorithm
- The adequacy of neural net training
- The relevance of the training samples

Drawbacks of neural-based trading systems include:

- Lack of explanatory capability
- Difficulty of including structured knowledge
- Bias toward quantitative data

Some traders are interested in how neural net recommendations are derived. Managers may actually demand more from trading systems than from traders, expecting neural net models to confirm their "gut feelings." As the newness of the technology disappears, these demands may fade. Who cares how combustion engines work when driving a car? Nevertheless, neural nets' lack of confidence intervals and inability to trace outputs, or even to include structured knowledge, have prevented rapid acceptance of neural net technology in trading operations.

Systems employing fuzzy logic present an alternative. Trading systems designed on the basis of fuzzy logic allow the inclusion of

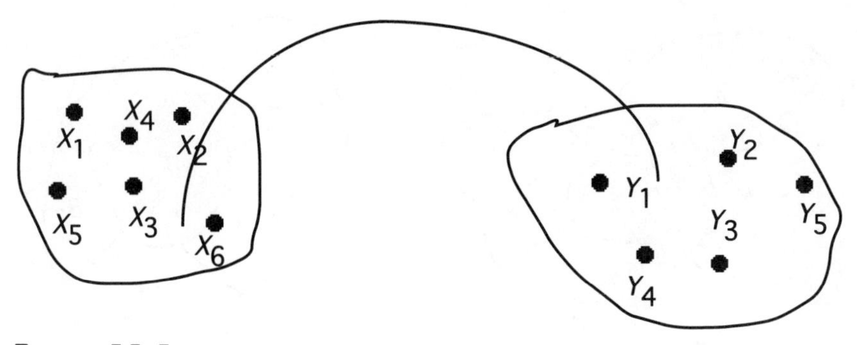

FIGURE 11.1
Neural network approach to mapping

trading rules provided by traders and are capable of explaining the trading recommendations made by the system. They also avoid over-reliance on quantitative data.

A simple model of a fuzzy trading system may consist of one or more "fuzzified" inputs (e.g., trend and volatility measures), one output variable (e.g., desired trading pattern), and a few fuzzy rules expressing the relationships between them. Membership functions are created, associations between inputs and outputs are defined in a fuzzy rule base, and fuzzy outputs are translated into crisp trading recommendations.

A fuzzy approach to the design of a trading system uses an adaptive clustering algorithm that infers fuzzy associations between inputs A_i and outputs B_i. This results in fewer rules than input antecedents. Figure 11.2 shows the mapping of fuzzy set A_i into fuzzy set B_i.

Only a partially filled rule matrix is required. Fuzzy rule bases can be designed by experts or estimated from trading samples. A fuzzy set encodes a structure that can be mapped as a minimal fuzzy association of part of the output space with part of the input space. A fuzzy associative memory, or FAM, is a mapping of fuzzy sets to fuzzy sets.

The first step in the design of a fuzzy trading model is to convert inputs into fuzzy representations or collections of membership functions. *Membership* defines the degree of adherence rather than the probability of an event. For example, a trader may have a 0.4 membership into the class of "experienced traders" while maintaining a 0.6 membership in the class of "average traders." In essence, contin-

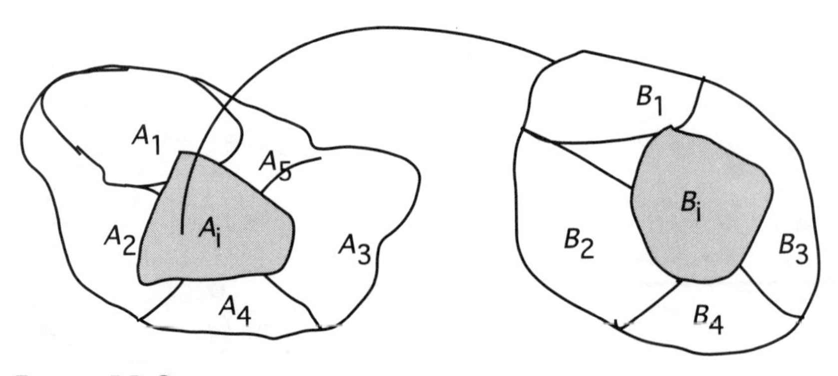

Figure 11.2
Fuzzy logic approach to mapping

uous inputs (e.g., experience divided into discrete subranges or classes, (such as novice, average, or experienced) may overlap. Within each class, the membership measures the degree of adherence to the particular attributes of a class.

These concepts can be applied to financial time series. For example, the trend of a security as measured by a moving average can be divided into various subranges or degrees of trending. To do this, one starts by finding the minimum and maximum of the trend over a historical period. The range of values defined by this minimum and maximum is called the *universe of discourse*. This range can be divided into trend subranges, such as big negative, small negative, no trend, small positive, or big positive. The starting, center, and end values can then be computed for each subrange. For example, the subrange "average" can be encoded as $\{1,0,-5,0,+3,1,+7,0\}$, where the odd-numbered items are the points of a membership function and the even-numbered items are the starting, center, and end points of the particular subrange. Figure 11.3 illustrates the concept of a membership function. The main advantage of fuzzifying a time series is the elimination of noise, which allows one to deal with a few subranges or classes rather than with minute changes.

Fuzzified time series are used as inputs in fuzzy trading systems instead of the original values. In neural networks, this technique is commonly known as a 1-of-*N* code. One method for constructing fuzzy membership sets has been described by Klimasauskas (1992). Several other methods for fuzzifying time series are discussed in Chapter 13.

Basic Math with Fuzzy Sets

In the previous section we discussed how fuzzy sets eliminate the need for precisely defining the change from one state to another. The fuzzification of time series has profound implications for the design of trading models. Based on fuzzified inputs and outputs, trading rules can be identified that embed imprecise ideas; thus, changes in market indicators and relationships between changing indicators can be captured in general terms. To appreciate the power and flexibility of fuzzy logic, we will first review some basic math techniques for dealing with fuzzy sets (Kandel 1986).

Basic fuzzy math techniques are essential for the design of a fuzzy trading system. Let's assume that the trend (input variable *A*) and the

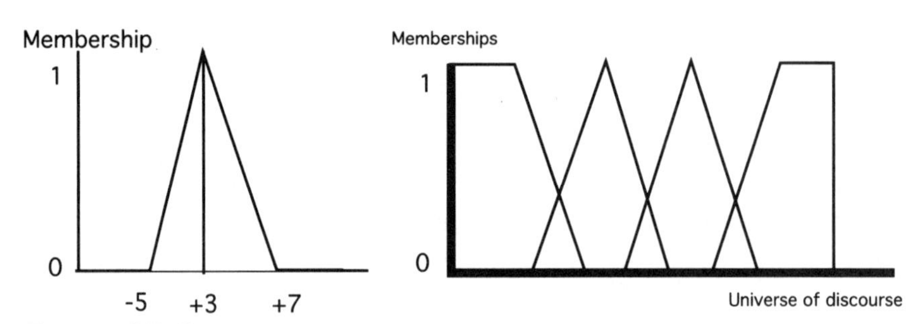

FIGURE 11.3
Membership functions and the universe of discourse

volatility (input variable B) of a security have been translated into fuzzy sets. Examples of a membership from each are $A^1 = \{0.3\ 0.7\ 0.5\}$ and $B^2 = \{0.4\ 0.0\ 1.0\}$, where the superscript indicates the membership segments.

Essential math operators are the intersection, union, and complement between two sets. The *intersection* of fuzzy sets A and B can be obtained by taking the pairwise minimum of all the elements, or:

$$m_{A \cap B} = \min(m_A, m_B), \text{ i.e. } \{0.3\ 0.0\ 0.5\}$$

The *union* of the fuzzy sets can be obtained by taking the pairwise maximum of all elements, or:

$$m_{A \cup B} = \max(m_A, m_B), \text{ i.e. } \{0.4\ 0.7\ 1.0\}$$

The *complement of fuzzy set A* can be obtained by reversal of the order of all elements, or:

$$m_{A^C} = 1 - m_A, \text{ i.e. } \{0.7\ 0.3\ 0.5\}$$

Next let's define overlap and underlap, both of which are important in defining the ambiguity or entropy in a fuzzy set. The *overlap* of fuzzy set A is the intersection of set A with its own complement, or:

$$A \cap A^c, \text{ or } \{0.3\ 0.3\ 0.5\}$$

The *underlap* of fuzzy set A is the union of that set with its own complement, or:

$$A \cup A^c, \text{ or } \{0.7\ 0.7\ 0.5\}$$

The midpoint of the fuzzy set representation is the point where its own overlap equals its underlap, or:

$$A = A \cap A^c = A \cup A^c = A^c$$

In order to define the ambiguity, or fuzziness, of a fuzzy set, we first must measure its size. The size of fuzzy set A is equal to the sum of all the fit values of A, or:

$$M(A) = \Sigma\, m_A(X_i), \text{ or } 1.5$$

The size of A, or the *fuzzy Hamming distance*, can be computed for multiple X_i. To measure the ambiguity of a fuzzy set we must compute the *fuzzy entropy*. The fuzzy entropy of A, $E(A)$, is the fuzzy hamming distance of the intersection of A with its own complement, or $M(A) \cap A^c)$, divided by the hamming distance of the union of A with its complement, or $M(A \cup A^c)$; thus:

$$E(A) = \frac{M(A \cap A^c)}{M(A \cup A^c)}$$

The fuzzy entropy varies from 0 to 1 and is highest at the midpoint. For example, if a trading recommendation can vary from very short to very long, or -1 to $+1$, and the fuzzy composite output resulting from the firing of one or more rules into the set is {0.3 0.7 0.5}, then the entropy of this trading recommendation is 0.579, which is derived from:

$$E(A) = \frac{M(A \cap A^c)}{M(A \cup A^c)} = \frac{1.1}{1.9} = 0.579$$

where

$$M(A \cap A^c) = \Sigma\, \{0.3\ 0.7\ 0.5\} \cap \{0.7\ 0.3\ 0.5\} = 1.1$$

and

$$M(A \cup A^c) = \Sigma\, \{0.3\ 0.7\ 0.5\} \cup \{0.7\ 0.3\ 0.5\} = 1.9$$

For completeness, we must also define *fuzzy subsethood* and *entropy subsethood*. The fuzzy subsethood estimates the degree to which sets contain subsets; the entropy subsethood estimates the degree of fuzziness or ambiguity as the degree to which a superset $A \cup A^c$ is a subset of its own subset $A \cap A^c$.

Fuzzy entropy and its related concepts are useful for estimating the degree of ambiguity in a trading recommendation. For example, if the fuzzy entropy of the trading recommendation is small, the position size should be set to large; if the fuzzy entropy is large, the position size should be kept small.

The concept of fuzzy entropy can also be applied to neural networks. Neural nets compute activation levels or weights of interconnections between processing elements. With proper normalization, these weights range from 0 to 1 and thus represent fuzzy sets. Fuzzy entropy can be computed from the weights of the interconnections between processing elements achieved after every *N* training cycles. Since neural network training is the evolution of a dynamic system from a random set of weights to a set of weights that achieve a certain fit, fuzzy entropy of the weights changes evolves to either a fixed point, a limit cycle, or a strange attractor of a chaotic system. These concepts will be explained in Chapter 15. Thus, fuzzy entropy provides a basis for optimizing the training of neural nets, independent of testing net performance against a set of test data samples. A neural net that evolves to a fixed point has memorized the training examples; a neural net whose weights have maximum entropy has optimum generalization capability.

Creating a Fuzzy Rule Base

Fuzzy rules are constructed from the membership functions of inputs and outputs. These rules provide the connections between antecedents and consequences. Fuzzified inputs and outputs facilitate rule extraction and generalizations. In the case of a trading system, rules can be obtained from (1) one or more expert traders, (2) historical data, or (3) a combination of both approaches.

Let's start with trader advice. Assume that a trader defines a trading rule whereby a long position is established when a price trend is equal to or exceeds a particular value x and the volatility is equal to or below a certain value *y*. A mechanical expression of this rule could be

IF the trend => x
AND the volatility is <= *y*
THEN the LONG position should be z

where x, *y*, and z are parameters provided by the trader or estimated based on simulations. A fuzzy version of the same would be

IF the trend of x is RISING RAPIDLY
and the volatility of x is LOW

Then GO LONG
or

IF the current position is LONG, INCREASE LONG POSITION SOME

which is a more natural way of expressing a trading rule and is less prone to changes. Thus, fuzzy rules are expressed in English with a syntax similar to the mechanical rules:

IF ⟨fuzzy proposition⟩,
THEN ⟨fuzzy proposition⟩

or in the case of multiple antecedents,

IF ⟨fuzzy proposition 1⟩

AND (or OR) ⟨fuzzy proposition 2⟩
THEN ⟨fuzzy proposition⟩

where a fuzzy proposition is of the form "x is *Y*" or "x is not *Y*"; x being one of the original scalar variables (e.g., the trend of a security) and *Y* being a fuzzy set associated with the variable (e.g., negative big).

The specificity of the mechanical rule dictates frequent updates. Rule updates pose significant problems in terms of performance evaluation; the performance of systems whose parameters change frequently cannot be aggregated. The fuzzy implementation, on the other hand, implies that rule firing will depend on fuzzy sets (in this case, the trend and volatility of a security) and that a fuzzy rule ties the inputs to output properties.

A group of fuzzy rules forms a *fuzzy associative memory*. When a set of input values is read, each rule that has any truth in its premise will be fired. Generally, the number of rules a system requires is related to the number of control variables. Thus, if the output of a trading system (e.g., recommended trading position) is a single variable influenced by two inputs that are fuzzified into five segments, each having a membership function with five elements, then 25 possible input combinations are possible. This suggests that 25 rules would be required. In most cases, it may be possible to use fewer rules. A simplified version of such a fuzzy rule base is shown in Figure 11.4.

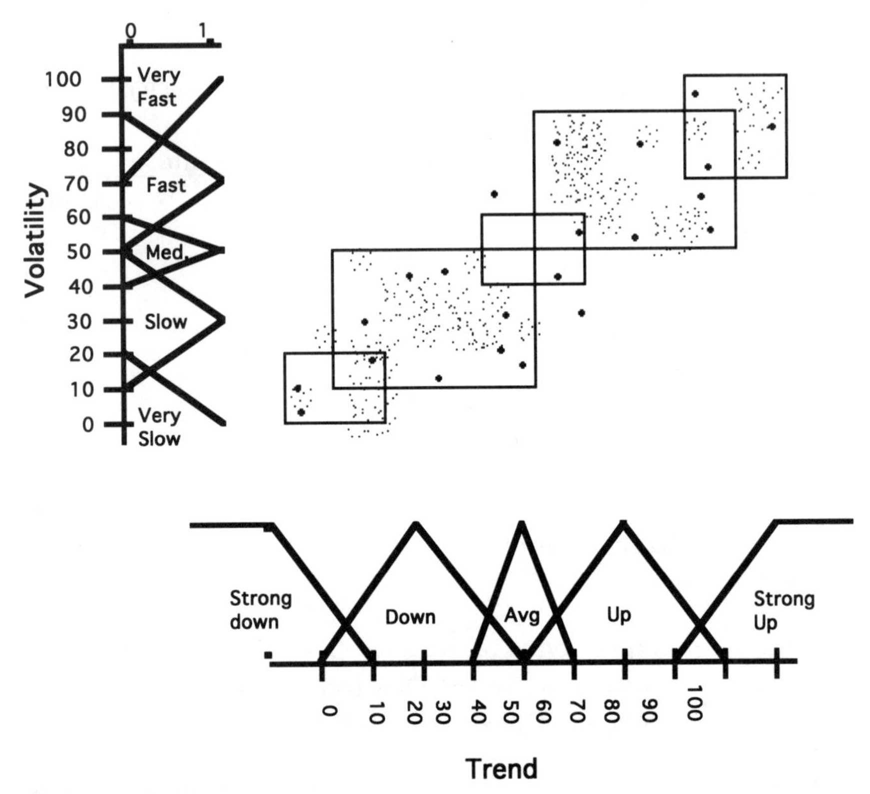

FIGURE 11.4
Extracting fuzzy rules

The rules or fuzzy associations represent knowledge that may be important in order for the system to be able to respond to all possible combinations of inputs. The incorporation of structured knowledge into a fuzzy rule base by traders overcomes one of the main disadvantages of neural-based trading models.

Fuzzy associative memories can also be automatically extracted from a historical data set. The FUNCTION FAM_Bank on the CD-ROM that can be ordered with this book provides an automated method of finding fuzzy relationships between pairs of fuzzy inputs and outputs. The FAM_Bank script requires at minimum one pair of fuzzy sets, *X* and *Y*, and can be used to extract fuzzy relationships between *X* and *Y* as inputs and *Z* as output. The script reads the fuzzy sets of *X*, *Y*, and *Z* (stored in column_*X*, column_*Y*, and column_*Z*, with Mbs1, Mbs2, and Mbs3, respectively, as the number of fuzzy sets) and creates a series of FAM matrices that provide the fuzzy relationships between X and Z, Y and Z, and the combined X and Y to Z relations. It also provides the frequencies of the intersections between the fuzzy segments of each of these variables.

For example, applying FUNCTION FAM_Bank to some 125 samples of trend and volatility measures of a security and the corresponding desired trading position, (whereby each input and output was fuzzified into five memberships), produced the set of FAM matrices shown in Table 11.1. FAMs provide the fuzzy relationships between each of the inputs and the output variable. The cells in these matrices represent the membership intersections, or *clipped fuzzy sets*, between the fuzzy sets of the trend, volatility, and recommended trading positions. The composite of these sets of FAM matrices is shown in Table 11.2.

A three-dimensional bar representation of the composite matrix for all fuzzy sets on the trading positions is shown in Figure 11.5. The height of each bar in this figure provides the degree of membership or activation level in neural network terms for various combinations of fuzzy inputs with fuzzy output. A three-dimensional surface representation of the same is shown in Figure 11.6, which is like a control surface of a trading system.

The fuzzy rule bases that are extracted automatically from historical data can complement the trading rules provided by traders. The performance of such systems depends on the initial inputs, the definition of the membership functions, and the rule extraction itself. Kasko (1992) has demonstrated how differential competitive learning (DCL), supervised competitive learning (SCL), and unsupervised competitive learning (USL) can be used to extract fuzzy rule bases from a training data set. Chapters 12 and 13 will discuss alternative methods for automating the extraction of fuzzy rules. Other research in this area involves the use of genetic algorithms to optimize the performance of fuzzy trading systems by minimizing fuzzy complexity and optimizing the shape of the membership functions. Chapter 14 will discuss combinations of genetic algorithms and fuzzy logic.

Defuzzifying Results

Thus far we have demonstrated how to fuzzify a time series and how to formulate or extract fuzzy rules from historical data. The use of

TABLE 11.1 *FAM Matrices of Trend, Volatility, and Trading Positions*

FAM Matrix 1	*Z*1 NL	*Z*2 NM	*Z*3 A	*Z*4 PM	*Z*5 PL
*X*1 NL	1	0.78	0.98	0.28	0.28
*X*2 NM	0.14	0.81	0.83	0.89	0.93
*X*3 A	0.31	0.8	0.92	0.9	0.86
*X*4 PM	1	0.83	0.76	0.98	0.99
*X*5 PL	0	0	0	0.85	1

FAM Matrix 2	*Z*1 NL	*Z*2 NM	*Z*3 A	*Z*4 PM	*Z*5 PL
*Y*1 NL	1	0.47	0.16	0	0
*Y*2 NM	0.47	1	0.61	0	0
*Y*3 A	0.16	0.61	1	0	0
*Y*4 PM	0	0	0	1	0.73
*Y*5 PL	0	0	0	0.73	1

FAM *Z*=1	*X*1 NL	*X*2 NM	*X*3 A	*X*4 PM	*X*5 PL
*Y*1 NL	1	0.14	0.31	1	0
*Y*2 NM	0.47	0.14	0.31	0.47	0
*Y*3 A	0.16	0.14	0.16	0.16	0
*Y*4 PM	0	0	0	0	0
*Y*5 PL	0	0	0	0	0

FAM *Z*=2	*X*1 NL	*X*2 NM	*X*3 A	*X*4 PM	*X*5 PL
*Y*1 NL	0.47	0.47	0.47	0.47	0
*Y*2 NM	0.78	0.81	0.8	0.83	0
*Y*3 A	0.61	0.61	0.61	0.61	0
*Y*4 PM	0	0	0	0	0
*Y*5 PL	0	0	0	0	0

FAM *Z*=3	*X*1 NL	*X*2 NM	*X*3 A	*X*4 PM	*X*5 PL
*Y*1 NL	0.16	0.16	0.16	0.16	0
*Y*2 NM	0.61	0.61	0.61	0.61	0
*Y*3 A	0.98	0.83	0.92	0.76	0
*Y*4 PM	0	0	0	0	0
*Y*5 PL	0	0	0	0	0

FAM *Z*=4	*X*1 NL	*X*2 NM	*X*3 A	*X*4 PM	*X*5 PL
*Y*1 NL	0	0	0	0	0
*Y*2 NM	0	0	0	0	0
*Y*3 A	0	0	0	0	0
*Y*4 PM	0.28	0.89	0.9	0.98	0.85
*Y*5 PL	0.28	0.73	0.73	0.73	0.73

FAM *Z*=5	*X*1 NL	*X*2 NM	*X*3 A	*X*4 PM	*X*5 PL
*Y*1 NL	0	0	0	0	0
*Y*2 NM	0	0	0	0	0
*Y*3 A	0	0	0	0	0
*Y*4 PM	0.28	0.73	0.73	0.73	0.73
*Y*5 PL	0.28	0.93	0.86	0.99	1

TABLE 11.2 *FAM for the Combined Influence of Trend (**X**) and Volatility (**Y**) on Trading Position (**Z**)*

FAM for all *Z*	*X1* NL	*X2* NM	*X3* A	*X4* PM	*X5* PL
Y1 NL	1	0.14	0.31	1	0
Y2 NM	0.47	0.14	0.31	0.47	0
Y3 A	0.98	0.83	0.92	0.76	0
Y4 PM	0.28	0.89	0.9	0.98	0.85
Y5 PL	0.28	0.93	0.86	0.99	1

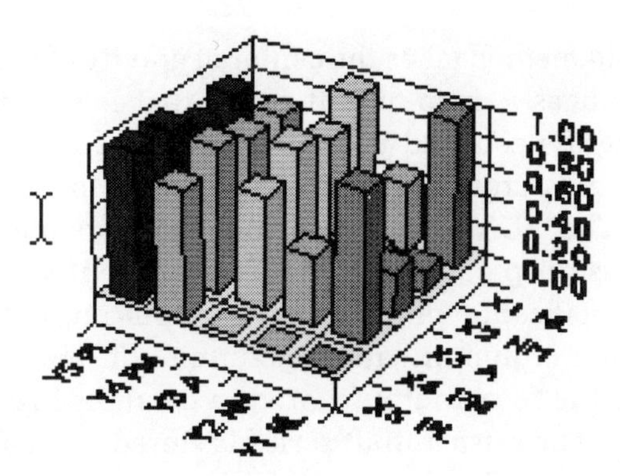

FIGURE 11.5
Composite FAM for trend, volatility, and trading positions

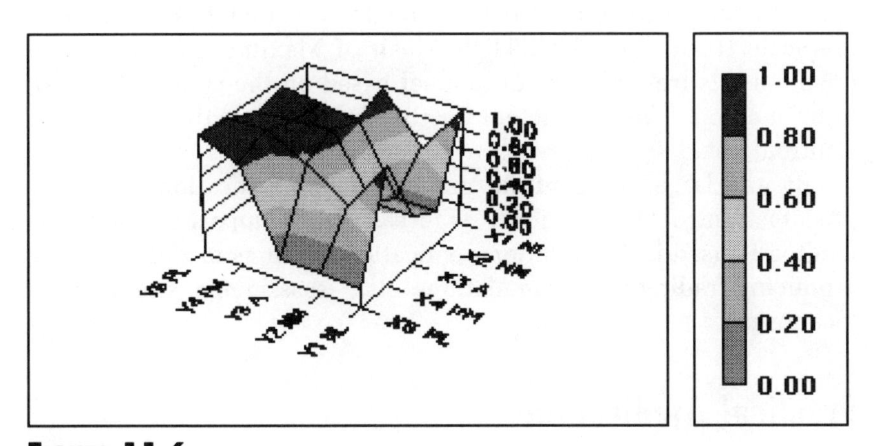

FIGURE 11.6
Control surface of a trading system based on trend and volatility information

fuzzy rule bases produces fuzzy outputs. For example, some combinations of trend and volatility indicators in Figure 11.4 produced recommendations for medium, long, or short positions. These fuzzy output recommendations must be translated back into crisp outputs in order to be put to use. The last section of this chapter will discuss *defuzzification* methods for translating fuzzy outputs into crisp values.

There are several methods for converting a fuzzy output into a crisp output. The two most common are the composite moment or centroid method and the composite maximum (mean of maxima) method.

The *centroid method* takes the center of gravity of the fuzzy output space and produces a crisp output result that is sensitive to all the rules. Applying the centroid method results in an output that tends to move smoothly across the control surface or the output universe of discourse. The FUNCTION Centroid_fnc (provided on the CD-ROM) computes crisp outputs from a fuzzy output set. The function requires the specification of the original output range, which is computed from the minimum and maximum of the values in the Z_column, and the fuzzy output stored in columns fuzzy_start_col to fuzzy_end _col. The crisp output series is stored in rscol. The FUNCTION Centroid_fnc calls the FUNCTION Centroid, which uses the Simpson method to evaluate the area under the fuzzy set. (Function Centroid_v2 is a simplified version of Centroid_fnc in which the user provides the minimum and maximum range of the output set.)

The *composite maximum method* produces a result that is sensitive to the truth produced by a single rule that has the highest predicate truth. The FUNCTION Mean_of_Maximum-fnc computes crisp outputs from a fuzzy output set based on the composite maximum method. The parameters for this function are the same as for Centroid_fnc.

By and large, the centroid method of defuzzification is used in process control applications, whereas the composite maximum method is used in information applications such as models aimed at producing trading recommendations, risk assessments, and asset allocations.

Practical Applications

This chapter has outlined the basic concepts and techniques of fuzzy logic for designing trading systems using a simple example with two inputs and a single output. More complex examples of fuzzy systems will be discussed in Chapters 12, 13, and 14 (see also LIFE 1991/92; Sugeno et al. 1992; Taber 1992; Wang 1991).

Wong et al. (1991, 1992) have developed fuzzy neural systems for stock selection. Yuize et al. (1991) have applied fuzzy reasoning to a decision-support system for foreign exchange trading.

Fuzzy portfolio management systems have been used by Yamaichi Securities since the late eighties. These systems have outperformed the Japanese stock index by 6% per year over a period of three years. Yamaichi offers several unit trusts listed in Tokyo based on fuzzy portfolio management. These unit trusts use the Yamaichi Dealing Expert System (Y-DES), which has been described as (Kaneko 1992):

> "a user friendly decision-making rule-based developing system. Users can easily define rules, simple and complicated, and test them against real market data to find promising rule sets. Y-DES is a sophisticated financial decision support system putting up-to-date computer science technology such as Artificial Intelligence, Fuzzy Theory, and Neural Networks altogether. . . . Discussions are [ongoing] to incorporate newly emerging Genetic Algorithms to enhance self-adjusting capability and reduce rule making and confirming work load."

When is it appropriate to use fuzzy logic? In the case of trading systems, it is appropriate to use fuzzy logic when several of the input variables are continuous, when mathematical models of the interaction between inputs and desired trading strategies do not exist, or when elaborate expert systems for trading decisions are complex or difficult to evaluate in real-time operation. High ambient noise levels in market signals must be dealt with. Expert traders are available, but seldom have time to formulate a classical expert system. Rules underlying financial system behavior either are not well known or are ill or poorly defined. Fuzzy logic can improve the effectiveness of neural nets for trading by incorporating structured knowledge about financial markets, including rules provided by traders, and explaining how the output or trading recommendations were derived. They are also inherently more stable.

References

Deboeck, G., H. Green, M. Yoda, and G. S. Jang. 1992. "Design Principles for Neural and Fuzzy Trading Systems." *International Joint Conference on Neural Networks* (Beijing).

Kandel, A. 1986. *Fuzzy Mathematical Techniques with Applications*. Reading, MA: Addison-Wesley.

Kaneko, B. 1992. Yamaichi Securities Inc. Private correspondence.

Kasko, B. 1992. *Neural Networks and Fuzzy Systems*. New York: Prentice Hall.

Klimasauskas, C. C. 1992. "Hybrid Fuzzy Encodings for Improved Backpropagation." *Advanced Technology for Developers* (Sept.).

LIFE (Laboratory for International Fuzzy Engineering). 1991/92. "Fuzzy Engineering toward Human Friendly Systems." In *Proceedings of the First International Fuzzy Engineering Symposium* (Yokohama), 13-15 Nov. 1991; *Proceedings of the Second International Conference on Fuzzy Logic and Neural Networks* (Tizuka), 17-22 July 1992.

Sugeno, M., et al. 1992. *Fuzzy Systems Theory and Its Applications*. Tokyo Institute of Technology.

Taber, R. 1992. "Fuzzy Entropy." *Advanced Technology for Developers* (Sept.).

Wang, P. P. 1991. *Fuzzy Control Bibliography*. Durham, NC: Duke University, Department of Electrical Engineering.

Wong, F. S. and P. Z. Wang. 1991. "A Fuzzy Neural Network for FOREX Forecasting." In *Proceedings of the First International Fuzzy Engineering Symposium* (Yokohama), 13-15 Nov.

Wong, F. S., P. Z. Wang, T. H. Goh, and B. K. Quek. 1992. "Fuzzy Neural Systems for Stock Selection." *Financial Analyst Journal* (Jan.-Feb.).

Yuize, H., et al. 1991. "Decision Support System for Foreign Exchange Trading." In *Proceedings of the First International Fuzzy Engineering Symposium* (Yokohama), 13-15 Nov.

Note

1. An earlier version of this chapter was published under the title "Basic Techniques for Fuzzy Model Design" in the November 1992 issue of *Advanced Technology for Developers*.

CHAPTER 12

A Fuzzy System for Trading the Shanghai Stock Market

Zhongxing Ye and Liting Gu

Buddhism correlates with fuzzy engineering in countries with developing high-tech economies. I can state this sociological hypothesis in color.

Bart Kosko
Fuzzy Thinking, *1993*

Fuzzy logic is popular in both China and Japan. According to some sources there are more than 10,000 fuzzy logic researchers in China (compared to about 400 in the United States). In this chapter Professor Zhongxing Ye, vice chairman of the Department of Applied Mathematics at Shanghai Jiao Tong University, and student Liting Gu show how a hybrid neural net system can be enhanced with fuzzy logic to trade the Shanghai stock market.[1] Professor Zhongxing made a valuable contribution on this subject to the international panel on financial applications of neural nets, which I chaired at the International Joint Conference on Neural Nets held in Beijing in November 1993. This chapter is concise but rich in suggestions for the design of hybrid systems.

Introduction

Several systems based on neural networks for analyzing the stock market produce good results (Asakwa 1990; Kamijo 1990). Most use only a single index, such as closing price or moving average, as an input signal or use a single neural network to predict a time series. However, the stock market is a nonlinear, dynamic, chaotic system, and stock prices are affected not by a single event but by many complex factors arising from economic or political domains. It is difficult to discover the rules hiding in the stock market, because very little useful information is utilized. In this chapter we discuss the design of a hybrid system composed of an artificial neural network and a fuzzy decision system. It is used to forecast stock market trends based on fine-grained daily data without much computational cost. We have selected the Shanghai stock market as the simulating objective because it represents both an emerging market and a microcosm of Chinese economic reform.

The procedure of building this system included the following stages: data selection and preprocessing, data classification, fuzzy rules selection, and fuzzy decision and trend forecasting. Each of these stages will be discussed separately.

Data Selection, Preprocessing, and Classification

Data selection is very important in building a successful system and requires the active participation of domain experts. Raw daily data, such as closing price, opening price, and volume indicators, must be carefully collected.

Data preprocessing is another important stage. The purpose of preprocessing the raw data is to prompt the quality of pattern classification, to compress the residual information, to reduce the computational cost, and to facilitate subsequent processing.

Various preprocessing methods have been discussed in Chapter 2. We focus on preprocessing based on previous work, transforming continuous values within a range of 0 to 1 (or some other bounded closed interval). Some neural network models (e.g., ART1, a popular model regarded as a paradigm for machine learning) require binary input signals. Actually, this process is a quantization or encoding procedure. To reduce the quantization error, those divisions of the range where the data appear frequently should be compact. Divisions where the data seldom locate must be incompact so as to lessen the number of code words. However, quantization is not always necessary—for instance, it is not necessary when we use a back-propagation algorithm to train multilayer feed-forward networks where data have been processed by using the method reported by Deboeck (1992).

Data classification follows data preprocessing. Suppose a series of 30-day moving averages and 30-day relative strength indicators is used. As we all know, in the stock market prices can move up and down quite frequently within a trading day. *Predicting these values is meaningless.* Instead, attention should be paid to the trends of the series. The principle is to identify various trends of data over a fixed period. Here is a concrete example. Suppose we have a series of moving averages. The chosen training set contains patterns composed of 20 values selected consecutively from the series. These patterns are then classified into nine categories (see Figure 12.1), a very ambitious scheme. It is important to select patterns appropriately. Suppose the patterns can be described by:

$$x^t = (x_1^t, \ldots x_{20}^t) \in \text{class } 1,\ t = 1, 2, \ldots k \tag{12.1}$$

where k is the number of patterns that belong to class 1. Then,

$$x_i^* = \sum_{t=1}^{k} x_i^t \,/\, k \tag{12.2}$$

We obtain a new pattern $x^* = (x_1^*, \ldots x_{20}^*)$ called the center of class 1. Meanwhile, we define $d = \max_t x^t - x^*$. For any pattern x, if $x - x^* < d$, then x belongs to class 1 with membership = 1; if $2d > x - x^* > d$, then x belongs to class 1 with membership $= (2d - x - x^*)/d$. Otherwise, the membership is 0. We can similarly obtain the centers of the other eight classes and the corresponding membership functions.

We then build a three-layer forward network with 20 input neurons accepting signals sent from the above selected patterns, one hidden layer, and nine neurons outputting nine values representing the nine category memberships of input patterns. Thus, the training set

FIGURE 12.1
Nine categories of patterns

is composed of the patterns selected above, and the teaching signals are the corresponding nine membership values. The result is a simple fuzzy classifier.

However, certain problems involve more complex patterns that cannot be classified so easily. In such cases, the ART1 model may be a better choice. (Note that in practice the number of categories must be controlled when using ART1.) The best choice depends on the actual problem.

Fuzzy Rules Selection

We have discussed the classification of several important stock indicators. Unfortunately, each single indicator tells us only a little information, which is always unilateral. If we combine them and discover the rules of their effect on the stock market, we may be able to remedy this drawback to some degree. Fuzzy rules can be expressed as follows:

if pattern $x \in$ class s^x
AND pattern $y \in s^y$
THEN pattern $o \in$ class s^o

where $\in$ means "belongs to." Generally, x and y are selected from different kinds of data series; the output pattern, o, is usually a series of future closing prices (represented by the category).

Fuzzy rules typically are based on the opinions of experts or the experience of traders. Here, however, we suggest another method—one which depends on neural networks rather than on human thoughts. Having learned training patterns, the neural networks can abstract significant information from patterns and generalize them into several principles reflected by the distribution of the connective weights. We shall explain only the main idea; more details can be found in Kosko (1992).

Consider a product space $X * Y * O$ of the input space X, Y and the output space O. Suppose that the vector $p = (x,y,o)$ is a pattern in $X * Y * O$; it is then trained by using the following composite learning algorithm:

$$m_j\,(t+1) = m_j(t) + a\,(t)\,(p_j - m[t]) \quad \text{if } j \text{ is the winner,} \qquad (12.3)$$
$$m_j\,(t+1) = m_j\,(t) \quad \text{if } i \neq j$$

The weight vector m_j converges to fuzzy-matrix centroids in $X * Y * O$ (see Kosko 1992). More generally, they are used to estimate the density or the distribution of the fuzzy rules in $X * Y * O$. In practice, we select several most frequent rules as the rules of the terminal system.

Fuzzy Decision and Trend Forecasting

The structure of the hybrid system is shown in Figure 12.2. The system comprises two major parts: neural network classification and a fuzzy decision system. The input signal X^i is a vector whose components are series of data over a period selected consecutively from a type of stock indicator. The classifier output Y^i is also a vector, of which the jth component is the membership value of X^i belonging to class j. Y^i also serves as the input signal of the fuzzy decision system.

A fuzzy decision system is really a correlation-minimum inference procedure. For simplicity, suppose that $k = 2$ and $l = 9$, and that the following rule exists:

$$\text{If } x^i \in \text{class } s1_{j1} \text{ and } x^2 \in s2_{j2}, \text{ then } Z \in \text{class } s3_{j3} \tag{12.4}$$

and by the classifier the following classification results: Y^1, Y^2, where:

$$Y^1 = (m1_1, m1_2 \ldots m1_9),\ Y^2 = (m2_1, m2_2 \ldots m2_9) \tag{12.5}$$

$m1_j$ is the membership of X^1 belonging to class j, and $m2_j$ is the membership of X^2 belonging to class j. According to the rule expressed in Eq. 12.4, we can conclude that Z belongs to class $s3_{j3}$, with membership $m = \min(m1_{j1}, m2_{j2})$. Other rules are dealt with similarly (for details, see Kosko 1992).

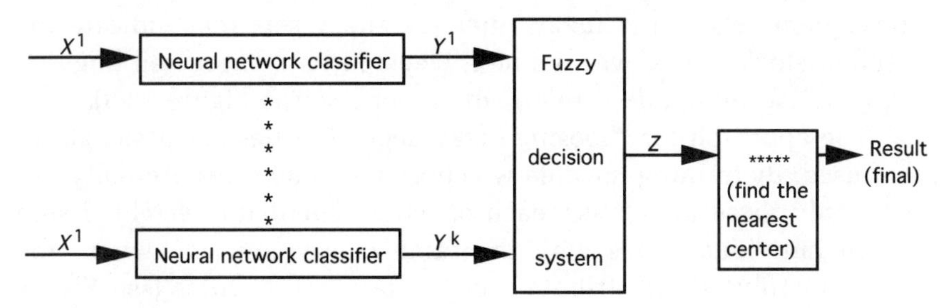

FIGURE 12.2
Hybrid cognition system

Fuzzy output vector Z is not the final result. It should be compared with nine class centers previously obtained by learning patterns and finding the class centers. Thus, the final result of the system is a sentence describing the category that forecasts the trends of the stock market or the corresponding action (buying, selling, or waiting) over the next period.

Empirical Results

To test this model, we selected the closing prices of individual stocks on the Shanghai stock market. The data used for training was from 21 May 1992 to 12 April 1993. Preprocessing the data produced a five-day moving average and a bias rate (the membership function according to the statistics of the data). Actually, we used the bias rate as an external input of the system. Another input was the Shanghai Stock Indicator. The objective was to detect the relationship between a single stock and the entire stock market.

The neural network was structured as follows: the first subnetwork was a three-layer net that classified moving averages with 20 input neurons, 30 hidden neurons, and nine output neurons. As discussed above, we selected a training set with 70 patterns, each composed of 20 values selected from a series of five-day moving averages, then classified them into nine categories by training according to the trend of the 20 values.

Since we do not have a distinct model to classify the Shanghai Stock Indicator, unsupervised learning was first used. A 15-dimensional binary vector encoded a 5-dimensional vector comprising five indicators selected consecutively; next an ART1 net with 15 inputs and 9 outputs classified this binary vector pattern adaptively. The consequent sets of the fuzzy rules are fuzzy sets that indicate the trend of stock prices over the next week and are expressed linguistically as *ascending*, *descending*, or *stationary* (see Figure 12.3).

The approach for choosing fuzzy decision rules discussed above was used. By training the data we chose about 30 rules. Actually, we used only three final rules, each of which equaled a weighted sum of the individual rules with the same consequent set, where the weights evolve the distribution of the original 30 rules (see Figure 12.4). The result of the fuzzy decision was a three-dimension vector where the component sets (ascending, descending, stationary) were

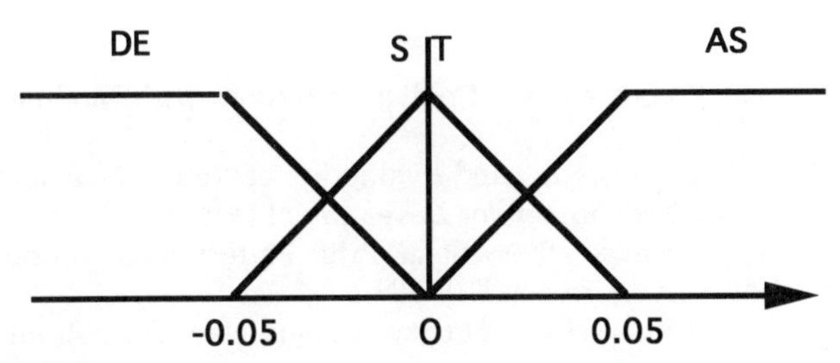

FIGURE 12.3
Consequent sets of fuzzy rules. AS, ascending; DE, descending; ST, stationary; *x = log (C2/C1), *where C1 is today's price and C2 is the average price over next week.*

fuzzy rules. The final procedure was really a defuzzification procedure using a maximum membership defuzzification scheme. The final result can be represented by three values:

- +1 means the trend is ascending.
- −1 means the trend is descending.
- 0 means the trend is stationary.

We selected 70 training patterns and 100 test patterns to test the performance of the system. Preliminary simulation results show that the rate of correct prediction is greater than 92% on the training set and greater than 74% on the testing set.

Future research will be oriented toward:

- Quicker convergence of networks using the back-propagation algorithm
- Using a network with other structures
- Accounting for more economic and political factors, and joining them with the fuzzy system

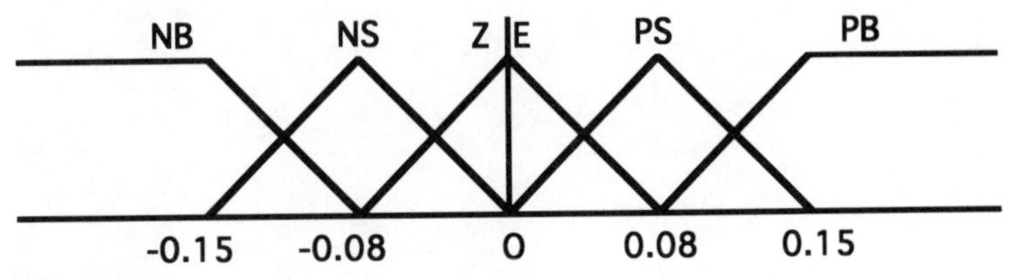

FIGURE 12.4
Membership function of bias rate. NB, negative big; NS, negative small; ZE, zero; PB, positive big; PS, positive small

References

Asakwa, T. K. 1990. "Stock Market Prediction System with Modular Neural Networks." *IJCNN* 1.

Deboeck, G. 1992. "Preprocessing and Evaluation of Neural Nets for Trading Stocks." *Advanced Technology for Developers* I (Aug.).

Kamijo, K., and T. Tanigawa. 1990. "Stock Price Pattern Recognition: A Current Neural Network Approach." *IJCNN* 1.

Kosko, B. 1992. *Neural Networks and Fuzzy Systems*. New York: Prentice-Hall.

Note

1. This work was supported by the Chinese Key Project of Fundamental Research, Climbing Project.

CHAPTER 13

Smart Trading with FRET

Dalila Benachenhou

What will it be like when everything you can do or think or create a smart machine can do better? Will we live all year on vacation? Will each generation . . . spend more time in virtual-reality cybersuits and cyberchairs? Will man end up an ultra-hightech couch potato?

Bart Kosko
Fuzzy Thinking, *1993*

The previous chapter illustrated how fuzzy logic can enhance a neural trading system. This chapter gives a concrete example of a fuzzy trading system. Dalila Benachenhou, a Ph.D. candidate in applied statistics at American University, describes a fuzzy rule extraction tool (FRET) that creates fuzzy membership functions, extracts fuzzy rules from input/output samples, and uses these in a fuzzy decision-support system. A fuzzy rule set derived from sample data is then used as a fuzzy expert system for trading. Preliminary results obtained using this system on U.S. Treasury securities are promising.

Fuzzy Systems for Trading

The basic concepts of fuzzy logic have been discussed in Chapter 11. The main advantages of fuzzy systems for trading are

1. A fuzzy system is more flexible than an expert system. Fewer rules or combinations of rules are needed to cover more possible outcomes. Unlike an expert system, fuzzy inferences can handle overlap or ambiguity between rules.

2. A fuzzy system can be adjusted over time: if the system doesn't perform properly, one can adjust the rules. In other words, a fuzzy system is more modular and amenable to modification than neural-based trading systems.
3. A fuzzy system is easier to comprehend than a neural network: fuzzy rules are expressed in English-like sentences using vague propositions and/or consequences that correspond more to the human way of thinking.

This chapter demonstrates a method for deploying fuzzy logic for the design of trading systems. We first review several methods for fuzzifying time series. Next we discuss in detail how to extract fuzzy rules from sample data, and how to use a fuzzy rule set to design a trading system. The fuzzy rule extraction tool (FRET) described in this chapter was written in an object-oriented programming environment (see the Appendix for details). FRET provides some of the qualities of more conventional fuzzy development tools, but it also permits more robust handling of nonstationary time series. The chapter concludes by demonstrating the usefulness of FRET for trading U.S. Treasury securities.

Fuzzifying a Time Series

Financial time series are difficult to analyze using conventional tools. We explored many fuzzy design tools and found the majority unsuitable for financial applications. Most do not produce membership functions or extract fuzzy rules, requiring the user to enter this information. Furthermore, financial series are usually nonstationary and thus require open or modifiable definitions of memberships and universes of discourse. The main reasons are that the sizes and shapes of membership functions are unknown and that the universe of discourse for many financial series are difficult to determine. Membership functions map the values of a time series into groups or segments that are labeled. Each real value is fitted between 0 and 1, expressing the degree of membership of the value in one or more groups.

Often it is not clear how many memberships or groups are necessary to map a financial series, or what the size of the memberships should be. To overcome these problems, we used a clustering method for extracting the size of membership functions from a time series.

A membership function cannot have a value larger than 1; the leftmost segment of the function should be nonincreasing, the rightmost segment nondecreasing. The middle segment can take on different shapes. Figure 13.1 provides an overview of common shapes for middle segments.

The first column in Figure 13.1 lists the name of the function. The second column defines the actual function that creates the memberships. The last column graphically represents the shape of the function and defines the parameters needed for the function to satisfy the definition of the function. For instance, an *arc tangent* is drawn if the scale factor s and the midpoint m are provided for a given membership function. A *sigmoid* is drawn if the position of its maximum c, minimum a, and midheight b are provided. Four if-statements must be met for the sigmoidal function. If the input x is larger than or equal to c, the function reaches a value of 1. If the input x is between a and b, the function is drawn by case 2. If the input x is between b and c, the function is drawn by case 3. Finally, if the input is smaller than or equal to a, the function reaches 0. All the functions represented map to a range between 0 and 1. Thus, there is no need to normalize the input and output data.

The possible shapes of the membership functions depend on the natural way of representing possible qualifiers. In time series, one should concentrate on convex functions and on nondecreasing or nonincreasing functions, as they are more natural for time-series data. Although we used the linear (triangular) shape, the Gaussian and sigmoidal functions probably are better choices. In the system described below, the shape of the membership functions is determined by the data. In fuzzy system applications, the most common membership function used is the linear triangle, being easier to implement and providing adequate results.

Next we'll discuss two methods of clustering data from a time series. One is by Klimasauskas (1992), the other by Bezdek (1981).[1] The Klimasauskas method assumes that the user already knows how many memberships are needed.

The method proposed by Klimasauskas has two steps: (1) form the clusters and (2) draw the membership functions representing each cluster. Hence, the range of the universes of discourse is extracted and the preliminary membership functions are formed. The sizes of the membership functions are derived using the k-mean clustering algorithm, which consists of the following steps:

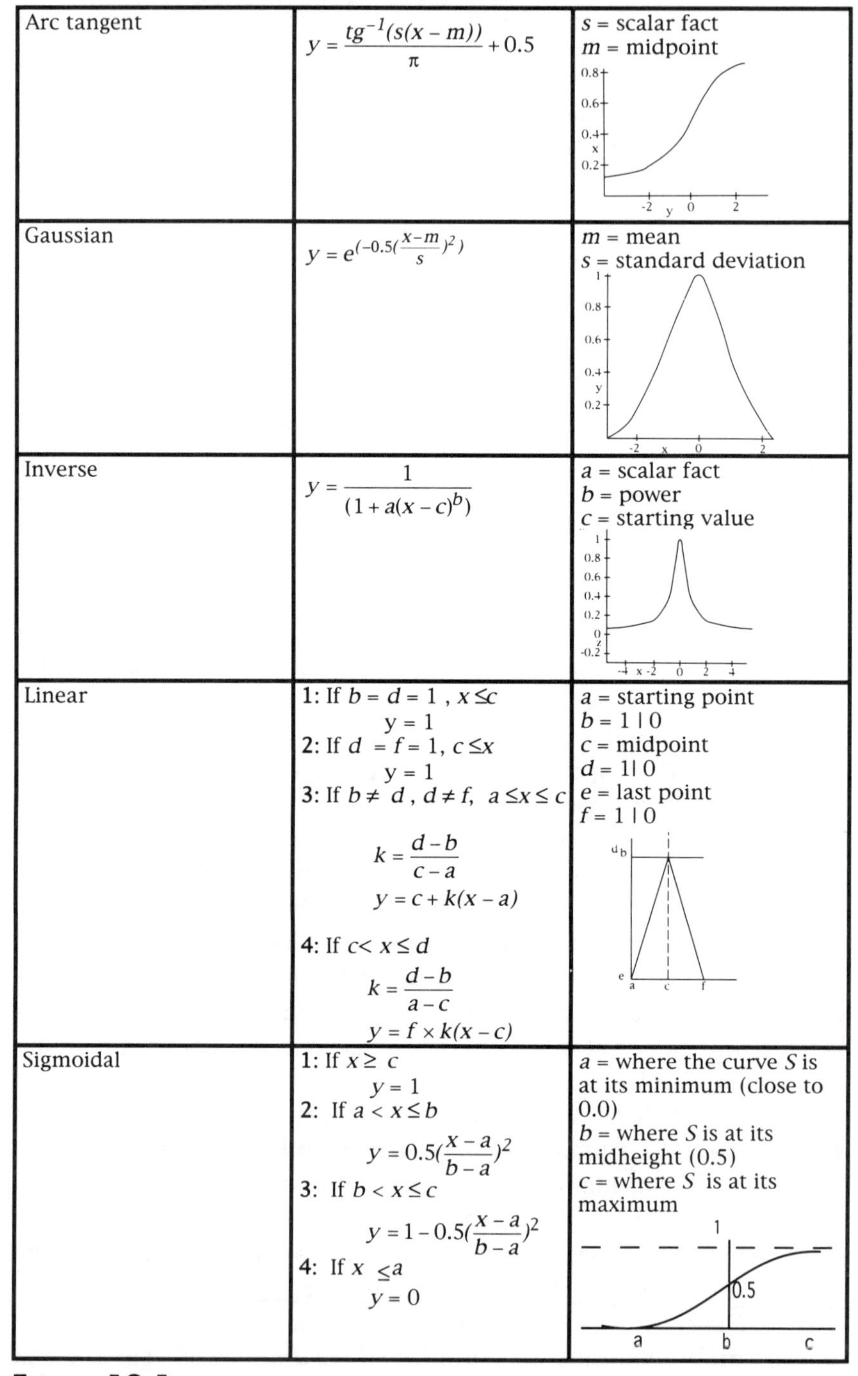

Arc tangent	$y = \frac{tg^{-1}(s(x-m))}{\pi} + 0.5$	s = scalar fact m = midpoint
Gaussian	$y = e^{(-0.5(\frac{x-m}{s})^2)}$	m = mean s = standard deviation
Inverse	$y = \frac{1}{(1 + a(x-c)^b)}$	a = scalar fact b = power c = starting value
Linear	1: If $b = d = 1$, $x \leq c$ $y = 1$ 2: If $d = f = 1$, $c \leq x$ $y = 1$ 3: If $b \neq d$, $d \neq f$, $a \leq x \leq c$ $k = \frac{d-b}{c-a}$ $y = c + k(x-a)$ 4: If $c < x \leq d$ $k = \frac{d-b}{a-c}$ $y = f \times k(x-c)$	a = starting point $b = 1 \mid 0$ c = midpoint $d = 1 \mid 0$ e = last point $f = 1 \mid 0$
Sigmoidal	1: If $x \geq c$ $y = 1$ 2: If $a < x \leq b$ $y = 0.5(\frac{x-a}{b-a})^2$ 3: If $b < x \leq c$ $y = 1 - 0.5(\frac{x-a}{b-a})^2$ 4: If $x \leq a$ $y = 0$	a = where the curve S is at its minimum (close to 0.0) b = where S is at its midheight (0.5) c = where S is at its maximum

FIGURE 13.1

Shapes of membership functions

1. Choose the number of classes C.
2. Sort the data.
3. Calculate the means of each class $\mu_1 \ldots \mu_c$, assuming that the classes are of equal size.
4. Classify each value in the series, x_k, using the square error function.
5. Recompute the means μ_i using the results of step 4.
6. If the μ_i are consistent, *stop*; otherwise, go to step 4.

Klimasauskas suggests the following approximation for the design of membership functions:

$C = 0.5 \times (S + E)$
$K = 2 \times (1 - M) / (E - S)$
$F_i = \max(0, 1 - K \times \text{abs}(X_i - C))$

where

M = the value of the fuzzy membership set at the boundaries
S = the leftmost boundary
E = the rightmost boundary
C = the center between the boundary
K = the scale factor
X_i = the input value
F_i = the fuzzy membership value

This approximation permits two adjacent membership functions to intersect at a 0.5 degree of membership. This method has two shortcomings. One is that the user must predetermine the number of memberships, and the second is the degree of overlap between the membership segments.

An alternative method for fuzzifying data is provided by Bezdek (1981). The concept is very simple: the method locates the number of clusters that best represents the data. The optimal number of clusters is the one where data from the same cluster are closer to one another, and data from different clusters are farther from one another. This is obtained by finding the local minimum of:

$$S(c) = \sum_{k=1}^{n} \sum_{i=1}^{c} (\mu_{ik})^m (\|X_k - \nu_i\|^2 - \|\nu_i - \bar{x}\|^2)$$

where

n = number of data to be clustered
c = number of clusters, $c \geq 2$
X_k = k^{th} data, usually vector
$\bar{x}$ = average of data
ν_i = vector expressing the center of the ith cluster
μ_{ik} = grade of kth data belonging to the ith cluster
m = adjustable weight

In other words, Bezdek's method tries to minimize the variance of the individual values x_k in each cluster and to maximize the variance between the clusters. Before $S(c)$ can be minimized, c clusters are formulated and their membership functions approximated. These are put in a matrix U of size n by c. This is done using the fuzzy c-mean (FCM) clustering method. The concept of FCM is to converge the membership functions contained in the matrix U to trapezoidal membership functions. The FCM algorithm has five steps:

1. c is fixed so that $2 \leq c \leq n$
2. $U^{(l)}$ is initialized so that $\sum_{i=1}^{c} \mu_{ik} = 1$

where

$l = 0,1,2...$
$i = 1,2...c$
k is the kth input

3. The fuzzy cluster centers ν_i are calculated:

$$\nu_i = \frac{\sum_{k=1}^{n} (\mu_{ik})^m X_k}{\sum_{k=1}^{n} (\mu_{ik})} \quad 1 \leq i \leq c$$

4. $U^{(l)}$ is updated:

if $d_{ik} = \| X_k - \nu_i \| \neq 0$ for all $1 \leq i \leq c$
then

$$\mu_{ik} = \frac{1}{\left[\sum_{j=1}^{c} (d_{ik}/d_{jk})^{2/(m-1)}\right]}$$

else

$$\mu_{ik} = 0 \text{ and } \sum_{i \in I_k} \mu_{ik} = 1$$

5. If $|U^{(l)} - U^{(l-1)}|$ is larger than some threshold, restart from step 3. Otherwise, *stop*.

This algorithm first calculates the centroid, ν_i, of each cluster (step 3). Second, it calculates the distance between each X_k and the centroids. Then it updates μ_{ik} if the distance between X_k and ν_i is different from zero (step 4). If X_k and ν_i are equal, μ_{ik} is set to zero, and its value is dispersed among the other μ_{ik}. It is this step that forms the trapezoidal membership functions. The last step checks whether U has converged. If it has not, the process is repeated.

If the process is repeated, S(c) is calculated and compared to S(c + 1). If S(c) is larger than S(c + 1), c is incremented by 1, and the process restarts at step 2. Otherwise, the S(c + 1) has reached the local minimum and the process stops.

This method is superior to Klimasauskas' because $\sum_{i=1}^{c} \mu_{ik}$ is always kept equal to 1. This means that, in most cases, a value more probably will belong to only one membership. This property makes the FCM more valuable for creating membership functions.

Formulating or Extracting Fuzzy Rules

While most traders manage to trade profitably—experienced traders achieve about 55% profitable trades—few are aware of the profitable trading rules they deploy on a systematic basis or know how to formulate them. If rules cannot be formulated in a coherent, machine-usable way, then a more automated method of extracting rules from data is essential. Statistical analysis, neural networks, and fuzzy logic can be deployed for extracting rules from sample data. Part One of this book showed how neural nets can be used to design trading systems. Part Two showed how trading rules can be extracted using genetic algorithms. This section discusses several methods for extracting rules from sample data using fuzzy logic.

Method 1

Kasko (1992) has demonstrated how fuzzy associative memories (FAMs) can be used to extract fuzzy rules (see Chapter 11). FAMs map input variables to an output variable(s), and the resultant rules can be presented in a matrix. Extensive experimentation with FAMs has shown that they fail to provide adequate rules for the memberships created by Klimasauskas, however. The reason is that $\sum_{i \in I_k} \mu_{ik} \geq 1$ causes a value to belong with high probability to two or more memberships. Using Bezdek's algorithm, we found FAMs to be quite valuable.

Method 2

Another method for extracting fuzzy rules was developed by Araki et al. (1991). Araki's method extracts rules for crisp outputs. The main concept is to minimize the inference error $D(t)$, defined as

$$\frac{1 \sum_{p=1}^{N_d} | y_p^* - y_p^r |}{N}$$

and the change of the inference error, defined as $\Delta D = D(t) - D(t - 1)$. This is achieved by creating new membership functions, and thus new rules. The new rules are weighted and their consequent parts are calculated. This process is repeated until $D(t)$ and ΔD are smaller than a threshold.

The steps of this algorithm are as follows:

1. Create three membership functions for two inputs, making sure that the membership functions intersect at 0.5 degree of membership. As there are three membership functions for each input, the number of possible rules is nine. The rules are represented by the elliptical symbol in Figure 13.2(a).

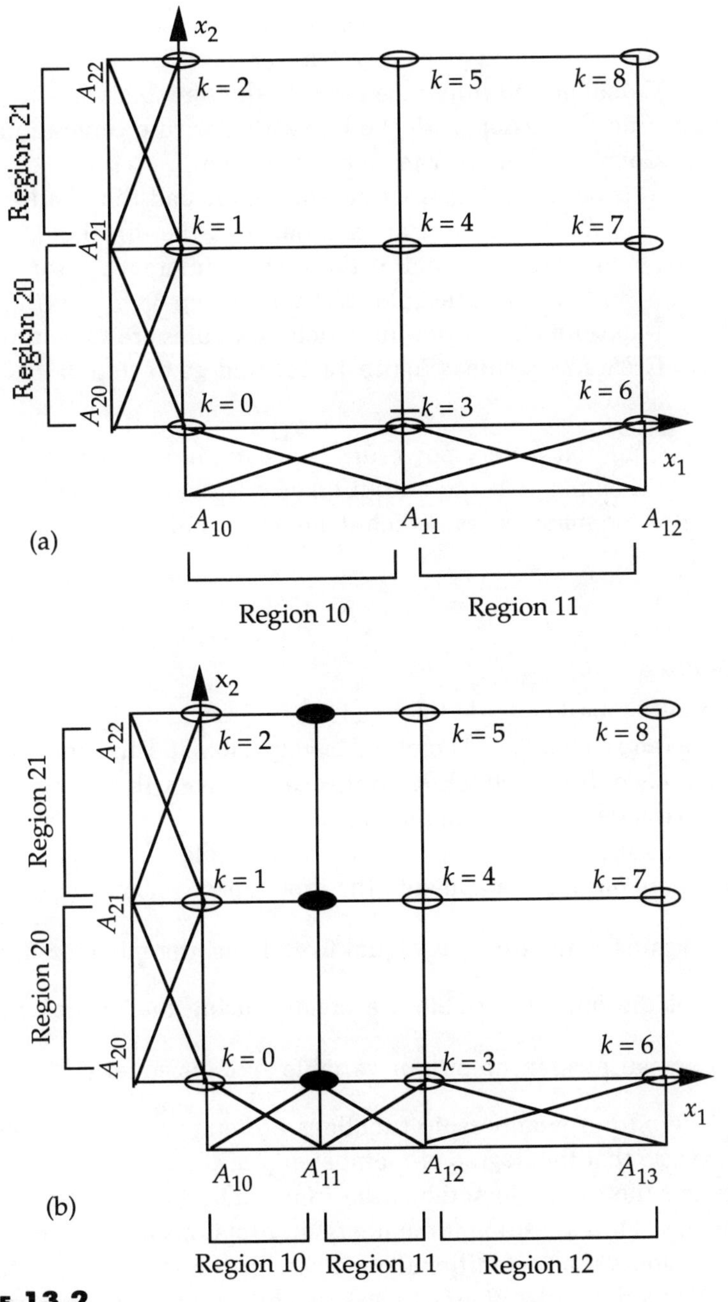

FIGURE 13.2
Fuzzy model creation using Araki's method. (a) Initial fuzzy model. (b) Fuzzy model after the region with the highest inference error is divided in two

2. Generate the real numbers of the consequence parts of the new rules, and place a weight on the rules.
3. Calculate the inference error in the new regions.
4. Find the region with the highest error rule generation.
5. Compare this error to some threshold.
   If both the highest inference error and the change of the inference error are smaller than the threshold, *stop*.
   On the other hand, if the highest inference error is larger than the threshold, go back to step 6.
6. Determine the region in which new rules are to be produced.
7. Create new rules (Figure 13.2b), and go to step 2.

This method is very powerful. However, one major shortcoming is that it assumes that the output takes crisp values. Another is that it creates too many rules, which defies the whole purpose of a fuzzy system.

Method 3

Yet another method for extracting fuzzy rules was designed by Sugeno and Takahiro (1993). This method is very sophisticated and considers all aspects of fuzzy modeling. In this section we will outline only the rule extraction part of the method.

The concept in Sugeno's method is simple and elegant. Unlike FAMs, the output is mapped to the input space. In addition, only the output must maintain $\sum_{i=1}^{c} \mu_{ik}$ equal to 1. Thus, the membership functions of the output variable are created using the Bezdek method, which ensures that the output variable $\sum_{i=1}^{c} \mu_{ik}$ is equal to 1.

The output membership functions are mapped to the input data cluster, so that the degree of membership of the ith inputs and output data are the same. Sometimes the input data do not form a convex function. Thus, additional membership functions are created for the given inputs. Each additional membership function causes the addition of a new rule. Figure 13.3 shows how two rules are created. In the examples presented by Sugeno, no more than six rules were created.

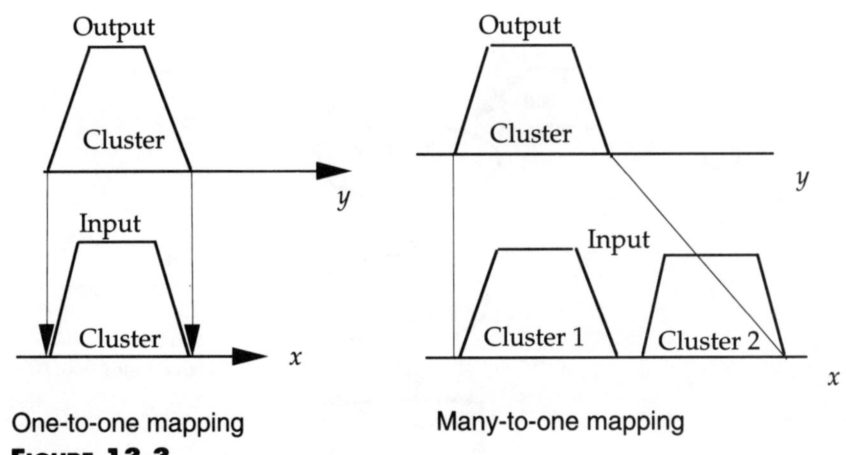

FIGURE 13.3
Possible mapping of the output to the input space

Of the several methods presented for automated extraction of fuzzy rules, the most promising is Sugeno's. One of its advantages is that FAM matrices need not be created. Thus, contradictory rules cannot be generated. In addition, it takes less memory, as there are no FAMs or extra rules to be stored. Furthermore, some authors have suggested that FAMs can be adapted in real time. This can be done by changing the "confidence coefficient" matching the rule, or by changing the size of the membership functions. From the literature, we have found that an adaptive fuzzy system increases the performance of time-series systems (Cox 1992, 1993; Kasko (1992).

How Do Fuzzy Systems Work?

In previous sections we described the preprocessing of input and output variables to extract membership functions as well as methods for automated extraction of fuzzy rules. Now we turn to the actual working of a fuzzy system. We will use a simple example (already mentioned in Chapter 11) that has two inputs (a trend and a volatility indicator) and one output (a measure of future profit potential). An overview of the method described here is provided in Figure 13.4, which includes the following steps:

Step 1. The values corresponding to the trend and volatility indicators are fuzzified. The fuzzification uses the triangular function (linear)

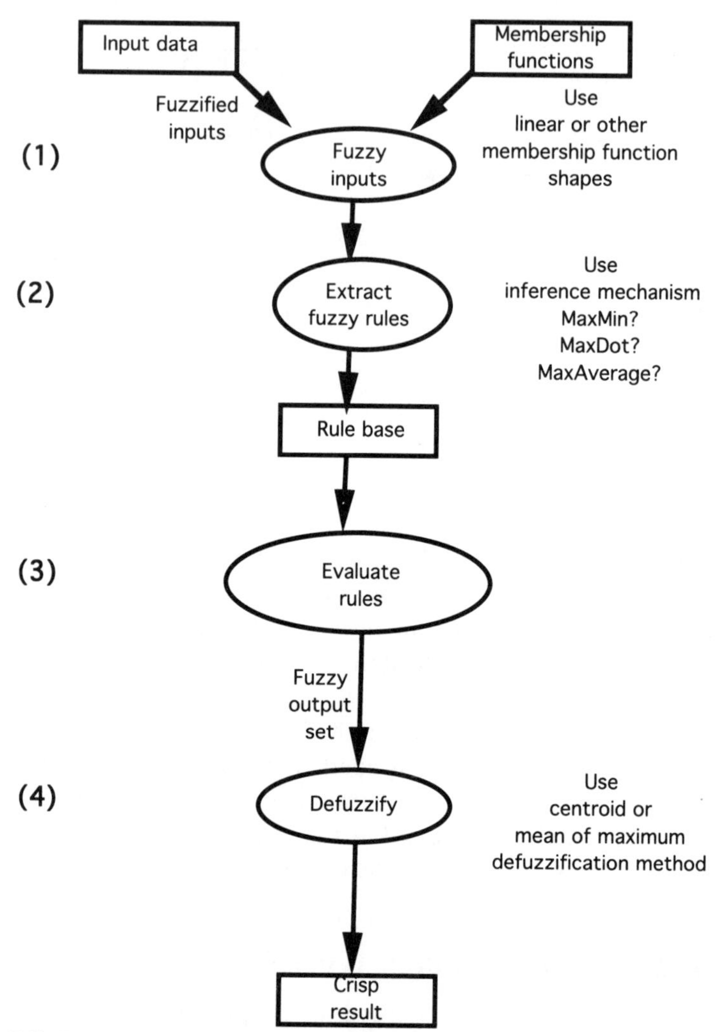

FIGURE 13.4
Fuzzy system development methodology

to map the trend and volatility to a range between 0 and 1. Thus, there is no need to normalize the inputs.

Step 2. Fuzzy rules are extracted using one of the methods described above.

Step 3. In this process only the rules containing the membership functions of the trend and volatility are executed (fired). For instance, if the values entered correspond to the membership high or medium for trend, and medium for volatility, only the rules "IF trend is high AND volatility is medium" and "IF trend is medium AND volatility is medium" will fire. Due to overlap between memberships, many rules can fire at the same time, and different memberships of the output can be chosen. The combination of the possible outputs is achieved through one of the three types of inference mechanisms: MaxMin (Figure 13.5), MaxDot (Figure 13.6), and MaxAverage. An example using MaxMin follows:

Step 3.1. If we are using AND in a rule (e.g., if x_1 is . . . AND x_2 is . . . then C is . . .), then:

$$c_1 = \min(\text{Medium}(x_1), \text{Low}(x_2))$$

If the operator is OR, then:

$$c_i = \max(\text{Medium}(x_1), \text{Medium}(x_2))$$

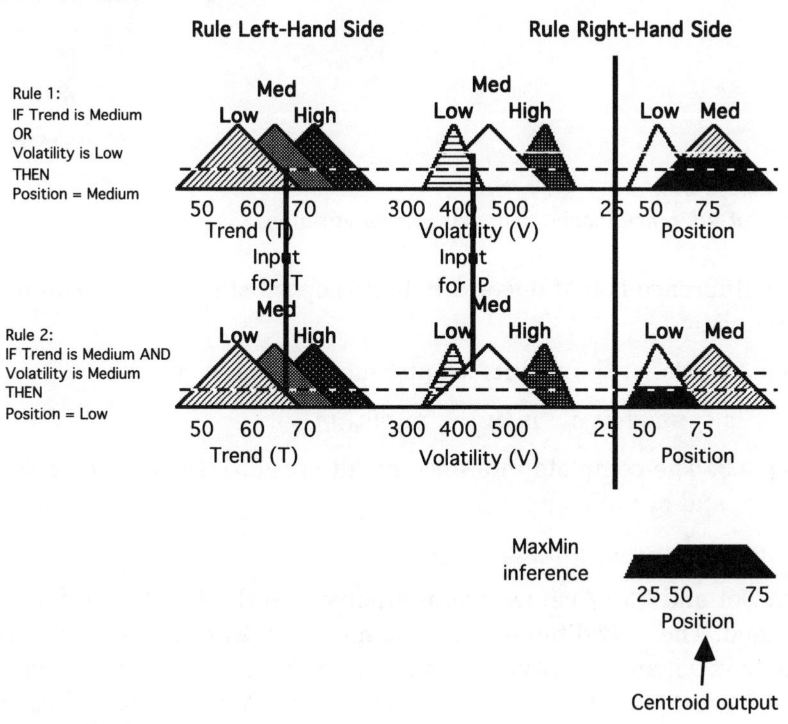

FIGURE 13.5
MaxMin inference mechanism using crisp variables

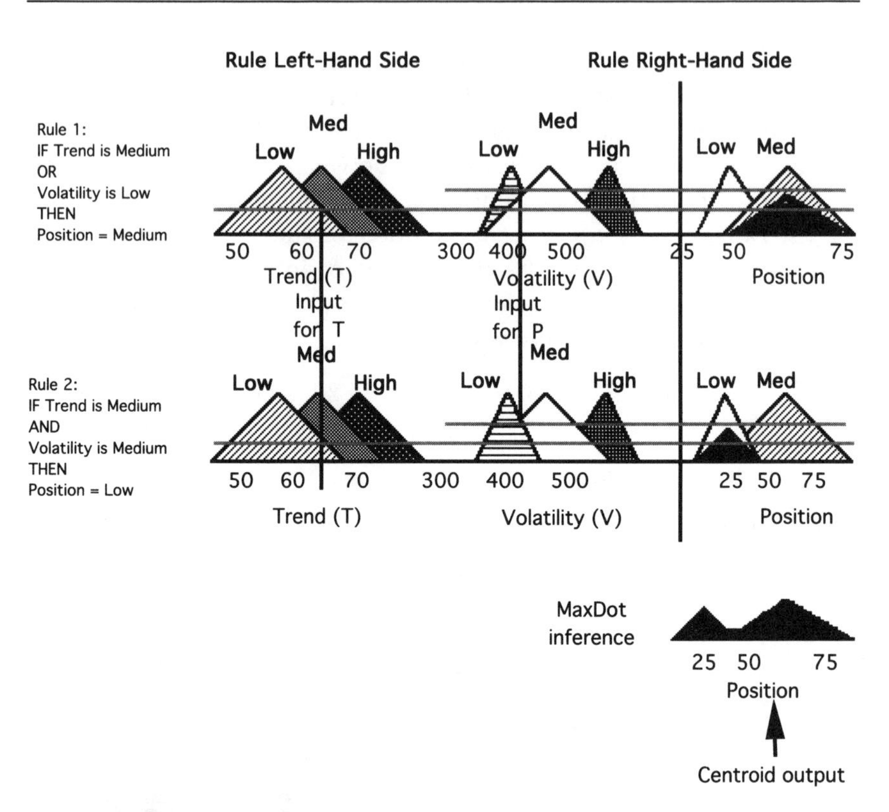

FIGURE 13.6
MaxDot inference mechanism using crisp variables

The difference is that one evaluation is optimistic, while the other is pessimistic.

Step 3.2. We let the results of inference for the ith rule be

$$y \text{ is } c_i \cdot m_i \text{ but } c_i \cdot m_i\,(y) = \min(c_i, m_i(y))$$

Step 3.3. The complete inference result (output) is constructed from $c_1 \cdot m_1$ and $c_2 \cdot m_2$:

$$m^* = \max\,(c_1 \cdot m_1, c_2 \cdot m_2)$$

MaxDot and MaxAverage are alternatives to the MaxMin inference method. The only difference between the MaxMin inference method and MaxDot and MaxAverage is that in step 3.2, instead of taking the minimum MaxDot takes the product, while MaxAverage takes the average between c_i and $m_i(y)$ (see Figure 13.7).

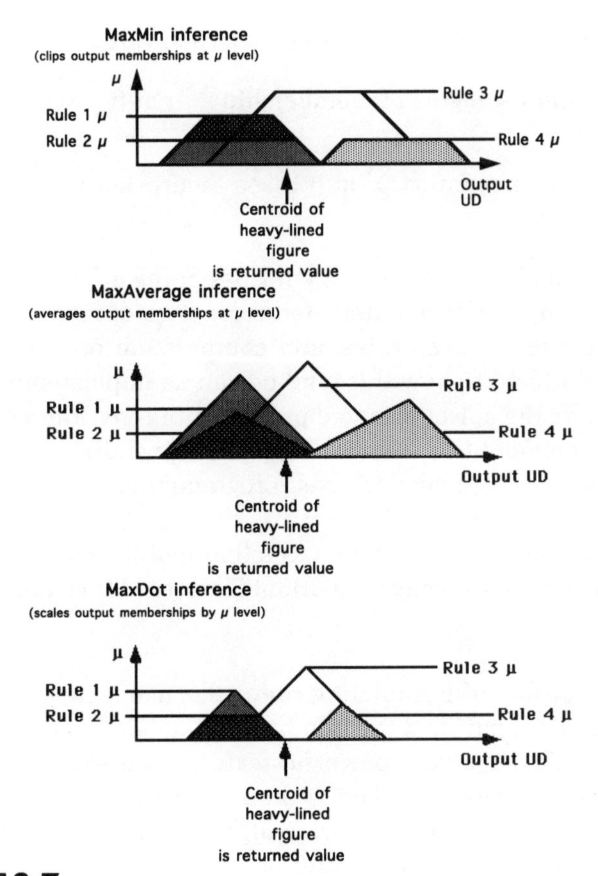

FIGURE 13.7
Inference mechanisms and defuzzification methods

Step 4. To obtain crisp outputs, either the centroid (center of gravity) or mean of maxima method is used. The mean of maxima finds the average of a series of numbers, while the centroid finds the center of gravity of a series of numbers. As one can see from Figure 13.7, the mean of maxima method does not consider membership functions, even though these functions represent the "weight" that each number in the universe of discourse contributes. This feature makes mean of maxima more prone to erroneous results, and thus less robust. The centroid method is a better approximation. It finds the proper crisp value, representing a list of values. It achieves its results by taking the ratio of the sum of the product between each value on the universe of discourse and its degree of membership $\sum_{j=1}^{p} y_j m_b(y_j)$, and the sum of the degree of memberships $\sum_{j=1}^{p} m_b(y)_j$ (see Figure 13.8).

In sum, the above methodology for designing a fuzzy system includes fuzzifying the data, automated extraction of fuzzy rules from sample data, use of fuzzy rules, and computation of crisp outputs. Next we will illustrate how this method can be implemented in software. We chose the object-oriented programming environment of Prograph™ to implement this methodology. The Appendix to this chapter elaborates on why an object-oriented programming environment was chosen.

Using Prograph, a fuzzy rule extraction tool (FRET) was created in about one month! The most important features of this tool are listed below:

1. It is capable of formulating triangular membership functions based on inputs and the number of membership functions desired. This is a very powerful feature, as it enables the clustering of time-series data without user intervention.
2. It uses the k-mean clustering algorithm to decide how large each membership set should be.
3. It determines the shape of the membership functions or better approximates the shape based on Bezdek's method.
4. It provides all fuzzification and defuzzification methods (see Figures 13.1 and 13.8). We chose convex functions for fuzzification, MaxMin for inference, and centroid and mean of maxima (Figure 13.8) for defuzzification.
5. It utilizes the standard menu objects for opening, saving and printing of files, so as to make the system easy to use.
6. It includes a performance menu from which a file containing the fuzzy information system can be read and displayed.
7. It allows the modified fuzzy system to be saved in a format that allows the results to be opened in a spreadsheet without having to perform any type of transformation.
8. It allows creation, modification, and/or deletion of universes of discourse and memberships. Rules can be modified on the

Mean of maxima	$h = \frac{max(y_j) - min(y_j)}{n}$ $result = min(y_j) + \frac{\sum_{k=1}^{n}(k.h)}{n}$	n=number of y_j = interval size of n of y_j
Centroid	$result = \frac{\sum_{j=1}^{p} y_j m_B(y_j)}{\sum_{j=1}^{p} m_B(y_j)}$	$m_B(y_j)$= membership value for y_j y_j= exact value corresponding to $m_B(y_j)$

FIGURE 13.8
Defuzzication functions

fly; likewise new membership functions, universes of discourse and memberships can be created. Rules can be modified on the fly; new can be created. The deletion of a membership function causes the deletion of the rules containing it. When a variable has been set as an output, it cannot be changed to an input. Furthermore, universes of discourse cannot be deleted. All these conditions ensure that the consistency of the fuzzy system is maintained. Figure 13.9 shows the screens illustrating a number of these capabilities.

Letting FRET Trade

To test the system, we used as inputs the 10- and 20-day trends of 10-year U.S. Treasury notes. We followed Bezdek's model, which suggests four memberships for the output, and we also decided to use four membership functions for the input variables. The input membership functions were created using part of Bezdek's model. The membership functions were labeled as follows: NB, NS, PS, PB.[2]

Figure 13.10 shows the membership functions of the inputs and outputs. They look like smooth, convex functions. The middle membership functions seem to be symmetrical in the case of the input data, while skewed in the case of the output data. The membership functions also seem to be somewhat different in size. Finally, two adjacent membership functions always meet at a 0.5 degree of mem-

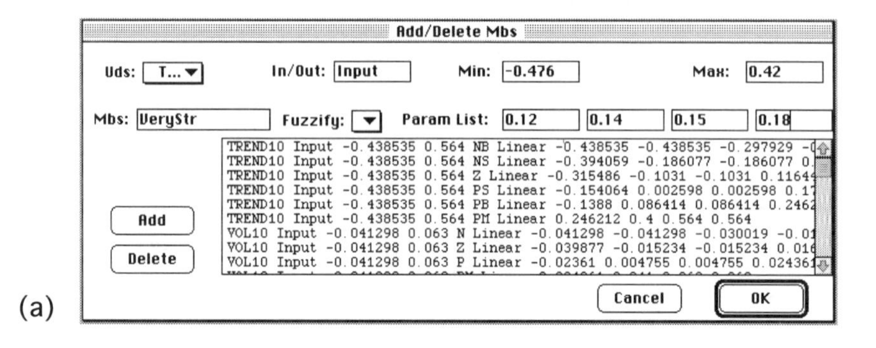

(a)

(b)

(c)

FIGURE 13.9

Sample screens from FRET. (a) Add or delete membership functions. (b) Modify the universe of discourse and the membership functions. (c) Modify rules

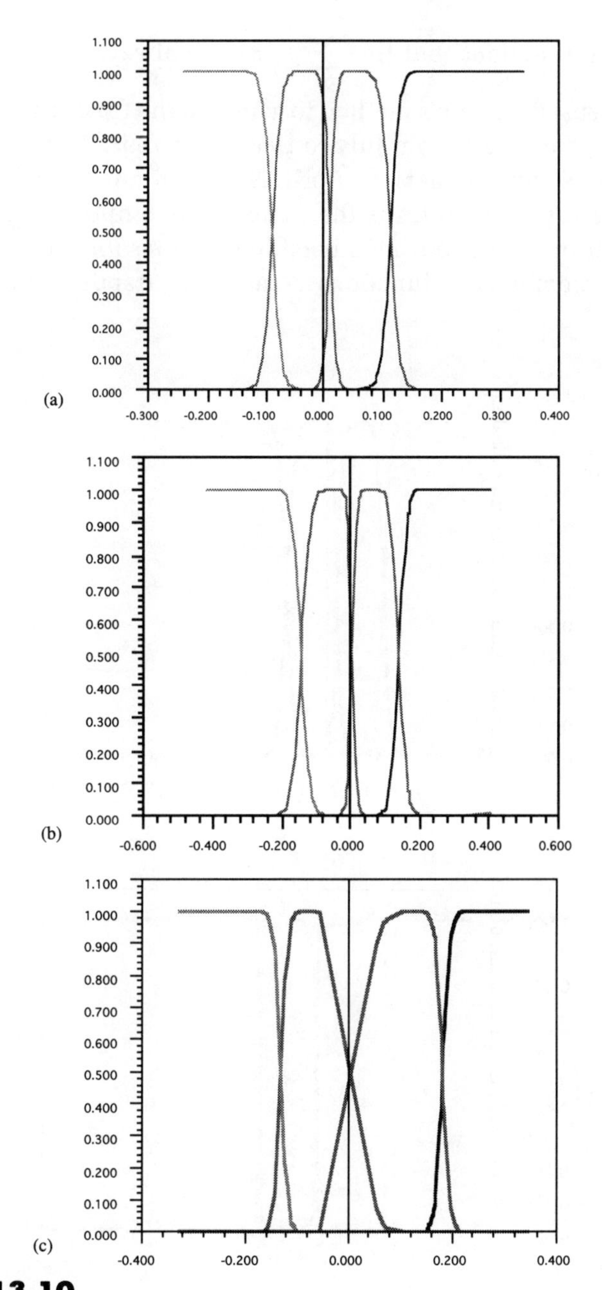

FIGURE 13.10

Membership functions. (a) 10-day trend. (b) 20-day trend. (c) Desired output bership. This shows empirically that Bezdek's model not only derives membership functions, but that $\sum_{i=1}^{c} \mu_{ik} \leq 1$ is always true.

We also used Bezdek's method to generate the fuzzy rules. A FAM was derived from data from July to December 1992. The system was tested on data from January to April 1993. The number of memberships of the input variables is the same as the output variable. Bezdek's algorithm found four membership functions for the output variable. The membership functions created are trapezoidlike (Figure 13.11).

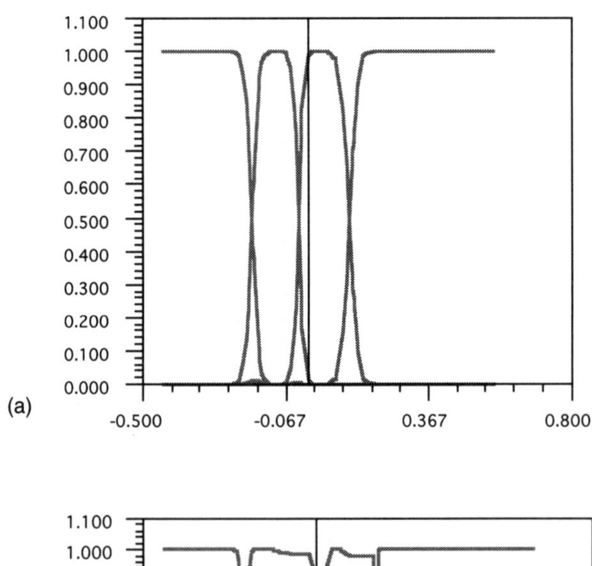

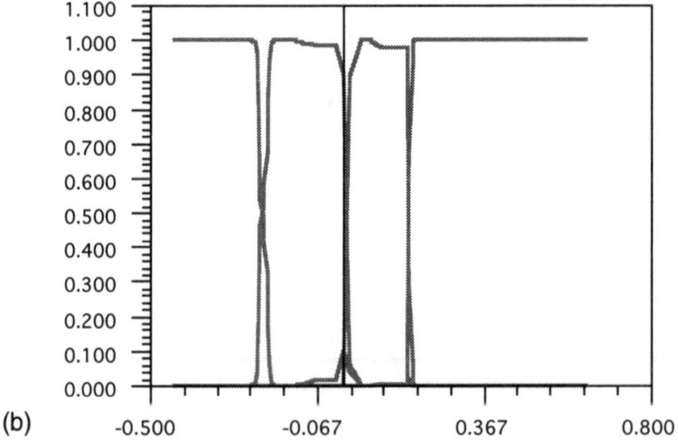

FIGURE 13.11
Membership functions. (a) 10-day trend. (b) Desired output

We used the FAM to extract a total of 11 rules. Using this fuzzy rule set on out-of-sample data produced the following financial results (in basis points comparable to the evaluation results provided in Chapter 6):

Trading days	125
Number of trades	6
Number of profitable trades	3
Percentage of winning trades	50%
Average P&L winning trade	22.4
Average P&L losing trade	−4.8
Average P&L winning/losing trade	4.6
Average P&L trade	8.8
Largest gain achieved	65.6
Largest loss achieved	−8.8
Maximum drawdown	36.7
Total cumulative P&L	117.3
Average P&L/maximum drawdown	3.2

Six trades were made, of which only three were profitable; however, the ratio of winning versus losing trades was 4.6. In addition, the largest positive trade was seven times larger than the largest loss, which was only −8.8 basis points. Furthermore, the other two profitable trades contributed more than half to the total cumulative P&L. The maximum drawdown was −36.7 basis points. The total cumulative P&L of this fuzzy rule-based system on six months of out-of-sample data was 117.3 basis points, or three times larger than the maximum drawdown.

We trained the system on the second six months of 1992, and tested it on January to April 1993. We used the same number of membership functions as in the previous example and used the FAM to extract the rules. Eleven rules were again extracted. The result using the FAM was as follows:

Trading days	74
Number of trades	6
Number of profitable trades	5
Percentage of winning trades	83%
Average P&L winning trade	5.6
Average P&L losing trade	−9.7

Average P&L trade	0.5
Largest gain achieved	10.8
Largest loss achieved	−9.7
Maximum drawdown	−9.7
Total cumulative P&L	12.6
Average P&L/maximum drawdown	1.3

The results here are significantly different from those obtained in the previous example. The model had only one losing trade, compared to three previously. The total cumulative P&L over four months was 12.6 basis points, just a little bit larger than the maximum drawdown. It should be noted the largest gain achieved was 1.2 basis points larger than the largest loss. In addition, unlike the previous example, the other profitable rules contributed to the total cumulative P&L only 1.8 basis points. Why were these results less successful? One reason is that the number of memberships allocated to the output and input variables were too small.

FAMs can yield good rules if there is an appropriate number of membership functions that don't overlap too much. Otherwise, membership functions overlap to the point where rules cannot be extracted. In addition, FAMs can produce good results if enough input variables are selected. It is obvious that too few input variables were used to be able to select appropriate positions in the market.

These examples show that automated extraction of fuzzy rules can produce similar or better results than those achieved with the neural network models described in Chapter 6. This is possible because the membership functions do not significantly overlap. We also found that the results obtained using the Bezdek method were better than those obtained using the Klimasauskas method. Klimasauskas' method for fuzzifying the data made rule extraction difficult. In the latter case, we had to use Kasko's FAM approach and our intuition to extract fuzzy rules.

Conclusions

This chapter has demonstrated how fuzzy logic can be used to develop fuzzy trading systems. We have shown the concepts and methods needed to build a fuzzy trading system and have provided an illustration. FRET, a fuzzy rule extraction tool designed in an object-oriented programming environment, is a prototype tool that is more powerful and flexible than current fuzzy shells on the market. It is particularly geared to handle financial data and has substantial capabilities for designing fuzzy trading models. While the object-oriented environment permitted development of this tool in record time, much remains to be done in terms of improvement of rule extraction, adaptation of the membership functions, and dynamic adjustment of fuzzy rule sets.

Appendix: An Object-Oriented Approach to Building FRET

The system described in this chapter was written in Prograph, an object-oriented programming lauguage for the Macintosh. An overview of the system organization under Prograph is provided in Figure 13.12.

Object-oriented programming, unlike procedural programming, focuses on the data. It is built on the notion of a type of data abstraction in which a function is implemented as an entity or object which responds to certain requests. Thus, an object supplying a service can change internally without the user's knowledge.

The reason for choosing an object-oriented programming tool is that the software is easier to test, maintain, refine, and extend. Hence, it is faster to implement than traditional programming tools. This advantage is especially felt when creating large and complex software systems. Traditional development tools do not provide the appropriate level of abstraction to permit reuse while providing a unifying model to integrate the development phases. This usually means a long and painful implementation. Using Prograph, the entire implementation of a system described above took about a month. Thus, it is clear that object-oriented programming can help reduce the development time of a system, especially if the desired system fits within this design philosophy.[3]

The limitations of the fuzzy development tools currently on the market motivated us to create our own system. This tool permits manual modification of universes of discourse, membership functions, and fuzzy rules. In addition, given input and output sample data it can automatically extract fuzzy rules. Inferences from fuzzy rules then produce crisp outputs when applied to new data. All the methods

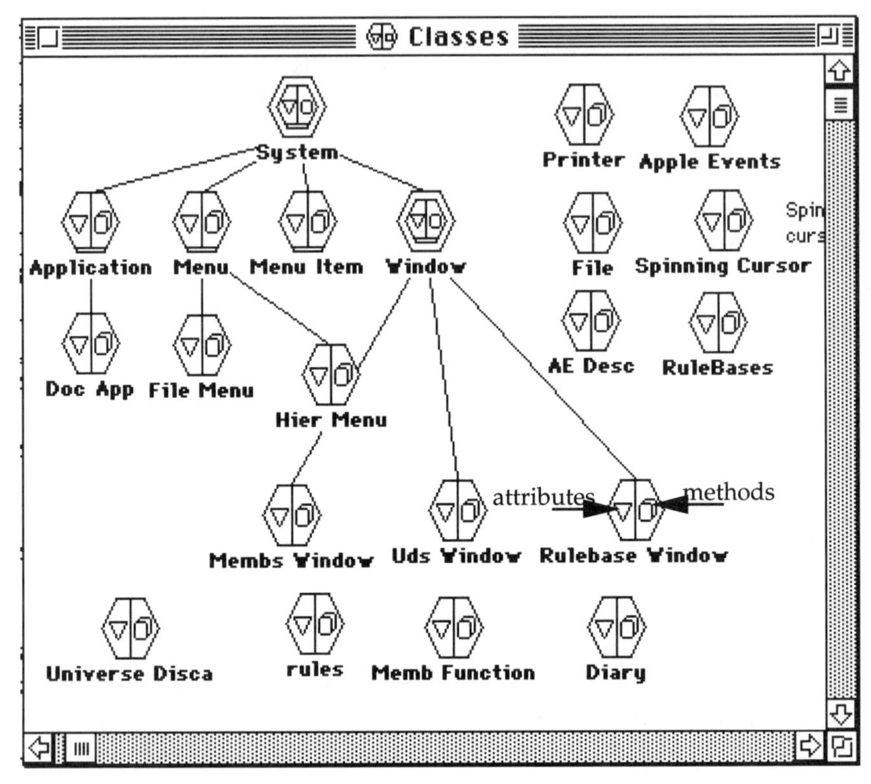

FIGURE 13.12
View of the fuzzy system organization in Prograph

discussed in this chapter, including inference methods and defuzzification methods, are provided.

The remainder of this Appendix shows how object-oriented technology was used to design FRET. Rules, universes of discourse, and membership functions are abstractions which in the object-oriented world form classes. Classes are sets of operations and data structures that characterize objects (TGS Systems 1992).

Classes are characterized by their attributes and by their methods. Attributes represent the features of the class, whereas methods are a library of behaviors of the class. Methods are like subroutines. For instance, the class *membership function* has the following attributes: labels (*name*), shapes (*fuzzy func*), estimated values (*functions*), and parameters (*parameter list*). The methods or behaviors of this class membership function are the routines for creating various shapes of membership functions, such as linear, Gaussian, or arc tangent. They are available only to the objects belonging to this class. *They are encapsulated.* This last feature permits errors to be isolated and facilitates debugging. More importantly, it preserves the integrity of the system as a whole. Figure 13.13 shows the attributes and methods of the object class called "membership function."

This example shows that an object-oriented programming environment provides a structure (just like fuzzy logic has a universe of discourse, membership functions, and fuzzy rules) for creating object classes, with attributes and methods, and without concern for space allocation or dangling pointers, which are required in structured programming in C or other languages.

Prograph, unlike C++ (the object-oriented version of C), uses a procedural or data-flow programming technique. A data-flow language

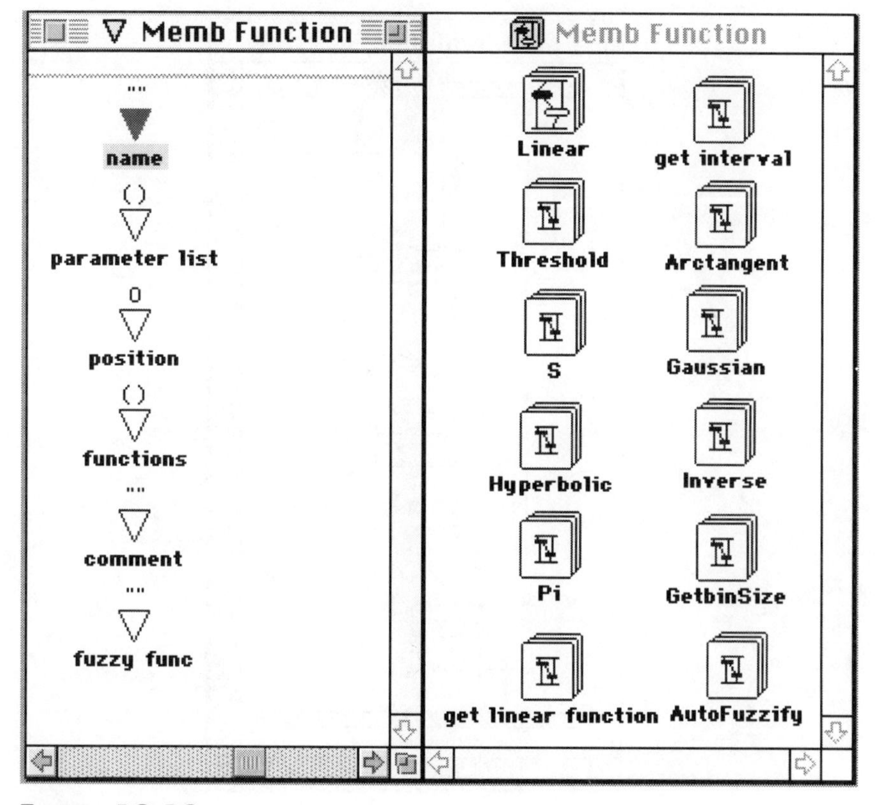

FIGURE 13.13
Attributes and methods of the class called "membership function"

is a graphical language based on the notion of data flowing from one function to another or directly supporting such data flow, independent of the explicit temporal effects of procedural programming. Contrary to procedural languages, where the data wait to be called by a procedure or a function, data flow activates an instruction as soon as all the required input data arrive. In addition, in data-flow programming the events are not sequentially ordered, but concurrent. As an operation is executed, the resultant data become available to other instructions. In addition, many operations can execute (fire) at the same time. Thus, data-flow programming, unlike procedural programming, does not create a bottleneck between the computer's control unit and its memory.

Figure 13.14 shows an example of data-flow programming. It illustrates a method that calculates the Gaussian function. The top bar

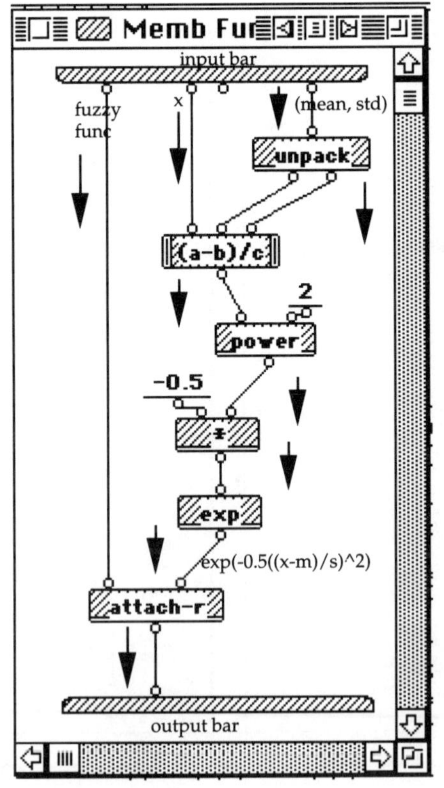

FIGURE 13.14
Example of data-flow programming

is for input, and the bottom bar is for output. The circles on the input bar represent the input parameters. In this example, four input parameters are passed to the method; however, only those with a line extending from them are used: a list (which can be thought of as a dynamic array), the input value x, and the mean and the standard deviation.

The lines represent the paths the data take. Apparently, the first instruction to execute (fire) is *unpack*, which extracts tokens from an input list. The second instruction is $(a - b)/c$. The third instruction is the multiplication of the result by -0.5. The fourth instruction is the exponential of the result. The final instruction is *attach-r*, which attaches the answer to the resulting list. The result is connected to the output bar, making the output data available to other methods. The result produced is a list of fuzzified inputs.

As the programming is graphical, it is easy to debug. Although this is a simple example, the programming language permits recursion and access to the Macintosh toolbox. Furthermore, a good debugger is available (TGS Systems 1992).

References

Araki, S., et al. 1991. "A Self-Generating Method of Fuzzy Inference Rules." *Fuzzy Engineering toward Human Friendly Systems*, Vol. 2. International Fuzzy Systems Engineering.

Bezdek, J. C. 1981. *Pattern Recognition with Fuzzy Objective Function Algorithm*. New York: Plenum Press.

Brubaker, D. 1993. "Togai Infralogic's TILShell." *Huntington Technical Brief* (Feb. 15).

Cox, E. 1992. "Fuzzy Fundamentals." *IEEE Spectrum* (Oct.)

Cox, E. 1993. "Adaptive Fuzzy Systems." *IEEE Spectrum* (Feb.)

Deboeck, G., H. Green, M. Yoda, and G. S. Jang. 1992. "Design Principles for Neural and Fuzzy Trading Systems." *International Joint Conference on Neural Networks* (Beijing).

Kasko, B. 1992. *Neural Networks and Fuzzy Systems*. New York: Prentice Hall.

Klimasauskas, C. C. 1992. "Hybrid Fuzzy Encodings for Improved Backpropagation Performance." *Advanced Technology for Developers* (Sept.).

Sugeno, M., and Y. Takahiro. 1993. "A Fuzzy-Logic-Based Approach to Qualitative Modeling." *IEEE Transactions on Fuzzy Systems*, Vol. 1. International Institute of Electronic Engineers.

TGS Systems Ltd. 1992. *Prograph Reference Manual*. Halifax, Canada: The Gunakara Sun System Ltd.

Williams, T. 1992. "Fuzzy Is Anything but Fuzzy." *Computer Design* (April).

Notes

1. Clustering methods range from the simplest, such as *k*-means, to the more complex, such as linear vector quantization (LVQ) or adaptive resonance theory (ART). These methods find the best representative clusters of the data; however, they do not create membership functions.
2. NB, negative big; NS, negative small; PS, positive small; PB, positive big.
3. The system we designed differs from many fuzzy shells on the market today. For instance, many fuzzy shells lack a simulator or graphical editor for memberships. C code can be neither incorporated nor extracted from the simulator. Most shells have a membership function editor that allows creation of evenly spaced membership functions; however, the user is limited to a small fixed number of membership functions per variables. Other limitations of existing shells are that parameters cannot be changed on the fly; there is no consideration that memberships can be evenly spaced (Brubaker 1993), and there is no automatic extraction of rules, although sophisticated rules can be created.

   Examples of fuzzy shells currently available are CubiCalc™ from Hyperlogic ($495 to $795) and Fuzzy Systems Manifold Editor™ from Fuzzy Systems Engineering ($295). Both have been written for Microsoft Windows. CubiCalc has pre- and postprocessing capability. Both permit rules to be expressed as IF/THEN statements. Furthermore, they permit the data to be displayed. However, the user still has to know what the rules are and to define the universe of discourse and the membership functions (Williams 1992). Fuzzy Systems Manifold Editor is more like an editor and can output a source-code written in C. In addition, it permits rules to be represented in a matrix format. A created system can have up to five input dimensions and two output dimensions with up to seven membership functions each.

   Another fuzzy development tool is TillShell™ from Togai Infralogic (around $4,500). TillShell runs on MS-DOS, Macintosh, and Unix workstations (Williams 1992). It has a simulator and permits the user to incorporate or extract C code. It permits the creation of evenly spaced membership functions, the creation of customized functions, and the flashing of C code. Unfortunately, it has the same limitations as CubiCalc and Fuzzy Systems Manifold Editor (Brubaker 1993).

CHAPTER 14

Hybrid Neural, Genetic, and Fuzzy Systems

Francis Wong and Clarence Tan

There are many ways to grow fuzzy systems from data. They all boil down to clustering data into rules. These make up the new field of adaptive fuzzy systems, or neuro-fuzzy systems, or fuzzy neural systems.

Bart Kosko
Fuzzy Thinking, *1993*

Several applications of neural networks, genetic algorithms, and fuzzy logic have been illustrated in previous chapters. In many cases, these techniques are improvements over traditional techniques. This chapter discusses the strengths and weaknesses of each of these techniques and outlines several hybridization methods for providing even better solutions. Francis Wong, the developer of NeuroForecaster™ (produced by NIBS Pte Ltd. in Singapore), and Clarence Tan, currently with Bond University in Australia, have collaborated as follows: first, Wong discusses the strengths and weaknesses of neural networks, genetic algorithms, and fuzzy logic; next, Wong and Tan define hybrid systems and provide illustrations.

Strengths and Weaknesses of Advanced Techniques

Advanced computing techniques such as neural networks, genetic algorithms, and fuzzy logic can be applied to a wide variety of applications in finance and economics. This section discusses the strengths and weaknesses of each of these techniques.

Neural Networks

Neural networks are multivariate, nonlinear analytical tools. As discussed in Part One, the most popular model used is a back-propagation (BP) network and its derivatives. A back-propagation network sends the input data through the hidden layer(s) to the output layer, determines the error, and back-propagates the error from the output to the input layer. The learning phase minimizes the global error by modifying the weights associated with the links. Other chapters have already pointed out that learning by a back-propagation network can be very slow.

Another model is a radial basis function (RBF) network in which each hidden node corresponds to a Gaussian function. Usually the training of the RBF networks is composed of two phases: the first applies self-organization to determine the center of the hidden units and to calculate the radii; the second phase trains the connections from the hidden layer to the output layer using the least-mean-square rule. This hybrid learning process speeds up the training enormously.

Neural networks have been severely criticized on the following grounds:

Lack of explanation facilities

Because of the complexity and interactions between the hidden nodes of a neural network, it is difficult to apply analytic techniques to understand how a decision is reached. Hence, we have to trust the output of the network blindly; this does not endear the neural network technique to fund managers, traders, or investors.

Design problems

Designing neural networks is often an onerous task. Several basic principles governing information processing in neural networks, such as parallel constraint satisfaction and distributed representation, are difficult to understand and even more difficult to exploit in the design of useful new network architectures. Complex distributed interactions among network layers usually make engineering techniques such as divide-and-conquer inapplicable. This complexity also precludes the concoction of analytic design methods. Furthermore, it is not possible to guarantee an optimal design solution when complex combinations of performance criteria (such as learning speed, compactness, generalization ability, and noise resistance) are given. As was pointed out in Chapters 2 and 6, in financial applications and particularly in the use of neural nets for the design of trading systems, there are many trade-offs among profitability, risk, trading frequency, transaction costs, and the like. So far, means of compromise have included relying on extensive human expertise and making elaborate simulations. As network applications continue to grow in number, size, and complexity, this human-engineering approach will gradually be replaced by better and more sophisticated software tools. More efficient, automated, and user-friendly tools are needed.

Difficulties in embedding existing knowledge

Unlike expert systems, where it is relatively easy to incorporate expert knowledge and any prior assumptions into the system, it is not easy to build a neural network with embedded a priori information about financial markets.

Learning problems

Due to inherent shortcomings in the learning algorithms for neural networks, the problem of finding a locally good solution instead of a globally optimal one is very real. Overfitting arises when one attempts to introduce "too many" explanatory variables, too many polynomial terms, or too many lags (training of data extending over too long a time period). In such cases, it is often found that while the additions improve the fit to the past data used for estimation purposes, forecasting performance on new data actually deteriorates, in some cases significantly. The standard treatment for this problem is to reduce the amount of data, to remove input variables, polynomial terms, or lags, or to apply some sort of sampling technique. Many methods for data filtering have been proposed, including the use of multiple regression of the "adjusted R-squared" rather than the simple R-squared measure of goodness-of-fit, the Box-Jenkins "parsimonious parameterization" methods, and Akaike or Schwarz "information criteria." After the data reduction process, the fitting (training) process is repeated, usually by seeking coefficients (weights) that minimize the sum of squared errors-of-fit over the data retained for estimation.

Genetic Algorithms

Genetic algorithms (GAs) are adaptive search procedures that can search large and complicated spaces, given certain conditions on the problem domain. Genetic algorithms tend to converge on solutions that are globally optimal or nearly so. As discussed in Chapter 8, GAs perform this search process in four stages: initialization, selection, crossover, and mutation. In the initialization stage, a population of genetic structures that are randomly distributed in the solution space is selected as the starting point of the search. After the initialization stage, each structure is evaluated using a user-defined fitness function and assigned a utility value. On the basis of their relative utility values, structures in the current population are selected for reproduction. A stochastic procedure ensures that the expected number of offspring associated with a given structure s is u (s)/u(P), where u(s) is the observed performance of s and u(P) is the average performance of all structures in the current population. Thus, structures with high performance may be chosen for replication several times, whereas poor-performing structures may not be chosen at all. In the absence of other mechanisms, such a selective process causes the best-performing structures in the initial population to occupy an increasingly larger proportion of the population over time.

The selected structures are recombined using crossover, which operates by swapping corresponding segments of a string representation of the parents. Crossover serves two complementary search functions. First, it provides new points for further testing of schemata already present in the population and, secondly, it introduces instances of new schema into the population. A schema is a subset of strings with similarities at certain positions. Generally, crossover draws only on the information present in the solutions of the current population in generating new solutions for evaluation. If specific information is missing (due to storage limitations or loss incurred during the selection process of a previous generation), then crossover is unable to produce new structures that contain this piece of information. A mutation operator, which arbitrarily alters one or more components of a selected structure, provides the means for introducing new information into the population. However, mutation functions as a background operator with a very low probability of application. The presence of mutation ensures that the probability of reaching any point in the search space is never zero.

Genetic algorithms have been increasingly applied in conjunction with neural networks (Wong and Tan [to be published]). An example of a hybrid system was discussed in Chapter 10. Two primary objectives of combining GAs and neural networks are neural net topology optimization and genetic training. In topology optimization, a genetic algorithm is used to select a topology (pattern of connectivity) for the neural network, which in turn is trained using a fixed training scheme such as back-propagation. In genetic training the learning of a neural network is formulated as a weights optimization problem, usually using the inverse-mean-square error as a fitness measure. In addition, GAs have been used to evolve optimal traders and in conjunction with fuzzy logic, which we will discuss later.

The primary drawback of genetic algorithms arises from their flexibility. The designer has to come up with encoding schemes that allow the GA to take advantage of the underlying building blocks, and to make sure that the evaluation function assigns meaningful fitness measures to the GA engine. In certain problems it is not clear as to how the evaluation function can be formulated so that the GA can produce an optimal solution. Another drawback is that GAs are computationally intensive, requiring much processing power. Genetic algorithms are readily amenable to parallel implementation, which should render them usable in real-time parallel environments.

Fuzzy Logic

As discussed in Chapter 11, fuzzy logic can quantify vague, qualitative ideas or judgments using computer-processible membership functions with values ranging from 0 to 1. In other words, fuzzy logic is a method for quantifying ambiguous information that conventional computers find difficult to handle. Fuzzy logic is an effective technology for understanding human speech, supporting human decision-making, and managing control processes. Fuzzy systems are applied to elevator banks (in order to minimize waiting times), to subway trains (for smoother, more comfortable stops), to many household appliances, to cameras and camcorders, and, as discussed in Chapters 12 and 13, to trading and portfolio management systems.

Fuzzy logic does not have learning capability. Hence, it is necessary to specify membership functions, select a fuzzy inference mechanism, define fuzzy rules, and apply a defuzzification method. As discussed earlier, many of these functions can be automated. Nevertheless, the optimization of a fuzzy model or system requires substantial effort in order to arrive at the optimal combination of membership functions/shapes, fuzzy rules, and fuzzy inference/defuzzification methods. There are as yet no commercially available tools that optimize these functions.

In sum, in a stand-alone mode, none of the techniques discussed so far is easy or optimal for any given problem. In the following section we will discuss hybrid systems—that is, combinations of two or more of these techniques.

Hybrid Systems

Hybridization of techniques can produce better systems. Several possible hybridization variations are shown in Table 14.1.

Neural networks are good at recognizing patterns from noisy, complex data; fuzzy systems are useful for decision-making where there is a great deal of uncertainty as well as vague phenomena; nonlinear dynamics and chaos theory (described in Chapter 15) provide a better understanding of the dynamics and evolution of financial time series and can help find periods of chaotic behavior in financial markets. Through hybridization of these techniques, more complex problems can be addressed and better financial applications can be built. A good introductory discussion of how these technologies can be combined is provided in Treleaven and Goonatilake (1992). In the following

TABLE 14.1 *Hybridization of Technologies*

Hybrid Combination	Advantages	Source
Neural network and genetic algorithm	GA creates better neural network (application: market timing adviser); GA selects relevant indicators.	Bornholdt 1992; Harp and Samad 1991; Montana and Dowls 1989; Wong et al. 1992b
Neural network and fuzzy logic	Neural network generates membership functions and fuzzy rules; fuzzy logic enhances neural net design.	—
Neural network and expert system	Neural network provides patterns; expert system provides rules (application: commodity trading model).	—
Fuzzy logic and genetic algorithm	GA optimizes fuzzy rule base and indicates reliability of forecasts.	Welstead 1993.

sections, several ways to hybridize these technologies for investment analysis are discussed.

Combining Neural Networks and Genetic Algorithms

As demonstrated in Chapter 10, the design of a neural network can be automated by combining two adaptive processes: genetic search through network architecture space, and back-propagation learning. Many aspects of neural network design can be optimized using GAs—for example, determining the optimal hidden layers and nodes, creating new indicators based on existing ones, selecting promising indications, and tuning the learning parameters (see Chapter 9). For purposes of illustration, we will discuss the application of GAs to the selection of indicators.

A common problem faced by neural network designers in forecasting economic time series is limited sample size (Wong 1990–91). The number of factors bearing on the task at hand is usually myriad; even experts have a hard time agreeing on the relative importance of these factors. The natural inclination is to use as many factors as possible, but this gives rise to the problem of overfitting. On the other hand, if the number of parameters in the model is small, the model will not necessarily represent the data but will be more likely to capture the regularity of the data and extrapolate (or interpolate) correctly. When the data are not too noisy, the optimal choice is the minimum-size model that represents the data.

In genetic algorithm, the problem of searching for minimal representative subsets of indicators is formulated as a bit-string optimization problem. Each indicator set is represented by a binary bit-string with N bits, where N is the total number of indicators. Each bit determines whether an indicator is used in the training phase of the neural network. A bit "1" indicates that an indicator is present in a subset, and a bit "0" indicates that the indicator does not belong to it. We will apply this paradigm to selection of input indicators for currency exchange rate forecasting.

The U.S./Singapore dollar monthly exchange rate was used to compare the performances of three different forecasting methods. The main objective was a six-month forecast based on available historical data. Different network architectures with different parameters were tried on the BP and the RBF models. To improve the results of back-propagation, we used GAs to select input indicators. A comparison of the performance of five subsets of input indicators produced by genetic algorithm with that of a network using all input indicators showed prediction accuracy to be better with subsets. Improvement ranged between 31% and 69%. In addition, training time was reduced drastically by 47% to 80% and the space complexity of the neural network was decreased by 50 to 83%. The smallest normalized mean-square prediction error was obtained with a subset of input attributes. One combination produced results that were 70% more accurate for a one-month-ahead forecast.

Genetic Fuzzy Systems

Learning capability can be provided to fuzzy logic systems by using a genetic algorithm to evolve the inference rules. The inference rules are encoded in a decision table and the fuzzy sets used are generic variables. A table is considered a genotype with chromosomes that are fuzzy set indicators over the output domain. The chromosomes are evaluated according to their behavior in response to the fuzzification, fuzzy inference, and defuzzification operations. The gene values reflect the fuzzy sets of the variables involved. A chromosome is formed from the decision table by going row-wise and producing a string of numbers from the code set. Standard crossover and mutation operators can act on these strings.

Combining Neural and Fuzzy Computing

The architecture shown in Figure 14.1 applies these various technologies to analyze input data, predetermine the predictability of a

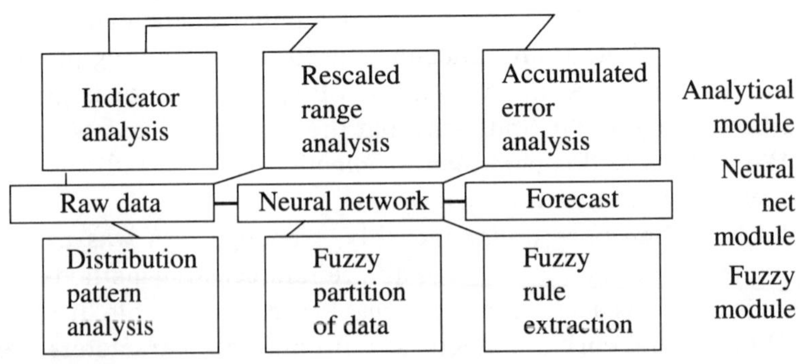

FIGURE 14.1
Architecture of a system combining neural and fuzzy computing

given time series, generate the fractal dimension, unveil the hidden cycles, determine the appropriate input window size and forecast horizon, determine the sizes of input and output layers of the neural net, decompose the problem to speed training, and achieve better forecasting accuracy.

The main problem noted with the back-propagation algorithm is the large amount of training time required for convergence. Although a number of training algorithms have been proposed to speed up the convergence, it seems unlikely that a significant reduction in training time can be achieved. Moreover, with too few hidden nodes, the network may not be trained at all. If just barely enough nodes are present the network may learn but it may not be robust when faced with noisy data and may not recognize new patterns. Furthermore, in order to capture the temporal information contained in time series, one may have to use a three-dimensional structure for better forecasting accuracy.

The phases of data preparation for modeling can be broadly classified into three areas: data specification and collection, data inspection, and data preprocessing. After the data are identified and collected, they must be examined for characteristics that may reveal important relationships between the target and the indicators. The distribution pattern of a variable provides valuable information about the existence of outliers, correlation of that variable with the target to be forecast, and whether enough data points have been gathered for training. A useful way to inspect the distribution pattern is shown in Figure 14.2.

The distribution pattern function often reveals the quality of the input data (e.g., the existence of outlying data or the insufficiency of training samples). The amount of training data is not very important for training, but the distribution pattern is. This function provides an easy way to visually inspect the distribution of the indicators versus the target in the input space. Each data point, or cluster of neighboring data points, represents a rule that describes the relationship between the indicator and the target. The input space is divided into five regions: VP (very positive), P (positive), M (moderate), N (negative), and VN (very negative) (see Figure 14.2).

A number of fuzzy if-then rules can be easily derived from this distribution pattern. For instance,

If Indicator 1 is VP, then Target is VN

If Indicator 1 is N, then Target is M
etc.

These fuzzy if-then rules can be used to decompose a given problem to achieve faster training and more accurate results. We proposed using a neural net structure as shown in Figure 14.3. Each of the

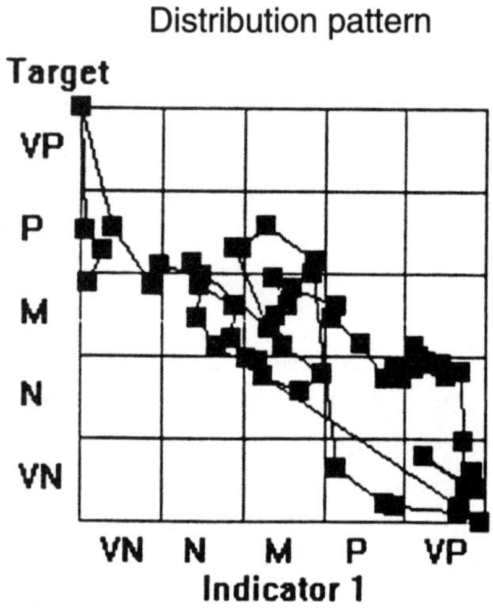

FIGURE 14.2
Distribution pattern of an indicator versus the target

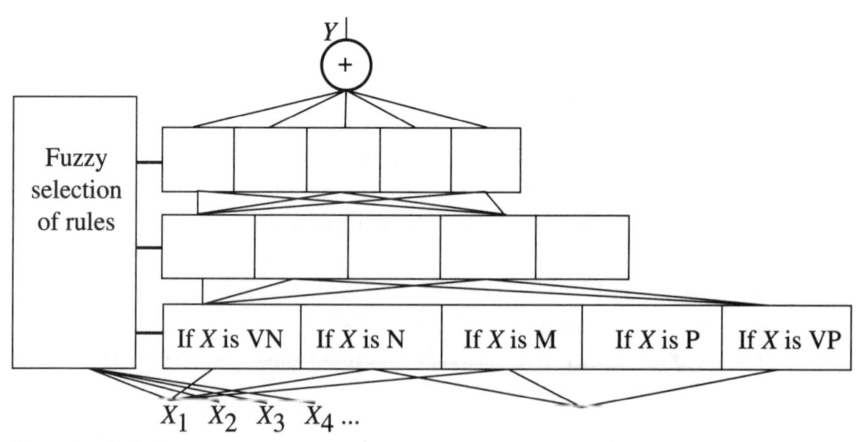

FIGURE 14.3
Selection of neurons for activation

hidden layers is partitioned to represent five fuzzy rules. The fuzzy selection module receives the current set of input vectors, determines the fuzzy partition, and selects one group of the neurons to activate. The mathematical description of the selection strategy can be found in Wong et al (1992). The basic idea is to decompose a given problem into subtasks via a divide-and-conquer strategy, without having to train the entire network for a given set of input vectors.

The fuzzy selection module partitions the input space into the five regions described earlier: VP, P, M, N, and VN. Because the output of the neurons is sigmoidal in nature, the input space is not divided crisply for either binary or continuous input values. This is similar to the membership function of a fuzzy set. The target value corresponding to the current input partition is used as the supervised data to train the network to generate the THEN part of the current partition. The output will be a value in one of the five regions.

If any of the regions is not well represented by the input data points, as illustrated in Figure 14.4, the network will not be able to learn the rule(s) describing that region and will fail if the test or future input data lie in that region.

One way to reduce training time is to activate and train the neurons selectively according to certain criteria. One such criterion is the current "situation" of the set of input data (Wong and Wang 1991; Wong et al. 1991) For example, if the current economic situation is

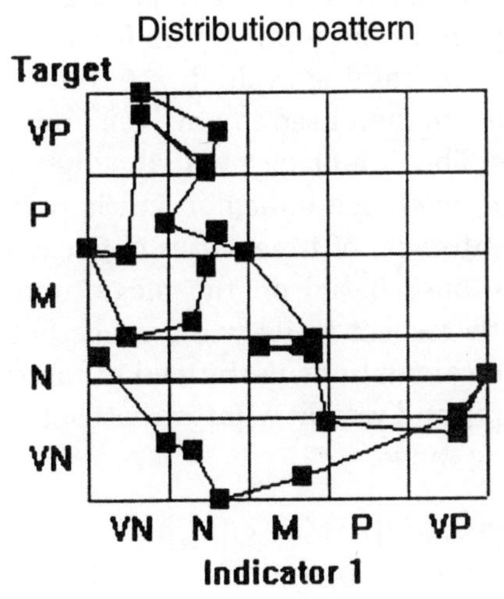

FIGURE 14.4
Absence of training data in the P (positive) region

very good, then only certain groups of neurons in the hidden layers (which are associated with the concept of "very good economic situation") need to be involved in the forward computation of the signals, as well as the back-propagation of error. One may use a clustering method to cluster the time series into several situations prior to the training. The algorithm we present performs the clustering of situations dynamically during the training process.

The neural network associates a specific output signal with the input data by transforming the weights stored in it into an explicit form using the causal index method. The causal indexes can only be calculated at the end of the training process when the system error has converged to an acceptable level, whereas the clustering of input data has to be performed during training. The FastProp algorithm (Wong 1991) uses the accumulated input error (AIE) index to obtain an instantaneous causal relationship between the input and the output, which is useful for finding the cluster to which the current set of input data belongs. The AIE index of the *i*th input at time *t* is calculated as follows:

$$AIE_i(t=T) = {}_{-t=0 \text{ to } T} \mid e_{it}[s=0] \mid$$

where *i* indexes over the neurons in the input layer ($s = 0$). The total influence of the *i*th input factor X_i on the output at time *t* can be expressed by $AIE_{i(t)}$, which is the causal relationship between the *i*th input and the output signals and is calculated based on the absolute errors back-propagated to the input layers for all the input data presented, over the entire training cycle ($t = 0$ to $t = T$).

The AIE indices are then used to rank the input factors according to their magnitudes. The input factor with the highest AIE index value is referred to as the *reference indicator* and is represented by using the index i^*. The entire set of time-series data can thus be clustered into several "situations" based on the magnitude of the reference input indicator with respect to its value range between $Max(x_{i^*}[0])$ and $Min(x_{i^*}[0]$. The calculations performed at the *s*th layer for input signal $x_j[s]$, error $e_j[s]$, and weight $w_{ji}[s]$ are restricted to those neurons within the following range:

$$F(x_{i^*t}[0]*N[s], Max(x_{i^*}[0]) - Min(x_{i^*}[0]))) \text{ to}$$
$$F(\_N x_{i^*t}[0]*N[s], (Max(x_{i^*}[0]) - Min(\mathbf{x}_{i^*}[0]))) + \_\mathbf{N}$$

where $x_{i^*t}[0]$ is the reference input indicator at time *t*, _*N* is a small positive integer (usually 1), and $N[s]$ is the total number of neurons in the sth layer.

With this selection scheme, only a small portion of all the neurons is involved in the computation at any instant. Therefore, training can be accomplished more quickly. Since this architecture is able to self-organize the input/output mapping relationship by learning, it is applicable to problems where unknown relationships exist. Moreover, it is capable of expressing any nonlinear relationship by virtue of its nonlinear structure, and there is no need to design membership functions that are difficult to implement in a multidimensional space using experience and initiation. It is also possible to extract "human comprehensible" rules from such a trained network using methods described by Towell and Shavlik (1991).

Rule-based Neural Trading Systems

Many trading systems are designed for only one objective: to make money. Financial institutions look at systematic methods of increasing rewards while reducing risks. Many of the trading systems used are entirely rule-based, utilizing buy/sell rules that incorporate trading signals generated from technical/statistical indicators such as moving averages, momentum stochastics, and relative strength index, or from chart formations such as head and shoulders, trend lines, triangles, wedges, and double top/bottom. This is hardly surprising, because many traders still assume that the market is efficient and that any system that is purported to make money will cease to do so if everyone begins using it.

The two major pitfalls of rule-based trading systems are the need for an expert to provide the trading rules and the inability to learn to adapt the rules to changing market conditions. The need for an expert is a major disadvantage in designing a trading system, as it is difficult to find an expert willing to impart his or her knowledge about trading. Furthermore, many successful traders are unable to explain the decision-making process that they undergo in making a trade. Indeed, many of them just put it down to a "gut feeling." This makes it very difficult for the knowledge engineer to derive the necessary rules for the inference engine.

The inability to adapt the rules means that a rule-based system may fail when market conditions change (e.g., from trending to non-trending), resulting in big trading losses. Different sets of rules may be needed for different market conditions, and since markets are dynamic, condition-determining parameters must be set. Such systems are also unable to learn and self-modify the rules/trading strategies from historical data. Many rule-based systems require frequent optimization of the technical indicators' parameters, which may result in curve-fitting.

Neural nets can be used to replace the human knowledge engineer in defining and finding the rules for the inference engine. An expert's trading record can be used to train the neural net to generate the inherent rules being used. A neural net can be taught profitable trading styles using historical data and then used to generate the required rules. It can also be presented with charts so that it can learn the necessary chart pattern formations that may provide profitable trading opportunities in the future. Furthermore, neural nets can also be presented with fundamental data, such as GNP, interest rates, inflation rates, unemployment rates, and so forth, and find the rules that relate these data to price movements.

Most rule-based trading systems are trend-following systems that respond to moving averages, momentum, and so on. Such systems work on the principle that the best profits are made from trending markets and that markets will follow a certain direction for a period of time. This type of system will fail in nontrending markets. Some systems also incorporate trend-reversal strategies by attempting to pick tops or bottoms through indicators that signal potential market reversals. A good system needs to have tight control over its exit rules, minimizing losses while maximizing gains. It must exit if the position taken is no longer favorable, even if this means taking a loss, and rides on the profits if the market starts trending in a favorable direction.

To improve on rule-based trading systems, a neural network can be added and utilized in the following tasks: generation of profitable trading rules; recognition of recurring chart patterns; incorporation of technical analysis; and incorporation of fundamental analysis. Each of these tasks will be discussed in detail.

Generation of profitable trading rules

A set of profitable trades is presented to the neural network. The neural net finds the inherent rules that lead to profitable trades. Once the rules are generated, they are incorporated into an expert system to produce buy/sell/hold signals depending on the new input data (see Figure 14.5).

Recognition of recurring chart patterns

The neural net is trained on chart formation patterns of the historical price data, such as double tops, triangles, and so on. It creates a pattern recognition algorithm that is incorporated into the expert system trading rules and its shell. It is then used to check new input data for the possible recurrence of chart patterns. Once it has identified a potential chart formation pattern, it attempts to complete the pattern and thus enable the system to predict the price movement of the security. By incorporating this new information with the trading rules, the system will then generate a buy/sell/hold signal (see Figure 14.6).

Incorporation of technical analysis

The neural net is trained on a set of data consisting of historical prices and the respective technical indicators (e.g., momentum, RSI, etc.).

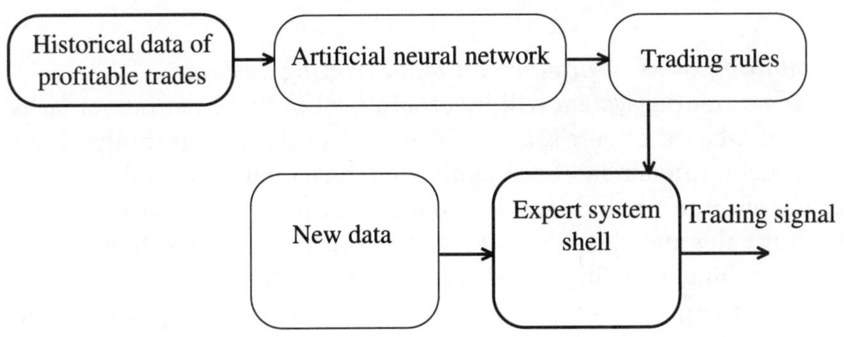

FIGURE 14.5
Incorporation of rules generated by a neural network into an expert system shell to produce trading signals

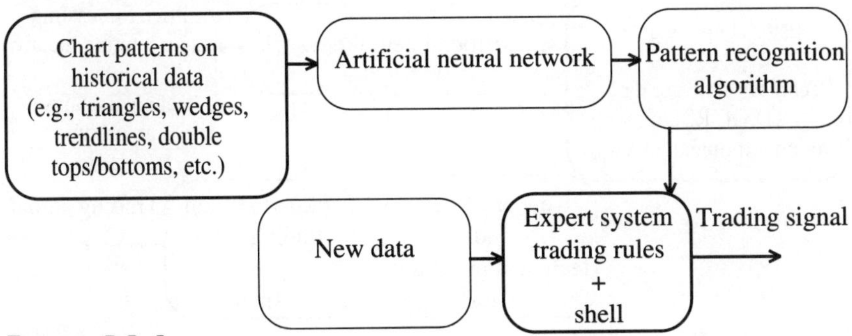

FIGURE 14.6
Incorporation of neural net pattern recognition into an expert system shell to generate trading signals

The neural net generates a price-forecasting algorithm based on the relationships between indicator and price. This is incorporated into the expert system trading rules and its shell, and a trading signal is generated when new data inputs are processed by the system (see Figure 14.7).

Incorporation of fundamental analysis

The neural net is trained on a set of data consisting of historical prices and the corresponding fundamental economic data (e.g., GNP, unemployment rate, inflation rate, interest rate, national debt, etc.). The neural net generates a price-forecasting algorithm based on the relationships between the fundamental economic data and price. This is incorporated into the expert system trading rules and its shell, and a trading signal is generated when new data inputs are processed by the system (see Figure 14.8).

Combining tasks to produce an ideal trading system

An ideal trading system will incorporate all of these neural net tasks in a rule-based expert system. Historical price data, technical indicator data, fundamental economic data, chart patterns, and different profitable trading strategies fed to a neural net will in turn generate trading rules and algorithms tying all the information together to generate profitable trading signals (see Figure 14.9).

The neural net can be evaluated and trained using profits and losses from the generated trades rather than squared forecast error. Trading strategies are presented to the neural net for evaluation and

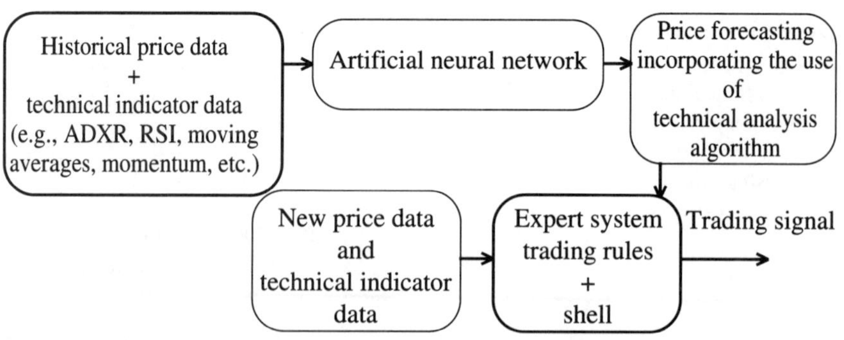

FIGURE 14.7

Incorporation of price forecasting via neural net technical analysis into an expert system shell to generate trading signals

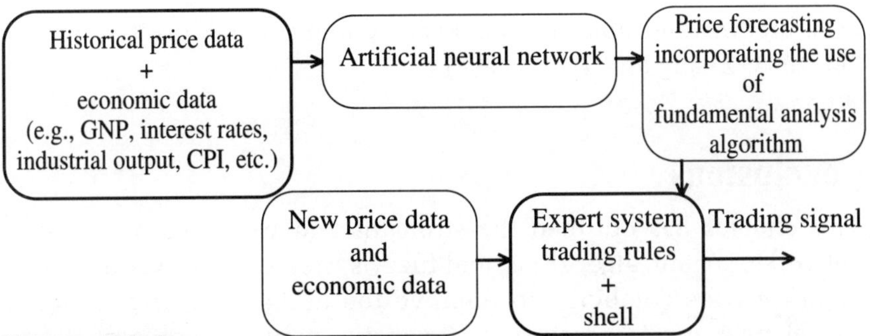

FIGURE 14.8

Incorporation of price forecasting via neural net fundamental analysis into an expert system shell to generate trading signals

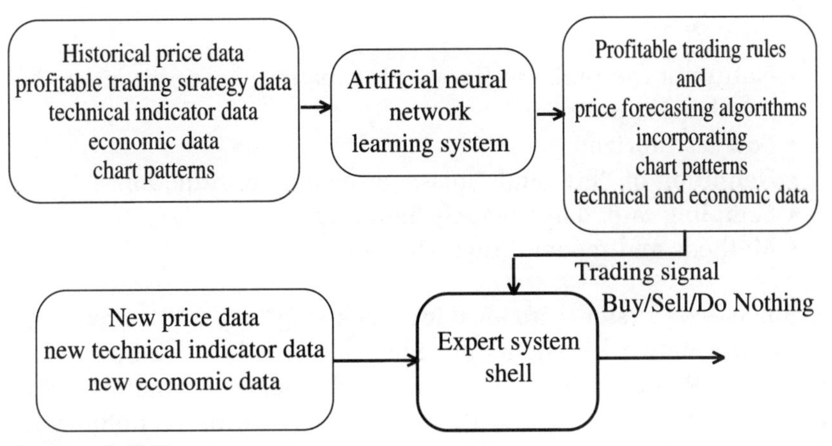

FIGURE 14.9

A system incorporating the neural net tasks illustrated in Figures 14.5 to 14.8

generation of profitable trading rules. Technical indicator data must be derived manually or through third-party technical analysis software. Ultimately, the system will be able to generate the technical indicator data automatically from a live price data input feed. Fundamental economic data may have to be normalized for use as training inputs. Periodic economic reports must be synchronized with price and technical indicator data.

The neural net uses an iterative (time horizon window) method of training so that the relationship of sets of data (rather than just a single set) to price can be uncovered. Implementation of the above modules will allow the neural net to complement the expert system in its trading decisions and enhance system profitability (Wong and Lee 1993; Zhang and Wong 1992).

Conclusions

This chapter has outlined the strengths and weaknesses of neural networks, genetic algorithms, and fuzzy systems. These various techniques can be combined to improve the design and/or training of neural nets and to create neural-genetic, neural-fuzzy, and fuzzy-genetic systems. Neural networks can enhance the performance of rule-based expert systems.

To achieve good results, several general factors must be considered:

- Nature of the problem (forecasting, pattern recognition, etc.)
- Available input data
- Forecast horizon
- Definition of "hit" and "miss," or evaluation function
- Sampling rate: daily/weekly/monthly
- Methods and model design strategy

Since fuzzy systems are ideal for processing "explicit" knowledge, and neural network systems are ideal for "implicit" knowledge, a natural trend is to combine these two technologies for better performance. Furthermore, genetic algorithms are powerful techniques for adaptive search of large, complicated spaces. Several research groups are currently investigating hybrid systems; the results are very encouraging.

References

Bornholdt, S. 1992. "General Asymmetric Neural Networks and Structure Design by Genetic Algorithms." *Neural Networks* 5.

Harp, S., and T. Samad. 1991. "Genetic Synthesis of Neural Network Architecture." In *The Handbook of Genetic Algorithms*. L. Davis, ed. New York: Van Nostrand Reinhold.

Montana, D., and L. Davis. 1989. "Training Feedforward Neural Networks Using Genetic Algorithms." In *Proceedings of the 11th International Joint Conference on Artificial Intelligence*.

Towell, G. G., and J. W. Shavlik. 1991. "The Extraction of Refined Rules from Knowledge-Based Neural Networks." *Machine Learning* (Aug.).

Treleaven, P., and S. Goonatilake. 1992. "Intelligent Financial Technologies." In *Proceedings of Parallel Problem Solving from Nature: Applications in Statistics and Economics*. EUROSTAT.

Welstead, S. 1993. "Financial Data Modeling with Genetically Optimized Fuzzy Systems. In *Proceedings of the 2nd Annual International Conference on Artificial Intelligence Applications on Wall Street* (New York).

Wong, F. 1990-91. "Time Series Forecasting Using Back-Propagation Neural Networks." *Neurocomputing* 2.

Wong, F. 1991. "FastProp: A Selective Training Algorithm for Fast Error Propagation." In *Proceedings of the International Joint Conference on Neural Networks* (Singapore).

Wong, F., and D. Lee. 1993. "A Hybrid Neural Network for Stock Selection." In *Proceedings of the 2nd Annual International Conference on Artificial Intelligence Applications on Wall Street* (New York).

Wong, F., and P. Y. Tan. To be published. "Neural Networks and Genetic Algorithm for Economic Forecasting."

Wong, F., and P. Z. Wang. 1991. "A Fuzzy Neural Network Approach for Forex Investment." *International Fuzzy Engineering Symposium* (Japan).

Wong, F., P. Z. Wang, and T. H. Goh. 1991. "Fuzzy Neural Systems for Decision Making." In *Proceedings of the International Joint Conference on Neural Networks* (Singapore).

Wong, F. S., P. Z. Wang, T. H. Goh, and B. K. Quek. 1992a. "A Fuzzy Neural System for Stock Selection." *Financial Analyst Journal* (Jan.-Feb.).

Wong, F., P. Tan, and X. Zhang. 1992b. "Neural Networks, Genetic Algorithms and Fuzzy Logic for Forecasting." In *Proceedings of the 3rd International Conference on Advanced Trading Applications on Wall Street and Worldwide* (New York).

Zhang, X., and F. Wong. 1992. "A Decision-Support Neural Network and Its Financial Applications." Technical Report, Institute of Systems Science, National University of Singapore.

PART FOUR

Nonlinear Dynamics and Chaos

Sui ji fing bian:

Nothing is more important than adapting to changing conditions. Both chaos and order can be found in capital markets. Nonlinear dynamics and chaos theory are new concepts that can improve our understanding of financial markets.

CHAPTER 15

Basic Concepts of Nonlinear Dynamics and Chaos Theory

Mark Embrechts

Confusion evolves into order spontaneously. What God really said was, "Let there be chaos."

Rosario M. Levins in* Hierarchy Theory *(H.H. Pattee, ed.), 1973

Part Four of this book discusses techniques for gathering intelligence from financial time series. Several books and articles have characterized financial markets as nonlinear, dynamic systems. Recent work on fractal market analysis demonstrates that some financial time series fit the paradigm of chaos models. Our first chapter on this subject is by Mark Embrechts, professor at Rensselaer Polytechnic Institute in New York, who provides a concise introduction to chaos theory. The second chapter provides a detailed overview of the procedures and methods for applying these concepts to financial time series. The concluding chapter in this section demonstrates the application of these concepts and techniques to foreign exchange, stock, and bond markets.

What Is a Fractal?

Benoit Mandelbrot introduced fractals in the 1970s to describe irregular shapes, such as mountains, clouds, wiggly lines, and coagulations of points. Since then, several books have been published that carefully define what exactly constitutes a fractal (Barnsley 1988; Mandelbrot 1983; Feder 1989; Peters 1991, 1994).

Geometric structures such as points, lines, and surfaces can be categorized according to their (topological) dimension as follows: 0 for a point, 1 for a line, and 2 for a surface. Shapes such as wiggly lines or circles have a topological dimension of 1, because they can be stretched into a straight line. Mathematicians have defined several other dimensions. One type is the *Hausdorff dimension*, which has an interesting characteristic in that it can also take noninteger, or fractional, values. The figures that Mandelbrot studied can be assigned such noninteger Hausdorff dimensions. These dimensions are not topological in nature, but can be interpreted as more generalized types of dimensions.

Fractals are figures with a noninteger Hausdorff dimension. Mandelbrot wrote: "A fractal is a set for which the Hausdorff dimension strictly exceeds the topological dimension." The Hausdorff dimension can be approximated by the *fractal dimension* or *correlation dimension*.

A fractal dimension is a measure of how much an n-dimensional space is occupied by an object. For example, a cotton ball has three physical dimensions. If one squeezes it with a press, it tries to occupy two dimensions but fails. It still has bulk in the third dimension. So the flattened cotton ball has a "dimension" somewhere between 2 and 3. Similarly, a point has one dimension, and a straight line has two dimensions. A wiggly line, which is a straight line squeezed together, has a dimension between 1 and 2.

A classical method to measure the length of a wiggly line is to use a ruler or a stick. However, the actual length of a wiggly line depends on the length of the measuring stick. Lewis Fry Richardson first observed this empirically in 1961, and his results were described by Mandelbrot under the heading, "How long is the coast of Britain?" The coastline of Britain (like that of many other countries) is a wiggly line. It would seem obvious that its length can be measured by counting how many times a measuring stick with a known length can be fitted along the coastline and then multiplying this number the length of the measuring stick. However, the length of a wiggly line, L, actually is a function of the length of the measuring stick, δ, according to:

$$L = N(\delta) \times \delta$$

where $N(\delta)$ is the number of times that the measuring stick can be fit to the wiggly line. For a familiar Euclidean shape, the number of times the measuring stick can be fitted around the perimeter is inversely proportional to the length of the measuring stick, or:

$$N(\delta) \propto \frac{1}{\delta}$$

For a fractal curve, the measuring procedure turns out to be different. The number of times we can fit the measuring stick around the perimeter is not just proportional to $1/\delta$, but it is also proportional to $1/\partial^{D}$. The length can thus be expressed as:

$$L(\delta) \propto \delta^{1-D}$$

By plotting the length of a wiggly line versus the length of the measuring stick on a double logarithmic scale, one obtains roughly straight lines because:

$$\log[L(\delta)] = \text{constant} + (1\text{-}D)\log(\delta)$$

On such a double logarithmic scale these measurements lead to straight lines with a slope of $1 - D$. The proportionality constant D is a measure of the jaggedness or nonsmoothness of the wiggly line and corresponds to the *fractal dimension*. A more wiggly line has a higher fractal dimension D.[1]

Methods for Computing Fractal Dimensions

There are six distinct methods for computing fractal dimensions: rescaled range analysis, relative dispersion analysis, correlation analysis, Fourier analysis, maximum likelihood estimator analysis, and the "Higuchi" method. In addition, neural networks can provide good estimates of fractal dimensions for any signal, if trained on data derived from Fourier analysis.

To estimate the fractal dimension of a time series of financial data, the concept of the correlation dimension is often applied. This concept is an elegant way of estimating the fractal dimension of a set of n-dimensional data points. The method of the correlation dimension consists of centering a hypersphere about a point in hyperspace or phase space, letting the radius (r) of the hypersphere grow until all points are enclosed, and keeping track of the number of data points that are enclosed by the hypersphere as a function of the hypersphere radius. A phase space is shown in Figure 15.1. The slope of the line

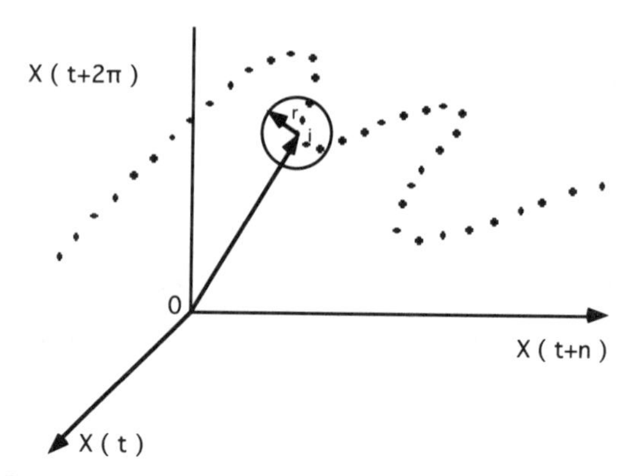

FIGURE 15.1
Phase space reconstruction of a signal X(t)

on a double logarithmic plot will then be an estimate of the fractal dimension of the set of data points.

The fractal dimension is a measure of the relation between the variance of the signal and the time scale. More succinctly, it provides a measure of the homogeneity of the signal. In addition, nearly all techniques use a power law to associate the statistic and the fractal dimension. Thus, by plotting on log-log scales and running a linear regression, one is usually able to estimate the fractal dimension from the slope of the regression line.

What Is Chaos?

Simple nonlinear maps or coupled differential equations can exhibit very complicated dynamics in a chaotic regime. These deterministic models can lead to seemingly random and highly irregular signals or trajectories. Chaotic systems such as the logistic map or the Lorenz system demonstrate typical characteristics. Several regimes, such as fixed points, limit cycles, and strange attractors, can be identified. Whether or not the regime is chaotic depends on the values of one or more system parameters.

As there is no commonly accepted definition of chaos, we can only characterize chaos as *a seemingly random and irregular signal generated by deterministic process with some additional properties.*[2] The first of these properties is that the deterministic process that generated a chaotic signal must show sensitivity to initial conditions.

A second property is that a chaotic signal is associated with a strange attractor (i.e., that the phase space shows a figure or distribution of points characterized by a fractal or noninteger dimension).

Finding chaos in financial data provides insight into the limits of predictability. Identification of chaos indicates that short-term predictions are possible in principle.

Neural nets can be trained in that case for predicting the time series. The fractal dimension is indicative of the minimum number of observations that must be shown to the neural net for a particular training pattern. The higher the dimension, the more training samples are necessary. Thus, the fractal dimension tells us how much data are necessary for training a neural net. Identifying chaos in financial data, however, is a complex process. Chapter 16 elaborates the techniques for identifying chaos in financial time series. A brief introduction to methods for detecting and describing chaos follows.

Methods for Detecting Chaos

Chaotic time-series analysis determines whether an irregular and seemingly random set of data points resulted from a chaotic process. A set of discrete data points that represent values of a signal as a function of time (e.g., daily stock quotes) is a *time series*. *A chaotic time series has a phase space trajectory that evolves toward a strange attractor.* A strange attractor is nothing more than a phase space trajectory that stays on a fractal set. The strange attractor occupies just a small fraction of the available phase space and is a fractal. As such, it can be characterized by a noninteger, or fractal dimension. Chaotic time-series analysis finds the dimension of this attractor. The higher the dimension, the more complex the underlying process that generated the time series and the more difficult it will be to train a neural net for predicting the time series.

In principle, a chaotic system can be modeled either by a number of coupled nonlinear first-order differential equations or by maps. The minimum number of differential equations is equal to the integer that *embeds* (i.e., is just larger than) the fractal dimension. The dimension of the phase space that spans this minimal number of differential equations is called the *embedding dimension*. The embedding dimension will always be larger than the fractal dimension of the strange attractor.

The most popular analysis technique for a chaotic time series is the *time delay method* (Packard et al. 1980; Takens 1981). In this method a series of equally spaced data is considered. Takens showed that a topological picture of the attractor in phase space can be reconstructed by the time delay method. Such a topological picture bears a reasonable resemblance to the original attractor, but certain directions have been stretched and deformed. The time delay method consists of choosing a proper delay time and reconstructing a set of n-dimensional vectors or n-tuples, where n is not known a priori. The components of the n-tuples form the directions of the new (topologically equivalent) phase space in which the attractor will be reconstructed by plotting the points corresponding to the n-tuples in this new phase space.

To illustrate this approach we will use the yen/dollar daily returns from 1971 to 1980. The returns are defined as the logarithm of today's exchange rate divided by the logarithm of yesterday's exchange rate. Using the procedure described above, the correlation dimension for various embeddings was calculated. About 2,000 10-tuples were generated from this data. There is no correlation between the daily returns, and daily returns were therefore considered without skipping any data. Figure 15.2(a) shows the correlation integral, $C(R)$, versus the radius, R, of the hyperspheres on a double logarithmic scale for various dimensions of the n-tuples. The uppermost curve corresponds to dimension $n = 1$, while the lowest curve was obtained assuming $n = 10$. We can distinguish three regions in each of the curves in Figure 15.2(a) with increasing values for increasing R:

1. For small values of log (R), the curves are not very straight. This can be due to a variety of reasons.
2. The region with radii varying between 0.7 and 3.0 seems to be quite linear. This is the region that will be explored for estimating the fractal dimension as the slope of the log of the correlation integral versus the logarithm of R.
3. When the radius of the hyperspheres increases, it eventually becomes as large as the size of the attractor itself. In that case all the points of the attractor are contained within the hypersphere. The correlation integral, being nothing more than a measure for the fraction of points within a hypersphere of a certain size, will therefore saturate to unity.

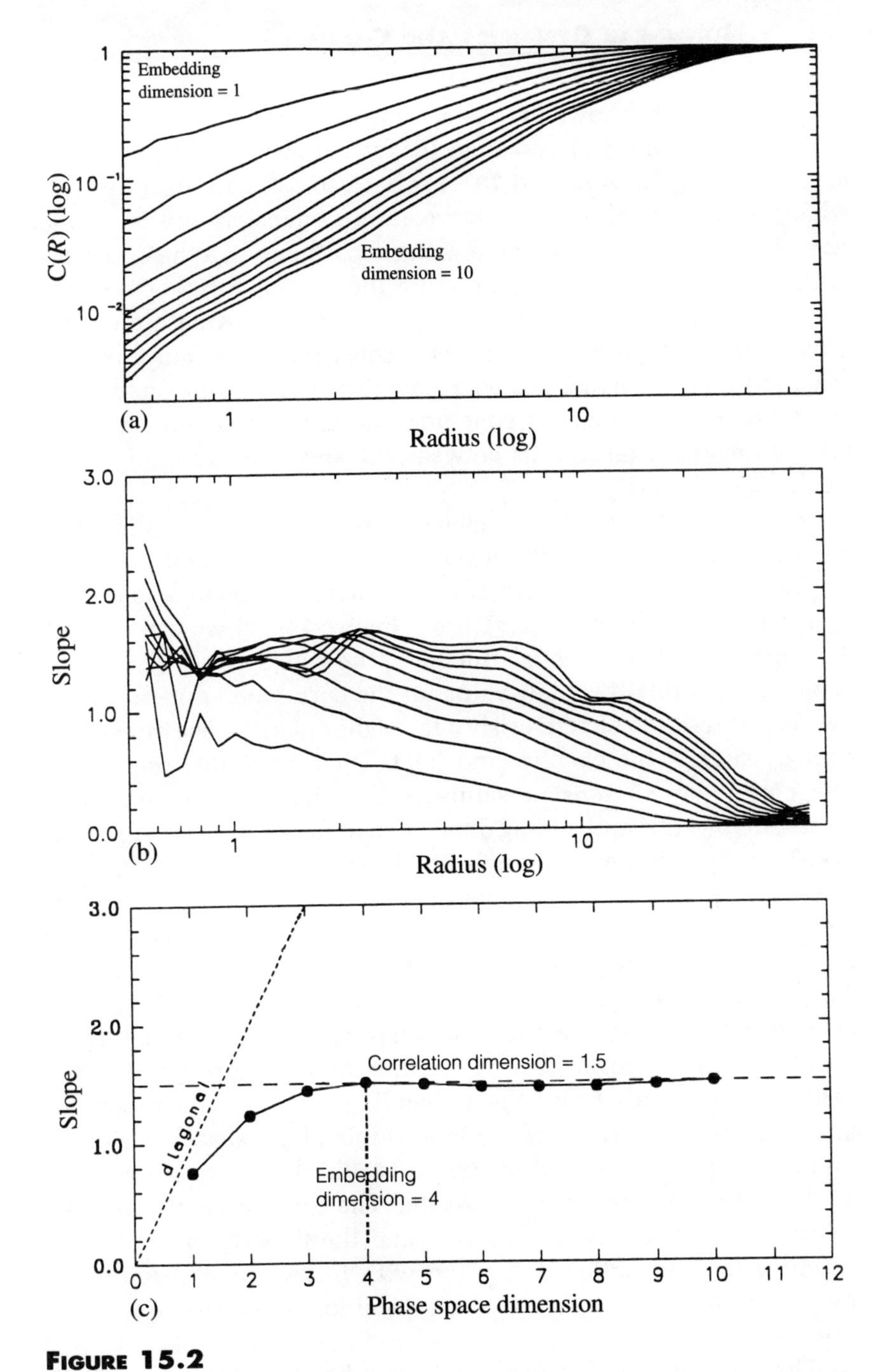

FIGURE 15.2
Calculation of correlation dimension for various embeddings for yen/dollar returns (1971–1980). (a) Log [C(R)] versus log (R). (b) Slope of log [C(R)] versus log (R). (c) As function of dimension of n-tuples

Figure 15.2(b) shows the instantaneous slope of the logarithm of the correlation integral versus log(R) for R ranging between 0.5 and 50, and n ranging from 1 to 10. It is hard to estimate from this plot what the proper value of the correlation dimension will be. Notice how the curves tend to coincide with each other when the dimension of the phase space (n-tuples) grows for the region with R between 0.7 and 3. For smaller values of R the plot shows a sawtooth tendency. This sawtooth type of structure is a consequence of too fine a bin spacing on the one hand, and poor statistics in this region on the other hand. Figure 15.2(b) lets us conclude that the correlation dimension for the linear region (radius between 0.7 and 3) is between 1.4 and 1.6.

Figure 15.2(c) provides the global least-mean-squares estimate for the slope of the curve of the logarithm of the correlation integral in the linear region as a function of n. This slope was evaluated directly from the data of Figure 15.2(a). From Figure 15.2(c) we can estimate the dimension of the chaotic attractor of the yen/dollar exchange rate to be 1.5. This relatively low value for the correlation dimension and the clear saturation characteristics for higher phase space dimensions strongly suggest the chaotic (and fractal) nature of this time series. The correlation dimension saturates for phase space dimensions greater than 4. The embedding dimension of the attractor is thus equal to 4. The strong indication for chaotic behavior implies that the yen/dollar rate is strictly predictable in the short term. The embedding dimension of 4 indicates that when training a neural net for this time series, one would have to train on three successive daily returns for the yen/dollar exchange rate in order to predict the fourth value.

Figure 15.3 illustrates the time delay method for estimating the correlation dimension of the yen/dollar returns during the period 1980 to 1992. In this figure the embedding dimension increases linearly with the dimension of the embedding phase space. The correlation dimension estimated in Figure 15.3(b) does not saturate, and this time series is therefore not chaotic. The correlation dimension is infinity (or at least extremely large), and the linearly increasing behavior of the correlation dimension with increasing embedding dimension of the phase space is also typical for Brownian motion and Gaussian noise.

The yen/dollar exchange rates during the eighties are not chaotic because no saturating value for the correlation dimension can be found. The yen/dollar evolution during the eighties is not predictable

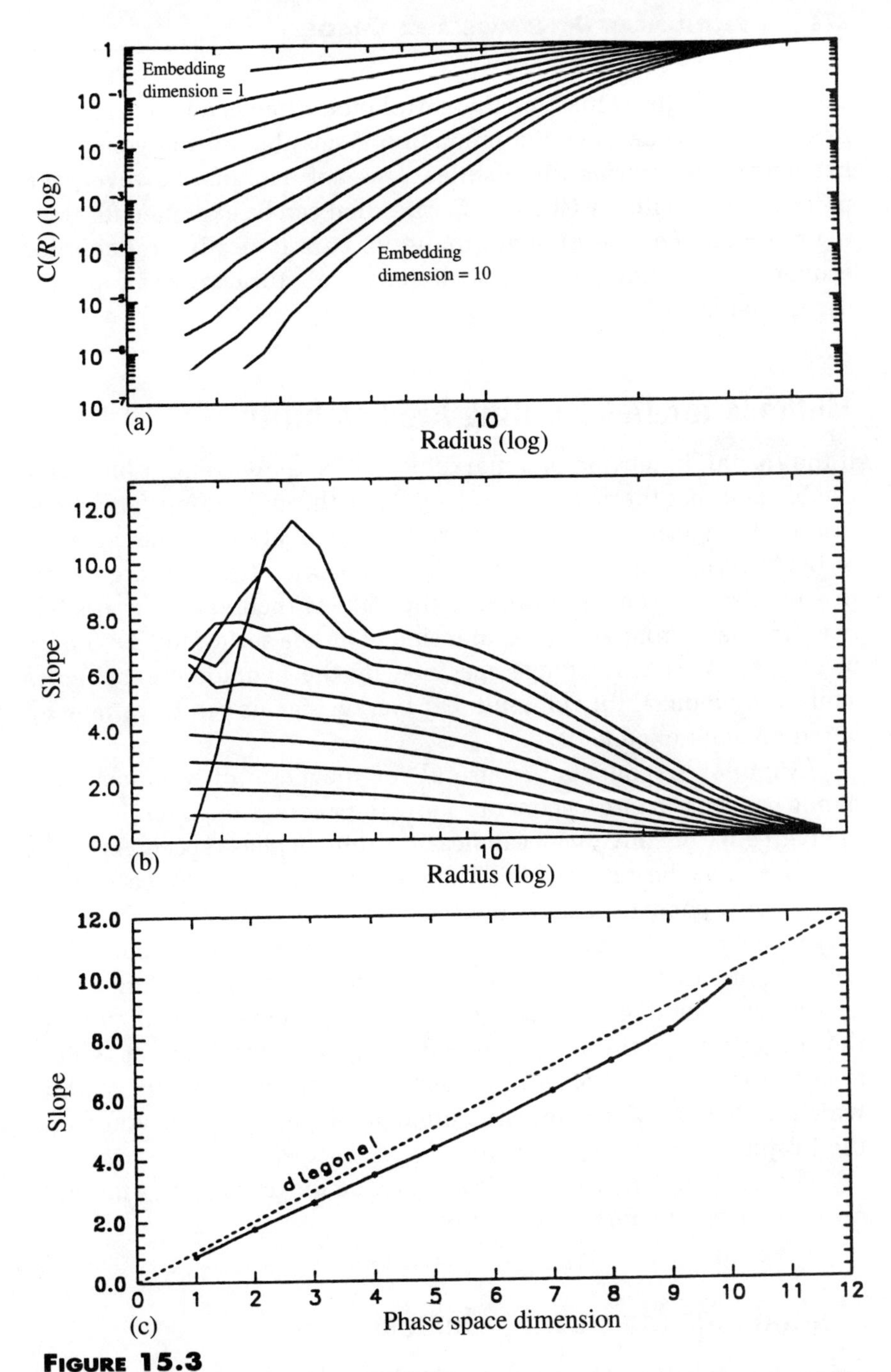

FIGURE 15.3

Calculation of correlation dimension for various embeddings for yen/dollar returns (1980–1992). (a) Log [C(R)] versus log (R). (b) Slope of log [C(R)] versus log (R). (c) As function of dimension of n-tuples because no finite values for the correlation dimension can be identified. The fact that no strict prediction is possible for the yen/dollar exchange rates during the eighties does not exclude, however, the presence of certain statistical properties that can be exploited for trading purposes. For the limited number of data typically available for economic indicators, only low-dimensional attractors can be recovered at best.

Methods for Determining Predictability

If the fractal dimension of a market signal is known, and it has been established that there is evidence of chaos through strange attractors in the phase space, then it is possible to estimate the predictability of the market signal. In other words, if chaotic attributes are found in a time series, then according to the Takens theorem[3] it is possible to extrapolate, from a single dimension of a time series, the dynamics of the series within a range specified by the *Lyapunov exponents.* Thus, the predictability of nonlinear market signals can be estimated based on such exponents.

Lyapunov exponents are useful for diagnosis of nonlinear dynamic systems. These exponents show the average exponential rates of divergence or convergence of nearby orbits in phase space. As such, they describe the rate at which close points diverge. Since nearby orbits correspond to nearly identical states, exponential orbital divergence means that systems whose initial differences are small will behave quite differently over time. Therefore, any system containing at least one positive Lyapunov exponent is defined as chaotic, with the magnitude of the exponent reflecting the time scale at which system dynamics become unpredictable. The time frame within which a chaotic system remains predictable can be estimated from the Lyapunov exponents as well.

The next section briefly reviews rescaled range analysis, a method for measuring memory in a time series.

Methods for Measuring Memory

Rescaled range analysis was first introduced by British scientist Harold Edwin Hurst (1880–1978), who studied the flood patterns of the Nile. Hurst introduced this statistical analysis method in order to design the optimal capacity for water reservoirs (Hurst et al. 1965). Mandelbrot rediscovered the work of Hurst and deserves the credit putting rescaled range analysis in a more general framework.[4]

Rescaled range analysis, also called *Hurst analysis*, is a statistical method to evaluate whether memory patterns are present in a series of data. Rescaled range analysis (which can be applied to any time series) provides the *Hurst coefficient*, *H*, which is a measure of the bias or trend in a time series. According to Peters (1994), *H* measures "the fractal dimension of the time trace by the fractal dimension $2-H$." It is also related to "the statistical self-similarity of the process, $1/H$, which measures the fractal dimension of the probability space." Rescaled range analysis is essential to generalize the concept of Brownian motion. Brownian motion cannot be characterized by a low-order strange attractor and therefore is not chaotic. A reconstruction of Brownian motion in phase space would lead to an attractor with an infinite correlation dimension. However, Brownian motion has certain fractal properties. It is just these (let us call them non-chaos-related) fractal properties that can be related to the Hurst coefficient.

Rescaled range analysis and estimation of the Hurst coefficient has been described at length by others. A summary follows. Rescaled range analysis follows the following relationships:

$$\frac{R_N}{S_N} = \left(\frac{N}{2}\right)^H$$

where N is the time window, *H* is the Hurst coefficient, and R_N/S_N is the dimensionless ratio of the average range divided by the standard deviation of the daily fluctuations over N. R_N is the difference between the maximum and the minimum of the cumulative deviations of C_i from the average for a particular *N*. Thus

$$R_N = \underset{1\leq k\leq N}{\text{Max}} \sum_{j=1}^{k} (X_i - \overline{X}_N) - \underset{1\leq k\leq N}{\text{Min}} \sum_{j=1}^{k} (X_i - \overline{X}_N)$$

The slope of the log R_N/S_N versus the log of *N* provides an estimate of the Hurst coefficient.

To derive the Hurst statistic, R_N/S_N is plotted on a log-log scale against different lag times. A regression line is then fitted over points greater than some lag, with 2-*H* being an estimate of the fractal dimension of the time series. The regression is carried out for all lags greater than a certain value, because Hurst's statistic is known to be sensitive to short-term dependencies. The Hurst statistic has been modified by Lo (1991) such that it is invariant to short-term dependence. Lo factored the weighted autocovariances into the scaling of the statistic; however, the selection of the weights is not definite, and guidelines are provided in the form of theorems.

This procedure for estimating the Hurst coefficient is illustrated for the yen/dollar returns during the 1970s and 1980s in Figure 15.4. If the data were related to a purely random process, the Hurst coefficient would be exactly equal to 0.5. Hurst found that for many natural phenomena the Hurst coefficient is actually about 0.73.[5] This Hurst relationship with $H = 0.5$ can actually be proven quite easily for a Gaussian distribution function. A Hurst coefficient different from 0.5 can be interpreted as a bias or memory effect. A Hurst coefficient larger than 0.5 indicates a memory effect, where there is a bias to enforce the current trend. This tendency is called *persistence.* A Hurst coefficient of less than 0.5 indicates a negative bias; this tends to oppose the ongoing trend and is called *antipersistence.*

Figure 15.4(a) illustrates the Hurst coefficient for the yen/dollar returns during the 1970s ($H = 0.64$), and Figure 15.4(b) illustrates the Hurst coefficient for the yen/dollar returns during the 1980s ($H = 0.62$). Note that even though the yen/dollar behavior is chaotic during the seventies, and not chaotic during the eighties, the Hurst coefficients are very similar for both regimes. The Hurst coefficient could be related to chaos or to fractional noise. It indicates a memory effect that can be exploited for obtaining a statistical edge for forecasting exchange rate returns. Further uses of the Hurst coefficient for financial time series will be illustrated in Chapter 17.

References

Baker, G. L., and J. P. Gollub. 1990. *Chaotic Dynamics: An Introduction.* Cambridge: Cambridge University Press.

Barnsley, M. 1988. *Fractals Everywhere.* New York: Academic Press.

Feder, J. 1989. *Fractals.* New York: Plenum Press.

Hurst, H. E., R. P. Black, and Y. M. Simaika. 1965. *Long-Term Storage: An Experimental Study.* London: Constable.

Lo, A. W. 1991. "Long-Term Memory in Stock Market Prices." *Econometrica* 59(5).

Mandelbrot, B. B. 1983. *The Fractal Geometry of Nature.* New York: W. H. Freeman.

Matsuba, I., et al. 1992. "Optimizing Multilayer Neural Networks Using Fractal Dimensions of Time-Series Data." In *Proceedings of the International Joint Conference on Neural Networks* (Beijing).

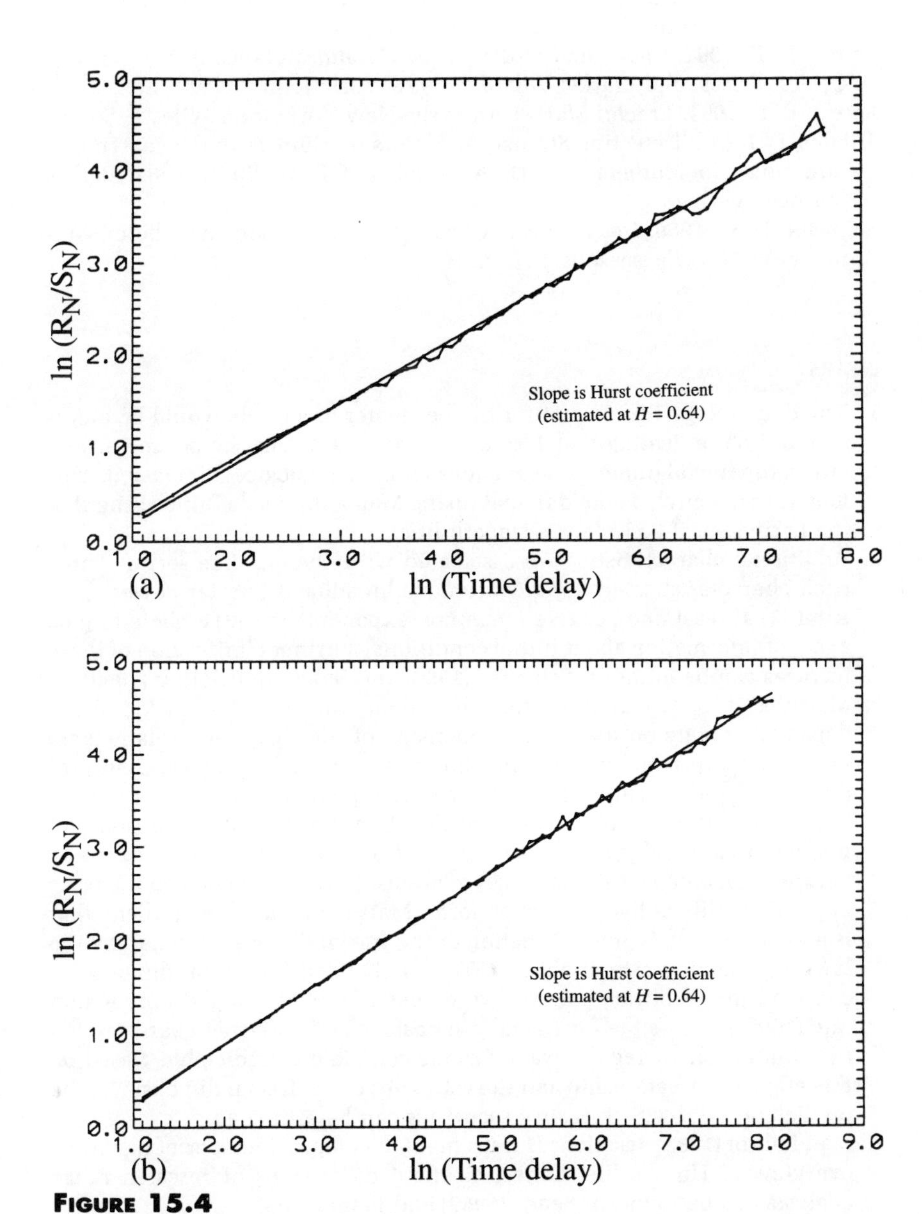

FIGURE 15.4

Estimation of the Hurst coefficient for yen/dollar returns. (a) 1971–1980. (b) 1980–1992

Packard, N., J. P. Crutchfield, J. D. Farmer, and R. S. Shaw. 1980. "Geometry from a Time Series." *Physical Review Letters* 45.

Peters, E. E. 1991. *Chaos and Order in the Capital Markets: A New View of Cycles, Prices, and Market Volatility.* New York: John Wiley & Sons.

Peters, E. E. 1994. *Fractal Market Analysis.* New York: John Wiley & Sons.

Takens, F. 1981. "Detecting Strange Attractors in Fluid Turbulence." In *Lecture Notes in Mathematics.* D. A. Rand and L. S. Young, eds. Berlin: Springer-Verlag.

Schuster, H. G. 1989. *Deterministic Chaos: An Introduction.* Weinheim, Germany: VCH Verlagsgesellschaft mbH.

Notes

1. The double logarithmic plot for the perimeter of a circle would be a horizontal line indicating that $1 - D = 0$, or $D = 1$. For the perimeter of a circle, the fractal dimension coincides with its topological dimension. This means that a circle is not a fractal (using Mandelbrot's definition) and that the perimeter of a circle is a smooth line.
2. Additional characteristics are associated with chaotic time series. Three such characteristics are (1) a continuous broadband Fourier power spectrum (2) at least one positive Lyapunov exponent, and (3) ergodicity (i.e., a loss of information about initial conditions). Further clarification of these features can be found in Schuster (1989) and Baker and Gollub (1990).
3. Matsuba et al. (1992) argue that the prediction rate of a market signal depends strongly on the fractal dimension of the signal. If a relationship between the prediction rate and the fractal dimension exists, then prediction reliability can be evaluated before a prediction is made. Matsuba has estimated an empirical relationship between fractal dimension and prediction rate. Using several time series from both economic indexes and natural phenomena, this empirical relationship was estimated on the basis of predictability of five distinct patterns. Matsuba found that predictability is an approximately linear function of the fractal dimension, whereby the lowest prediction rate of about 20% was obtained for fractal dimensions close to 2 (five arbitrary patterns were selected), and the highest prediction rate (100%) is possible for fractal dimensions of 1. A signal characterized by a limit cycle or regular wave form is completely predictable. Based on this empirical relationship and the estimates of the fractal dimensions, the predictability of U.S. Treasury securities can be estimated.
4. Mandelbrot (1983) describes Hurst's method and provides a brief historical overview of Hurst's life. Other excellent discussions of rescaled range analysis can be found in Feder (1989) and Peters (1991, 1994).
5. It is obvious from Hurst's experimentation that many natural phenomena do not adhere to Gaussian statistics. While this is an important finding, Hurst's conclusion might not be all that surprising. Why should nature follow Gaussian statistics in the first place? What is interesting in Hurst's study, however, is that the Hurst coefficient is more or less the same for

all natural phenomena. Nature definitely wants to adhere to certain statistical laws, and these laws are clearly not of the Gaussian type. There are several interpretations for Hurst's statistics. We refer the reader to the appropriate literature (Peters 1991, 1994). The concept of Brownian motion has been extended by Mandelbrot to fractal Brownian motion (fBm). Such fBm adheres to Hurst's statistics (Feder 1988). There is a correlation between the Hurst coefficient and certain fractal measures. Let us point out here that a fractal dimension can be associated with Brownian motion. Brownian motion is not a self-similar process but rather a self-affine process, where the time scale and the record scale have to be scaled by different parameters to generate a similar picture when rescaling the process. The interpretation of the fractal dimension is more complicated for such self-affine processes (Feder 1988). The various fractal dimensions that were introduced earlier will not be equivalent for self-affine fractals, and estimating the Hausdorff dimension can become quite challenging in such a case.

CHAPTER 16

Nonlinear Data Analysis Techniques

Ted Frison

Those skilled at the unorthodox are infinite as heaven and earth, inexhaustible as the great rivers. When they come to an end, they begin again, like the days and months; they die and are reborn like the four seasons.

Sun Tzu
The Art of War, (ca. first century B.C.)

Financial and economic time series are good candidates for chaotic analysis. Chaotic techniques can determine whether financial or economic time series are related and whether there is predictive value. As shown in Chapter 15, chaotic techniques can provide insights into the nature of financial or economic systems. These methods can also be used for model construction. In this chapter Ted Frison, president of Randle Inc., outlines nonlinear data analysis techniques for describing the nature of market signals and demonstrates how to compute the various measures defined in the previous chapter using chaotic signal processing (CSP).

Financial Time Series as Dynamic Systems

Financial and economic time series are signals from a system. Characteristics of these signals can be used to classify systems, monitor them, or possibly predict future behavior. Thus, we can analyze financial and economic time series in the same way signal-processing experts analyze signals from telephone circuits, television sets, radar emissions, or sounds radiated by submarines or dolphins!

Many signals from physical systems are chaotic. Chaotic signals appear to be random (noise), but they are in fact deterministic. These signals may be classified by invariant parameters that describe chaotic behavior, just as well-behaved signals can be classified by invariant Fourier coefficients (the FFT). To use these tools, we describe a new perspective of data analysis. The novel aspect of chaotic signal processing is that signals are analyzed in a time domain state space, rather than in a frequency domain. The new idea is that systems with only a few degrees of freedom can be chaotic and that it is possible to reconstruct and analyze chaotic behavior with scalar data collected in only one dimension.

The chaotic approach to data analysis treats both data and containments as *deterministic processes* in a *multidimensional state space*, also called a *phase space*. No assumptions are made about the behavior of the data. By deterministic processes we mean events that are not random and that are governed by some principles. By multidimensional state space we mean a geometric space of multiple dimensions where the value of each variable is plotted against its own axis.

Signals treated as *geometric objects* can be categorized by properties that are invariant. Systems analysis defines attributes of a system that are invariant and encompasses characterization, prediction, and control. Useful tools for characterization, prediction, and control are the *Fourier spectrum*, *state portraits*, and *difference maps*.

A periodic linear system can be described in terms of the Fourier spectrum—that is, a graph of how much energy is contained in a given frequency band. Since a nonlinear system can display multiperiod or aperiodic behavior, the utility of Fourier analysis diminishes as chaos is approached, and other methods to describe a system are needed.[1]

A state portrait is simply the values of all known variables at some point in time. The *trajectory* of a system is the change in the state variables over time. Plotting trajectories is a means of visualizing a dynamic system. A plot of the displacement versus velocity of a forced pendulum is one example of a trajectory.

A group of trajectories that originate from adjacent initial conditions is a *flow*. The point or set of points toward which the trajectory of a system tends in the long term is an *attractor*. An attractor may be a single point, such as a local minimum, or a closed curve for periodic behavior. The attractor may be such a complicated set of points that trajectory appears to wander randomly about the state space. The system dynamics are chaotic and said to have a "strange attractor."

In a chaotic system the future state of the system is critically dependent on its initial conditions. Close, but unresolvable, initial conditions lead to large divergences in trajectories. Therefore, information about prior states is lost as the system evolves, and long-term predictions by definition are doomed to failure. The inability to predict weather even a few days in advance is a dramatic example of man's futile grappling with a chaotic system. While long-term predictions of chaotic systems are not possible, it is sometimes possible to make short-term predictions—depending on the accuracy of the model, the precision of the variables, and how fast adjacent trajectories diverge, as measured by the Lyapunov exponents.

Lyapunov exponents are a means of classifying systems; they are independent of initial conditions. If at least one Lyapunov exponent is positive, close trajectories will diverge and the system is, by definition, chaotic.

A second necessary, but not sufficient, condition for identifying chaos is that information in data masquerades as broadband noise. Thus, linear methods for analyzing chaotic systems (such as the Fourier power spectrum) do not work. A signal can be contaminated by external broadband noise, which is indistinguishable from the true signal. Lack of understanding of "noise" prohibits development of simple models for noise rejection.

Finally, the *Hausdorff dimension* (or, more commonly, the *fractal dimension*) provides a measure of the extent to which the strange attractor fills the n-dimensional embedding space and is one means of classifying a signal.

The above brief overview of concepts from nonlinear dynamics and chaos theory provides an essential framework for absorbing the data analysis techniques described in this chapter. Application of nonlinear data analysis techniques to financial and economic market data is no different from identifying the properties of information in a text database or extracting signals from a spacecraft sensor.

Nonlinear Data Analysis Techniques

Acquire a String of Scalar Numbers

A time series of a single variable (acquired in one dimension) can be used to construct coordinates for a multivariate state space, because a financial time series is a nonlinear system where all variables are coupled. The behavior of any one variable has embedded in it full knowledge of the behavior of all other variables. The dynamical dimension of this system is the number of degrees of freedom and the number of Lyapunov exponents.

The ticker tape provides a sequence of prices or yields, x(1), x(3) . . . x(N), at times $t_0 + n\Delta t$. For example, a sample of daily opening prices of U.S. Treasury 10-year notes are shown at the top of Figure 16.1. Time delay vectors of this time series can be used to construct a d-dimensional phase space that serves as the coordinate system for capturing the attractor for the system (Eckmann and Ruelle 1985). The time delay vectors have the form:

$$y(n) = [x(n), x(n + T), x(n + 2T), \ldots x(n + (d - 1)T)]$$

Embedding the vectors $y(n)$ in the d-dimensional phase space to form a phase portrait is the fundamental operation that makes chaotic data analysis possible. The word "phase," as used by electronic engineers, may lead to some confusion. Our usage bears no relationship to signal phase. For the sake of clarity, we use the term "state space." To embed the time-lagged vectors in a state space a suitable time delay (sometimes called the time lag), T, must be found and a suitable number of dimensions must be identified. After the state space portrait is formed, the system can be analyzed.

This is a simple operation mathematically. Figure 16.1 arbitrarily shows three dimensions because that is the most that can be visu-

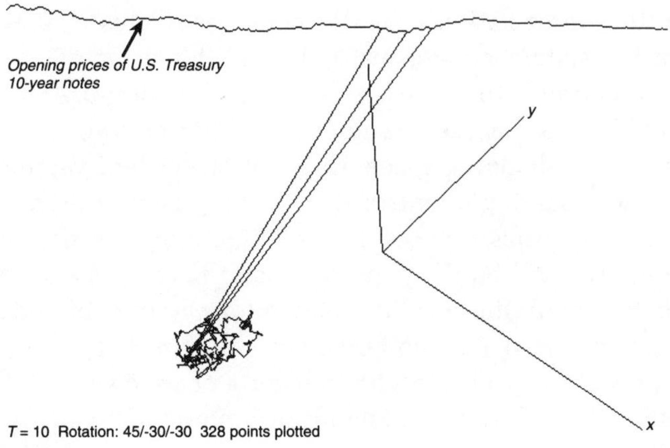

FIGURE 16.1

Phase space embedding of Treasury note data

alized. However, the theorem is valid for any number of dimensions. The phase space construction is performed by taking any value in the time series and calling it the x value. Usually, one likes to start with the first value in the series, but there is no requirement to do so.

Then, a second value is selected for the y coordinate. This number is T values from the first. Strictly speaking, T is a constant of time and is independent of the rate at which the data are sampled. In any time series the sample number can be used as a proxy for time.

The final value, $2T$ samples from the first, is selected to be the z value. This x,y,z vector is plotted in the geometric space (see Figure 16.1). We then move to the next sample, $x(n+1)$, and repeat the process.

Find a Suitable Time Lag, *T*

Recent approaches are to define T based on the idea of average mutual information (Fraser and Swinney 1986). The *average mutual information*, $I(T)$, is a prescription for selecting an appropriate time delay interval (T) for construction of the time-lagged vectors that will be used to build the attractor. $I(T)$ defines how much one learns about a datum by having knowledge of another datum. The mutual information of a system is the amount of knowledge (expressed as bits) that one can derive about two datums separated by the time lag, T.

Mutual information is a measure of general dependence and is loosely related to the idea of correlation functions and autocorrelation. However, correlation functions only measure linear dependence. Thus, mutual information provides better estimates of the time lag to use for the state space reconstruction of the attractor.

Determination of the time lag, T, is important because an optimum selection of T gives best separation of neighboring trajectories within the minimum embedding space. Calculation of the Lyapunov exponents relies on solving a matrix that is comprised of descriptions of how close trajectories diverge. If the trajectories are not separated, then the matrix will be ill conditioned and may not be solvable.

If T is too small, there is little new information contained in each subsequent datum. If T is too large, $x(n)$ and $x(n + T)$ will appear to be random with respect to each other for a chaotic system. In fact, T can be somewhat arbitrary for an infinite amount of uncorrupted data. The quality of the state space portrait depends on T, so it is desirable to select some value that is reasonable. One older method that is often used is to select the point where the autocorrelation of $x(n)$ and $x(n + T)$ first becomes zero.

The first local minimum of (T) determines an optimum value of T. For the exemplar data, $T_{opt} = 10$ (see Figure 16.2). For a chaotic system, as T increases past this point, ambiguities in the correlation between $x(n)$ and $x(n + T)$ arise—they start to appear to be random with respect to each other. The state space portrait begins to lose resolution and the fractal nature of the attractor starts to become blurred. Information is being lost. $I(T)$ for a truly random signal is zero for all time lags, except 0. Thus, mutual information provides one rationale of why chaotic signals appear to be noise.

One might also postulate that T_{opt} (10 days) is also an optimum trading horizon for these securities. For shorter periods of time the expected price of the security is so close to the current price that there is little information about more distant movements and hence little risk with a corresponding lack of return. Beyond this point, the future price is increasingly unrelated to the current price and the risk and expected return become random statistics if based on the current price.

Find the Minimum Embedding Dimension, *dE*

If a dynamic system can be described by n independent variables, then the full behavior of the system can be observed in an n-dimen-

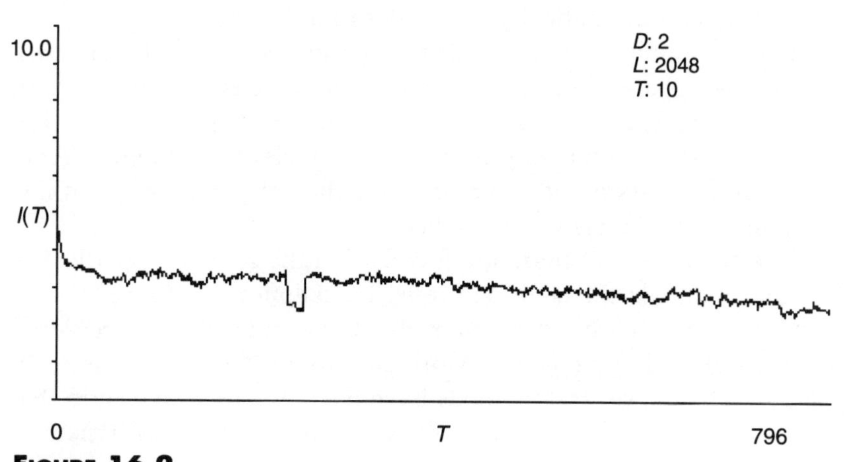

FIGURE 16.2
Average mutual information for Treasury notes

sional "state" space. However, the attractor of the system may be contained in a subset of the state space with dimension dA (sometimes called the fractal dimension), and may be described in a state space, d, that is much smaller than n. This minimum embedding dimension dE is, at most, the first integer greater than $2dA$; it may be less.

The dimension of the underlying dynamics, d, determines how many Lyapunov exponents are useful. Determination of the minimum embedding dimension, dE, is of practical interest because the computation burden rises dramatically as dimension increases. Further, noise fills all dimensions, so computations carried out in a higher-than-necessary dimension will be corrupted by noise. If dE is too small, the trajectory may cross itself, and neighbors at a point in this area may be indistinguishable in the lower dimension. Generally, by making $dE > 2dA$, self-intersections can be avoided (Mañé 1981; Takens 1981).

A new method of determining minimum embedding dimension has been developed (Kennel et al. 1992) that relies on the fact that as dimension is increased, attractors "unfold." Points on trajectories that appear close in dimension d may move to a distant region of the attractor in dimension $d + 1$. These are "false" neighbors in d.

The technique measures the percentage of false neighbors as d increases. Trajectories that are close in d are tallied, and the number of these trajectories that become widely separated in $d + 1$ are calculated. For a noiseless signal, the number of false neighbors becomes zero when the minimum embedding dimension dE is reached. A noisy signal drops off dramatically, but it does not become zero.

For the exemplar data, the first surprise is shown in Figure 16.3. These data apparently have an embedding dimension of 5! Further, the false neighbor count is zero for the higher dimensions. The implication is that if one were to develop a model to simulate the behavior of the market and to predict its behavior, a maximum of five independent variables would suffice.

The caveat is that there are few data in the set and regardless of the analysis tools selected, one must be skeptical of the results. In technical terms, there is discussion about how much data are required to ensure that all regions of the attractor are visited. As discussed in the next section, several methods have been proposed to ensure that for sparsely sampled data the results are caused by the true dynamics and not by insufficient data.

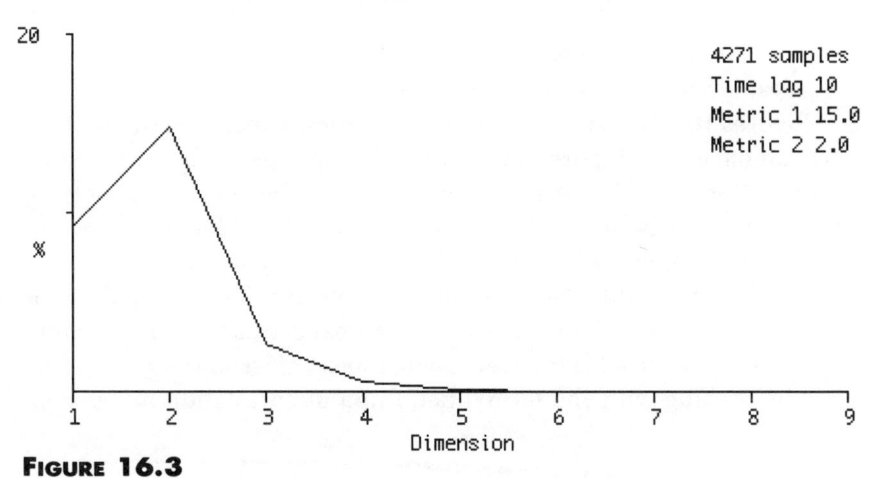

FIGURE 16.3
Embedding dimension results for Treasury notes

Embed the Time-Lagged Variables in the State Space

Once T and dE are determined, the time-lagged vectors are plotted in the state space to form the portrait. If the dimension is low enough, computer-generated graphics may be used to display the portrait. The demonstration software on the CD-ROM that can be ordered with this book has the system needed to display these data.

Compute the Fractal Dimension of the Attractor, *da*

The fractal dimension of the attractor (Hausdorff 1918), da, provides information on how much of the state space is filled by the system. One interpretation of da is that it measures how many degrees of freedom are significant. Another interpretation of the fractal dimension is that it provides a measure of how an object's bulk scales with its size: bulk = sizeda. Bulk that can be associated with volume and size is then interpreted as Euclidean distance. A plane, for example, has a dimension of 2 because the area = $d2$.

The fractal dimension of the attractor, da, may be estimated using Ruelle's approach by calculating the number of spheres or boxes, $N(r)$, of size r that capture all points as r approaches zero.

Grassberger and Procaccia (1983) defined a relatively easy approximation that may be done on a PC for high signal-to-noise ratio (SNR) signals. One major issue is the sensitivity of these calculations to signal SNR. The amount of data required to do the calculations may dramatically increase as SNR decreases.

Figures 16.4 and 16.5 show the results for the example time series. The top curves in Figure 16.4 are the histograms of the raw counts as a function of radius; the bottom curves are the slope of the log-log representation of the histogram. Usually, the actual log-log representation of the top curves is shown, but experience has shown the representation in Figure 16.4 to be more useful in selecting a radius for the final estimate. Figure 16.5 shows the final estimate of *da* for these data. The radius used is the well-defined peak close to the zero radius axis, in keeping with the formal definition of correlation dimension.

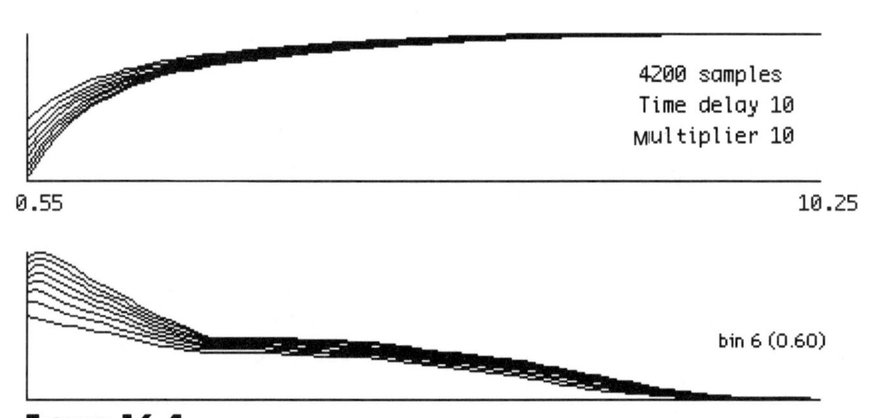

FIGURE 16.4
Histogram counts and local slope of log-log plot

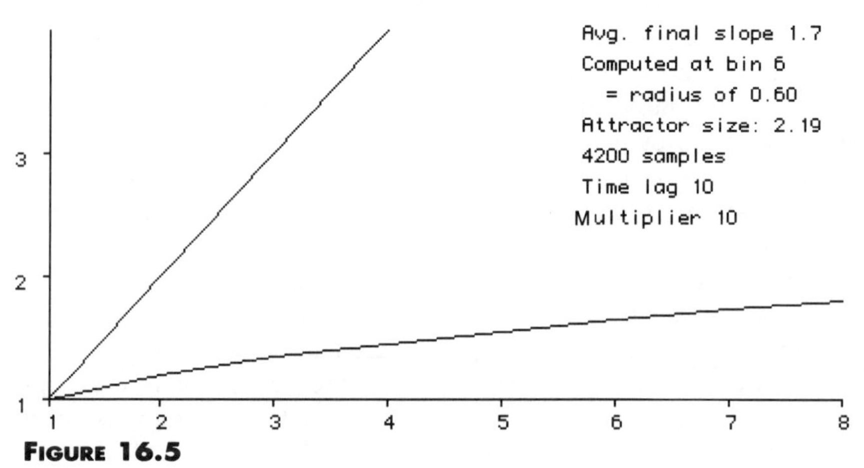

FIGURE 16.5
Final estimate of da *for Treasury notes*

The final estimate, $da = 1.7$, is the average of the last three dimensions plotted. The precise estimate is not important and is probably corrupted by the undersampling. The important point is that the results of these estimates are consistent with the embedding dimension (i.e., it's a good omen that $dA < dE$). The other interesting fact is that the correlation dimension is a noninteger value—a hallmark of chaos.

Compute the Global Lyapunov Exponents

The Lyapunov exponents describe the rate at which close points in the state space diverge. There is one exponent for each dimension. If one or more Lyapunov exponents is positive, the system is chaotic (Eckmann and Ruelle 1985). The Lyapunov exponents are invariant with respect to initial conditions. Therefore, they are another way of classifying a chaotic system.

All the Lyapunov exponents may be calculated from the Jacobian of the map by the QR decomposition technique discussed by Eckmann et al. (1986). The Lyapunov exponents are a measure of how quickly the trajectories of very close points in state space diverge. If the Lyapunov exponents are all zero or negative, the trajectories do not diverge and the system is stable. If one or more Lyapunov exponents is positive, the trajectories of near-neighbors diverge and the system is unstable, a requirement for chaos. Equally important is the relationship of the Lyapunov exponents and an ability to predict the behavior of a system. The more exponents one can correctly find, the better predictions will be (Abarbanel et al. 1989, 1992).

Compute the Local Lyapunov Exponents

Lyapunov exponents are a global invariant because they describe the effect of infinitesimal perturbations over infinite time (Abarbanel et al. 1991). "Local" Lyapunov exponents measure the divergence of trajectories in different regions of state space. The issue addressed by local Lyapunov exponents is predictability.

Figure 16.6 shows the Lyapunov exponent calculations when an embedding dimension of 6 is assumed and a local dimension of 4 is used. Two of the exponents are positive and two are negative. The positive exponents are one more indication of chaos. The fact that the forward and reverse exponents are identical (after we apply a sign

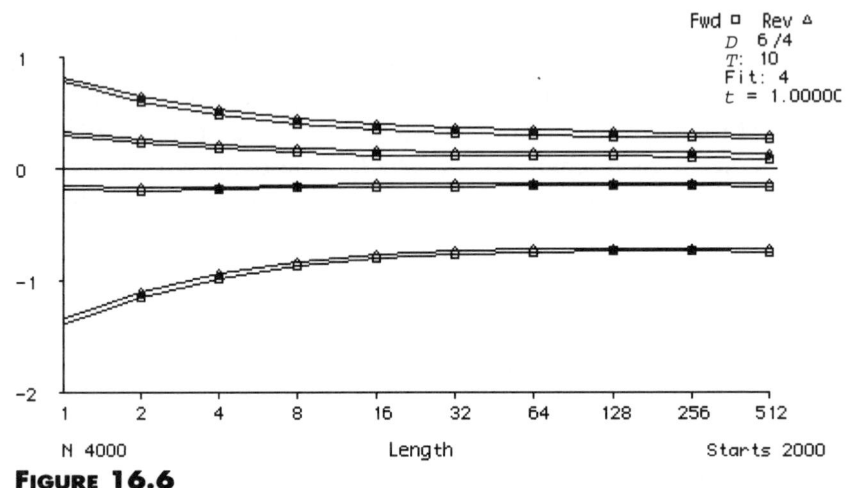

FIGURE 16.6
Lyapunov exponents for Treasury notes

change to the reverse exponents) for all four dimensions shows that our earlier calculations of embedding dimension were probably a function of the dynamics.

Using the above nonlinear dynamic data analysis methods, we can characterize U.S. Treasury 10-year notes as having the following characteristics:

Embedding dimension (degrees of freedom)	4
Optimum time delay	10 days
Correlation dimension	1.7
Lyapunov spectrum	2 positive exponents; largest is about 0.33%/day

What, then, are the prospects for predicting this time series? Figure 16.1 showed the reconstructed attractor for these data. Notice that it stays close to the line that defines x = y = z. In short, this instrument is not very volatile; price movements occur slowly. Closer inspection of Figure 16.1 shows that the points cluster in small "nodules" along this line. The system stays in these nodules for short periods of time before departing for the next nodule.

We have shown that the data of interest may be characterized using the methods of chaotic signal processing. But, given the low predictability, could these simply be random data, or random data riding some periodic wave? The answer is no. Three of the calculations tell us that these are not random data.

First, if the data were random, the average mutual information curve would lie close to the axis (indicating no predictability—just what is needed for random processes). In fact, many signals of great interest have less average mutual information than these data. Second, for noisy data, false neighbors continue to manifest themselves in higher dimensions. Finally, flawed as they are, the correlation dimension estimates behave like a dynamic system and not like noise.

The cautionary note is that even though the U.S. Treasury 10-year notes have been characterized with chaotic tools as being of low dimension, the limits on predictability seem severe and the lack of data should also provide some skepticism. The tools may be more useful in understanding the nature of the data and in model validation than they are in prediction.

Data Processing Guidelines

This section provides a practical guide for the processing and chaotic analysis of data using a system developed by Randle Inc. known as chaotic signal processing. A demonstration version of CSP can be found on the available CD-ROM, along with some interesting data. The procedural operations are discussed in the paragraphs that follow.

View the signal trace, FFT, or polyspectra. If these transforms are uninteresting, there is no reason to proceed with the more computation-intensive methods of chaos. The issues here are whether the data have a broadband spectra and are exponentially decreasing, and whether there are discernible patterns in the data trace. A exponentially decreasing pattern is commonly called $1/f$ spectra, because the Fourier components decrease as though divided by the frequency.

Make a preliminary estimate of the time delay, T. This is a chicken and egg problem because the computed optimum time delay changes with each assumed dimension (in general, for a chaotic system T_{opt} increases with increasing d). A reasonable estimate is needed to form the state vectors in each dimension. Experience has shown that a reasonable estimate of time delay in two dimensions is sufficient for determining the minimum embedding dimension, dE, using the method of false nearest neighbors in any dimension. The optimum value for the time delay is determined from the first local minima of the average mutual information for the signal. Precise estimates of the numerical value of $I(T)$ are not needed, because we are interested only in the first local minima of the time delay (T_{opt}). The average mutual information is used to find this point. This fast algorithm only works in two dimensions, but this is generally sufficient. The full algorithm may be used with no penalty other than computation time. If the accurate algorithm is used, T_{opt} for two dimensions provides a time delay that can be used to directly view the attractor. Because both algorithms provide a sample-size-constrained lower bound of $I(T)$, use the largest sample size possible and the fast algorithm. If precise numerical results are of interest, as when examining $I(T)$ between disparate data sets for leading indicators, then use the full algorithm.

It is worthwhile to take a look at the attractor in three dimensions. Visualization may provide clues to the nature of the system and/or defects. Signal collection poses many potential problems, and some of these are strikingly apparent in a visualization, especially if animated. Clipping and undersampling are an occupational hazard in signal processing, and these defects are immediately noticeable. If the signal has a dimension much over 3.5, features begin to become indistinct as a consequence of the embedding theorem (or perhaps in deference to it as a fundamental law).

Use the estimate of time delay to compute the minimum embedding dimension (dE) using false nearest neighbors. One can presuppose that any real signal of interest has an embedding dimension of at least 2. However, the computations for dimension 1 are reasonably fast and provide a baseline estimate. Estimating the upper limit is less clear. This is why we are doing the calculation.

False nearest-neighbor testing is robust and works well with remarkably few data—a necessity in financial applications. As an operational matter, we have found that 30,000 samples almost always yield the same estimate of embedding dimension as when much more data are used. A prescription for a lower limit on the amount of data has not been explored, but even 10,000 data should yield a reasonable estimate.

If insufficient data is a concern, then the method of Kennel and Isabelle (1992) may shed some light on whether the calculated *dE* is the undesirable artifact of a sparse attractor or the desired effect of system dynamics. This technique applies false nearest-neighbor testing to known random data with the same distribution and same number of samples. If the false neighbors in any dimension for the random number set are similar to the real data, then the results can be presumed to result from undersampling.

The correlation dimension, dA, comes next as a proxy for the fractal for Hausdorff dimension. Even with the comparative efficiencies of the Grassberger-Procaccia algorithm, these are time-consuming calculations. The computation penalty is significant in higher dimensions. In general, one would like to compute the correlation integral out to dimension 10, but $dE + 1$ may do in the interest of time. Computing the higher dimensions does provide a more satisfying final curve, so dimension 10 is a good target.

A practical limit is imposed by the amount of data available and their contamination. Calculations in higher dimensions require more data, on the order of 10 $d/2$, where d is the dimension of interest. Thus, for no-noise data in dimension 10, 100,000 samples are required.

A more practical matter is how well data populate the attractor and the resolution of the computations. The issue here is representation of numbers taken from a sensor and the number of samples. Sensors and analog-to-digital converters deliver a variety of numeric formats and ranges. Typically, an analog-to-digital converter may provide an integer number that represents some scaled and translated range of analog measurements at the sensor. The analog floating point value is now an integer. Although there is no reason that integer-valued floating point numbers cannot be processed, more signal-processing systems convert the integer numbers back to floating points and scale the data to another, perhaps original, range. Thus, a sensor output of 5.3333 mV, for example, may become 187 leaving the converter and revert back to 5.3300 in the signal processor.

For correlation dimension calculations, the histogram of distance between vectors is computed as:

$$C[r] = 1 \text{ if } | xi2 + xi2 | \leq r$$

Because r is used as an index, it must be an integer. There is an ambiguity associated with each vector in the resolution of the calculation. For example, 5.3300 may be indistinguishable from 5.6000. If the range of floating point values is small, the resolution is poor. The fine structure at small values of r is lost. However, the definition of the correlation integral requires good resolution for small radii.

The first application method is to use a multiplier to improve the resolution of close vectors. Scaling and translation of numeric values only moves the attractor and changes its size. The dynamics and invariant geometric properties are preserved. So, multiplying the floating point number by 10 or 100 increases the resolution of fine structure. A larger histogram is required, but there is no computation penalty because all radii must be computed anyway. The magnitude of the multiplier depends on the characteristics of the numbers. Using a multiplier of 10 on the example of 5.3300 increases the resolution by an order of magnitude. The difference between 5.33 and 5.60 becomes apparent. The first obvious consideration is, how far can we take this? For this example, a multiplier of 100 provides all the improvement possible.

However, there is another problem. The increase in resolution comes at the expense of population density at each radius, and more samples must be used if possible. Even so, the fluctuations between histograms can be large. Because the calculation depends on local slopes at small radii, these slopes can vary widely. One radius may have a sufficient count, while the next is half the count because insufficient data were available. It then becomes difficult to determine which radius to use for the final estimate. This effect can be mitigated by smoothing over a moving average. Experience shows that for moderately rough curves, a moving average over 10 points is adequate. Again, this is a situational problem. The trick is to use enough smoothing to mitigate uneven histograms but not so much as to completely corrupt the final estimate.

Fortunately, this compute-intensive analysis is of little operational value. After it is performed for research and characterization, it probably will not be used again.

The final relevant calculations are Lyapunov exponents. Global exponents characterize long-term behavior over the entire attractor. Local exponents characterize behavior as a function of length (time) and location (Figure 16.6). Normally, one looks at an aggregate of time periods over many locations to get the best estimate of the global exponents.

Experience has shown that the global estimate can usually be achieved within 1,000 samples, assuming the data are not grossly oversampled. We normally use a length of 2,048 (211) samples for the first try. If the curves are asymptotic, there is no need to look further. Note that the number of points used to characterize the attractor is much higher. Experience has also shown that 300 to 500 neighbors are sufficient to develop an accurate map. The "group factor" is used in the indexing of near points in state space. A value of 25 should suffice.

Calculation of the time-reversed Lyapunov exponents provides confirmation that the estimate of dE is correct and flushes the false exponents caused by using too high an embedding dimension (Parlitz 1992). These spurious exponents are caused by the fact that Taken's theorem states that the embedding dimension may be as high as $2dA + 1$, which possibly could lead to $dA + 1$ extraneous dimensions. Because the attractor is sparely populated (due to numerical precision or corruption) in the extraneous dimensions, false exponents are computed.

References

Abarbanel, H. D. I. 1989. "Prediction in Chaotic Nonlinear Systems: Time Series Analysis for Nonperiodic Evolution." In *Lectures at the NATO Advanced Research Workshop on Model Ecosystems and Their Changes* (Matrea, Italy). INLS preprint 1020.

Abarbanel, H. D. I. 1990. "Determining the Lyapunov Spectrum of a Dynamical System from Observed Data." Presented at *SIAM Conference on Dynamical Systems* (Orlando, FL).

Abarbanel, H. D. I., R. Brown, and M. Kennel. 1991. "Variation of Lyapunov Exponents on a Strange Attractor." *Journal of Nonlinear Science* 1.

Abarbanel, H. D. I., R. Brown, and M. Kennel. 1992. "Local Lyapunov Exponents Computed from Observed Data." *Journal of Nonlinear Science* (Sept.).

Eckmann, J. P. and D. Ruelle. 1985. "Ergodic Theory of Chaotic and Strange Attractors." *Reviews in Modern Physics* 57(3).

Eckmann, J. P., S. O. Kamphorst, D. Ruelle, and S. Ciliberto. 1986. *Physics Review A* 34.

Fraser, A. M., and H. L. Swinney. 1986. "Independent Coordinates for Strange Attractors from Mutual Information." *Physics Review A* 33 (Feb.).

Grassberger, P., and I. Procaccia. 1983. "Measuring the Strangeness of Strange Attractors." *Physica* 9D.

Hausdorff, F. 1918. "Dimension and Ausseres Mass." *Math Annalen* 79.

Kennel, M. B., and S. Isabelle. 1992. "Method to Distinguish Possible Chaos from Colored Noise and to Determine Embedding Parameters." *Physics Review A* 46 (Sept.)

Kennel, M. B., R. Brown, and H. D. I. Abarbanel. 1992. "Determining Embedding Construction." *Physics Review A* 45 (Mar.).

Mañé, R. 1981. In *Lecture Notes in Mathematics*. D. Rand and L. S. Young, eds. Berlin: Springer-Verlag.

Parlitz, U. 1992. "Identification of True and Spurious Lyapunov Exponents from Time Series." *International Journal of Bifurcation and Chaos* 2(1).

Takens, F. 1981. In *Lecture Notes in Mathematics*. D. Rand and L. S. Young, eds. Berlin: Springer-Verlag.

Note

1. Fourier analysis is inherently a one-dimensional tool; the dimensionality of many financial and economic processes is much higher. The waviest approach to market analysis is essentially Fourier analysis. Stating that there is a seven-year wave is the same as saying there is a predominant Fourier component with a wavelength of seven years.

CHAPTER 17

Nonlinear Dimensions of Foreign Exchange, Stock, and Bond Markets

Mark Embrechts, Masud Cader, and Guido J. Deboeck

Why do investors insist on the existence of cycles? Because periodicity is the most complicated orderly behavior they can imagine. When they see a complicated pattern of prices, they look for some periodicity wrapped in a little random noise.

James Gleick
Chaos: Making a New Science, 1987

Several authors have applied nonlinear dynamic techniques to financial markets. This chapter illustrates the concepts and techniques discussed in Chapters 15 and 16 and complements the analyses already available. The focus here is on foreign exchange, stock, and bond markets. The empirical results in this chapter underscore the importance of implementing intelligence-gathering techniques before engaging in the design of trading systems. They illustrate how short-term and long-term trading strategies can be formulated. Nonlinear dynamic analyses are important for determining feasibility as well as parameters for the design of trading systems. These techniques also have important consequences for the design of benchmarks as well as strategies for risk and portfolio management.

Introduction

Chaos analysis is based on the evaluation of the correlation dimension as a function of the embedding dimension and indicates whether a time series is random, chaotic, or correlated (Brock et al. 1991):

1. The time series is *random* when the correlation dimension remains roughly equal to the embedding dimension. There is no saturation of the correlation dimension with increasing embeddings. The returns follow a random walk process. In this case, neither short-term nor long-term predictions of the time series are possible.
2. The time series is *chaotic* when the correlation dimension clearly saturates with increasing embedding dimensions of the phase space. The phenomenon of chaos inherently implies that no long-term prediction is possible. However, it is possible to make short-term predictions.
3. The time series is *correlated* when the correlation dimension remains well below the value of the embedding dimension and keeps increasing with increasing embeddings without ever saturating. In that case it should be possible to make long-term predictions.

The purpose of a chaotic analysis is to indicate whether strict predictions of the time series are possible, whether these predictions will have a short-term or a long-term character, and whether forecasting will be easy or difficult. The purpose of a chaotic analysis is not to make these predictions but to help formulate prediction strategies that can be backed by fundamental analysis. Actual methodologies for predicting time series are not addressed; Part One of this book showed how to implement forecasting strategies using neural networks.

The results of chaos analysis are fundamentally different from rescaled range analysis. Whereas chaos analysis addresses the issue of strict predictability of a time series, rescaled range analysis addresses the issue of formulating long-term or short-term betting strategies with favorable statistical odds.

Rescaled range analysis reveals whether the data follow essentially a random walk or a biased random walk. In cases where data follow a random walk ($H = 0.5$), no successful trading strategies can be formulated. In cases where data follow a biased random walk, the

Hurst coefficient will be different from 0.5. If the Hurst coefficient is greater than 0.5 (e.g., $H = 0.6$), the data are enforcing the past trend or show *persistence*. Peters (1989) indicates that a Hurst coefficient of 0.6 means that there is a 60% probability that the data will continue in the ongoing trend. If the Hurst coefficient is less than 0.5, the data are essentially trend-reversing. A Hurst coefficient of 0.4 means that there is a 60% chance for trend reversal. This phenomenon is called *antipersistence*.

A Hurst analysis might result in one type of behavior in the short term and in a different type of behavior in the long term. A break in a Hurst plot indicates the cycles in the data. For example, if a Hurst analysis indicates a two-week nonperiodic cycle, this means that there is an average biweekly cycle, which may be 12 days during one run and 16 days the next time around. A Hurst analysis is superior to traditional techniques such as Fourier analysis for studying the cyclical behavior of a time series. Fourier analysis is very powerful when the cycles have a fixed length, but it fails to detect cycles with an average duration. The results from a Hurst analysis can be used to develop statistical betting strategies that tell to go with the trend or against the trend, as well as the time scale for development of these strategies. The revelation of average cycles in a Hurst analysis is also relevant for determining how many past data from a time series one has to consider for predicting the next point. While the results from a Hurst analysis are often consistent with the results of a chaos analysis, both paradigms apply in very different ways to trading: *chaos analysis addresses whether a time series of data is strictly predictable; Hurst analysis determines whether betting strategies with favorable statistical odds can be formulated.*

The remainder of this chapter illustrates chaos analysis and rescaled range analysis for:

- U.S. Federal Funds
- U.S. Treasury notes
- Swiss franc/U.S. dollar exchange rates
- The Japanese stock market (TOPIX)

Antipersistent Trend of U.S. Fed Fund Returns

The analysis in this section focuses on U.S. Fed Fund rates during two periods: from January 1975 to December 1991, and from August 1988 to December 1991.

U.S. Fed Funds (January 1975 to December 1991)

Figure 17.1 summarizes the results of chaos analysis and rescaled range analysis of U.S. daily Fed Fund rates during the period from January 1975 to December 1991. For these analyses 4,100 daily returns were calculated, where the return is defined as a scaling factor times the log(today's value/yesterday's value). Using returns rather than the raw Fed Fund data achieves three objectives:

1. Elimination of correlations inherently present in any random walk
2. Detrending
3. Reduction of the effects of nonstationarity

Figure 17.1(a) represents the daily Fed Fund rates; Figure 17.1(b) the daily returns. The distribution functions of the raw Fed Fund data and the Fed Fund returns are shown in Figures 17.1(c) and (d), respectively. Note that the distribution of the daily returns has a symmetrical shape and resembles a high peaked Gaussian distribution with fat tails. The results of the chaos analysis (Figure 17.1e) indicates a saturating tendency for the correlation dimension, leading to a fractal dimension of about 6. The embedding dimension (i.e., the dimension of the phase space for which saturation in the correlation dimension occurs) is 10.

These results of chaos analysis provide convincing evidence that the Fed Funds during 1975 to 1991 are actually chaotic and should therefore be strictly predictable in the short term. The embedding dimension of 10 indicates that 9 data must be shown to a neural net to predict the 10th data point of the time series. While short-term predictions are in principle possible for a chaotic time series, such predictions might prove difficult in practice because of the high value of the correlation dimension. Such a relatively high value for the correlation dimension requires more than 4,100 data points for accurate predictions.

Predicting chaotic time series might further be handicapped because:

1. Financial data are always corrupted by noise. While several authors suggest noise-filtering techniques, the process of filtering itself introduces parasitic correlations that might lead

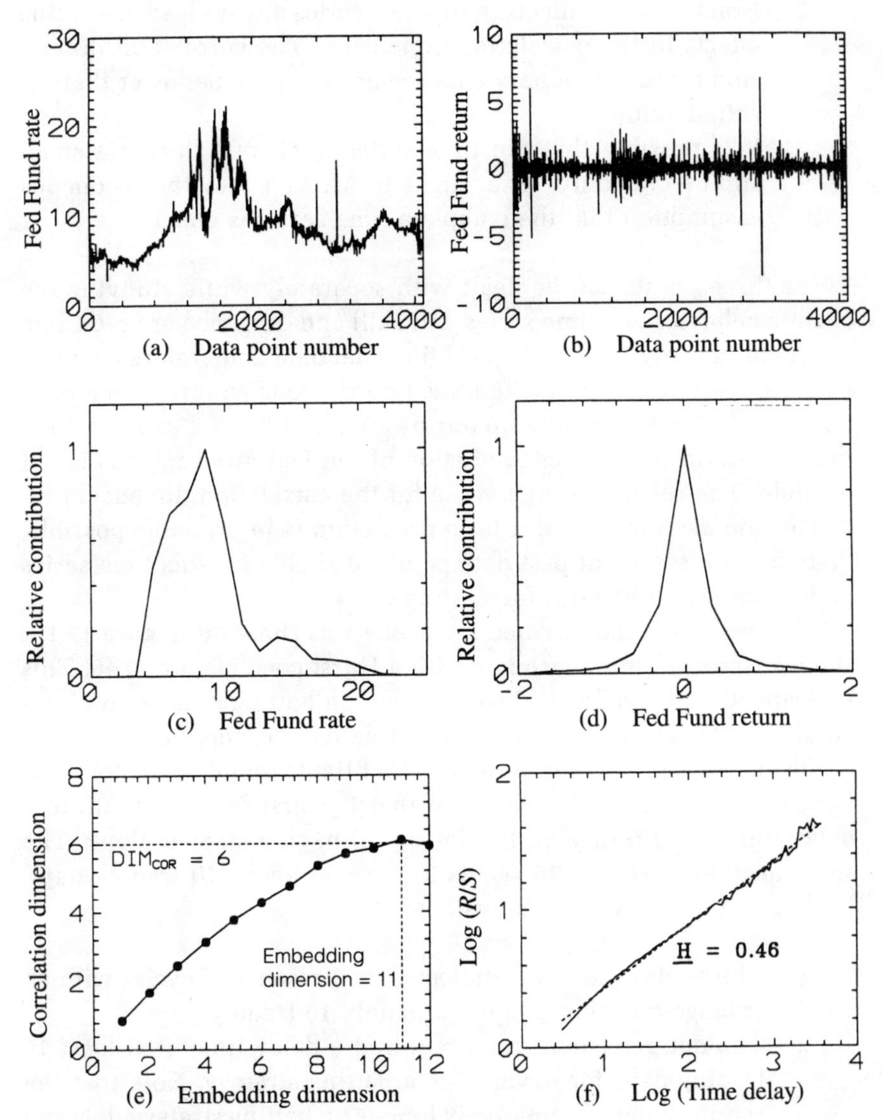

FIGURE 17.1

Results of chaos analysis and rescaled range analysis of U.S. daily Fed Fund rates from January 1975 to December 1991. (a) Daily Fed Fund rates. (b) Daily returns. (c) Distribution of raw data. (d) Distribution of daily returns. (e) Correlation dimension. (f) Hurst analysis to a seemingly low-order chaotic series, where no chaos would normally be present.

2. Nonstationary effects in the time series always lead to aliasing effects in the correlation dimension. The correlation dimension for nonstationary data seems always to be lower than its actual value.
3. It is possible that the time series is chaotic during a small subinterval only. One might be tricked into the erroneous assumption that the complete time series is chaotic.

These three points can be dealt with separately while studying the chaotic behavior of a time series, but skill and diligence are required.

A chaos analysis of real world financial data is therefore as much an art as a science. Because there is rather strong empirical evidence for chaos in the Fed Fund rate during 1975 to 1991, we can conclude that long-term time-series prediction of the Fed Fund returns is not possible. The relatively high value for the correlation dimension indicates the even when short-term prediction is in principle possible, there are not sufficient past data points available for this time series to do accurate short-term forecasting.

The results of the rescaled range analysis shown in Figure 17.1(f) show a trend of antipersistence with a Hurst coefficient of 0.46. This indicates that, by going against the trend, a betting strategy with favorable odds can be developed. The relative closeness of the Hurst coefficient to 0.5 indicates, however, that the favorable odds for coming out ahead are small. Note also that the first few points up to a delay time of 10 (unity on the log scale) have a steeper slope. The slope leading to $H = 0.46$ seems to cross at about 10 trading days, leading to the following conclusions:

1. The 10-day crossover indicates the presence of cycles with an average duration of approximately 10 trading days.
2. The antipersistent trend required a delay time of at least 10 trading days for developing a betting strategy. Note that the requirement of a relatively long-term betting strategy does not contradict the fact that no long-term predictions can be made. The betting strategy leads to favorable odds, but this is different from making long-term predictions.
3. To exploit the average cycle of 10 days, one could train a neural net with 10 input data to predict the next point. This

conclusion about how to proceed with training a neural net is actually consistent with the results from chaos analysis.

U.S. Fed Funds (August 1988 to December 1991)

The same analysis as in the previous section was repeated with Federal Fund data for the period from August 1988 to December 1991. The results of chaos analysis and rescaled range analysis are summarized in Figure 17.2. The chaos analysis (Figure 17.2c) does not show a saturating tendency for the correlation dimension with increasing embeddings. The fact that the correlation dimension remains well below the values of the dimension of the embedding space leads to the following conclusions:

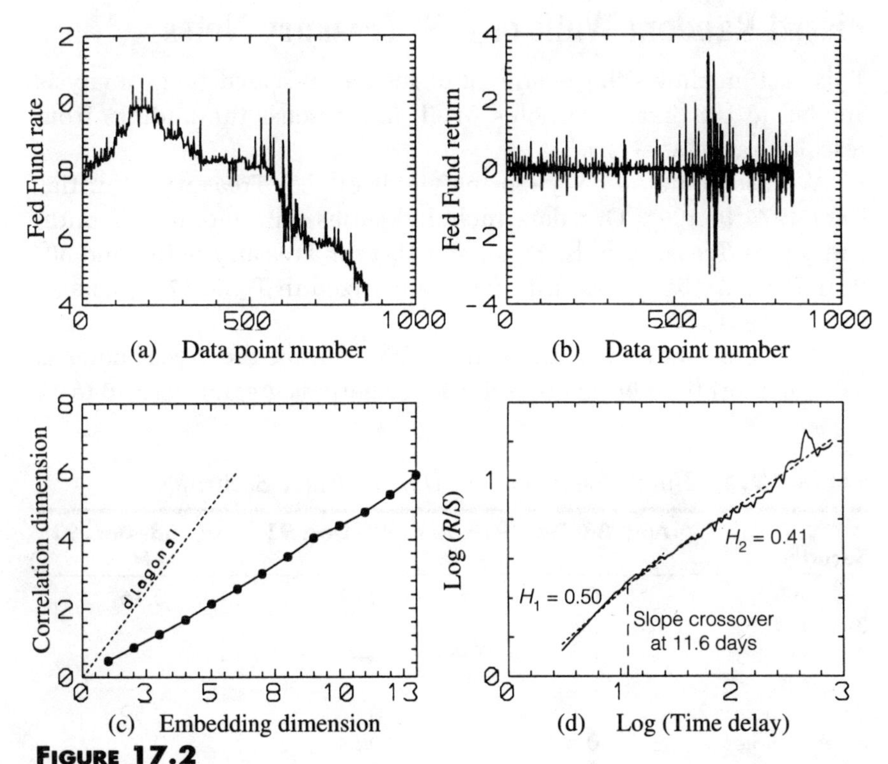

FIGURE 17.2

Results of chaos analysis and rescaled range analysis of U.S. daily Fed Fund rates from August 1988 to December 1991. (a) Daily Fed Fund rates. (b) Daily returns. (c) Correlation dimension. (d) Hurst analysis

- The time series is not chaotic.
- There is a certain correlation between the data of this time series. This correlation can in principle be exploited for long-term forecasting.

The Hurst analysis (Figure 17.2d) leads to conclusions similar to the analysis of the time series for the Fed Funds during 1975 to 1991. There is a long-term antipersistent trend. The short-term behavior is more or less a random walk ($H = 0.5$). The long-term slope crosses the short-term slope at 12 trading days, indicating an average cyclic length of 12 trading days. The relatively low value of the Hurst coefficient for larger time delays ($H = 0.41$) indicates that there is a relatively large likelihood that a trend is going to reverse itself.

Biased Random Walk of U.S. Treasury Notes

This section shows the results of chaos and rescaled range analysis applied to U.S. Treasury notes. We'll first discuss the findings from rescaled range analysis.

We used data on daily closing yields of U.S. Treasury securities from 1975 to 1991. Our data included series on 3-month, 6-month, and 1-year Treasury bills; 2-, 5-, and 10-year Treasury notes; and 30-year Treasury bonds. Results are summarized in Table 17.1 (see also Larrain 1991).

The table shows that the Hurst coefficient for the 2-year notes is 0.63 for short time horizons and 0.54 for periods greater than 10 to 12

TABLE 17.1 *Hurst Coefficients of U.S. Treasury Securities*

Security	Aug. 88–Dec. 91 H_1^a	Aug. 88–Dec. 91 H_2^a	Aug. 88–Dec. 91 H
Fed Funds	0.50	0.41	0.46
3-month T-bills	0.60	—	—
6-month T-bills	0.59	—	—
1-year T-bills	0.59	—	—
2-year T-notes	0.63	0.54	0.59
5-year T-notes	0.63	0.54	0.59
10-year T-notes	0.61	0.54	0.56
30-year T-bonds	0.54	—	—

(a) H_1, for approximately 10 to 12 trading days; H_2, for more than 10 to 12 trading days

days ($x = 2.48$). This suggests that there are persistent trends in the daily yield changes of 2-year notes, meaning that if the change is positive there is an approximately 63% chance that yield changes will remain positive within an average 12-day cycle.

The Hurst coefficients of all other Treasury securities (except 30-year bonds) range between 0.59 and 0.63 for short time horizons and 0.54 for horizons of more than 10 trading days. Thus, for all securities (except 30-year bonds) there are strong persistent trends in the daily yield changes. After 10 to 12 trading days, the Hurst coefficients for all Treasury securities show a tendency to Brownian motion (i.e., a random walk).

The Hurst coefficient of 30-year bonds for short as well as long time horizons is 0.54, which means that the movement of daily yield changes, even in the short term, is close to Brownian motion.

In sum, these empirical findings show that:

1. Daily yield changes of U.S. Treasuries do not follow a Gaussian distribution
2. Daily yield changes of U.S. Treasury notes follow a biased random walk
3. Two- to 10-year notes in particular have persistent trends
4. Daily yield changes of U.S. Treasury bonds are close to Brownian motion or a random walk
5. The correlations and covariances between yield changes are significant and thus should not be ignored in risk management of a portfolio of U.S. Treasury securities. This last point will be further elaborated in Chapter 18.

Predictability of Treasury Yields

In Chapter 16 the Lyapunov exponents were described as the rate at which close points in the state space diverge. They are a measure of how quickly the trajectories of very close points in state space diverge, and thus provide an indication of the predictability of a time series.

Lyapunov exponents for U.S. Treasuries were calculated assuming embedding dimensions of 2 and 6, and letting the evolution time vary from 7 to close to 400 days. All computations showed decreasing values for Lyapunov exponents, but even the smallest values remained positive—a further indication of the possibility of chaos and the predictability of the Treasury yields.

To estimate predictability, we then calculated the slope of a regression line drawn through the lowest values (e.g., evolution times greater than, say, 100) of the Lyapunov figures. For example, for 2-year notes we estimated the slope of the Lyapunov exponents to be about 0.2. This can be interpreted as the loss of predictive power at a rate of 0.2 bits per day. While in a "forward look" this figure provides an indication of predictive power, in a "backward look" it means that the system loses all memory of initial conditions at the rate of 0.2 bits per day.

Thus, on average, changes in market yields are no longer related or correlated after approximately 10 to 12 days. This interpretation of Lyapunov exponents is similar to the cycle found by *R/S* analysis. In *R/S* analysis, the crossover to random walk behavior implies that the long memory effect dissipates, or daily yield changes become independent. Hannula (1994) demonstrates how this information can be used to design specific trading systems.

Analysis of Swiss Franc/U.S. Dollar Returns

This section reports results of chaos and rescaled range analyses for the 30-minute and 1-minute closings of the Swiss franc/U.S. dollar exchange rates from September 1991 to February 1992, and from 3 February to 7 February 1992, respectively. Figure 17.3 summarizes the results of analysis of the 30-minute closing differences. Note that in this section the first-order differences rather than the returns were analyzed. This does not alter the conclusions. Only when a time series needs detrending (e.g., for the analysis of a stock market over several decades) are returns required.

The actual daily exchange rates and their daily differences are shown in Figures 17.3(a) and (b), respectively. The chaos analysis (Figure 17.3c) shows a correlation dimension that equals the embedding dimension and that does not saturate with increasing embeddings. This means that the Swiss franc/U.S. dollar exchange rates follow in the first instance a random walk or some other stochastic process and that making strict predictions for this time series is impossible for the long term as well as for the short term.

The rescaled range analysis (Figure 17.3d) shows two regimes with different slopes. For short time lags (less than a day) the Hurst coefficient is equal to 0.6, which indicates persistency. This is indicative

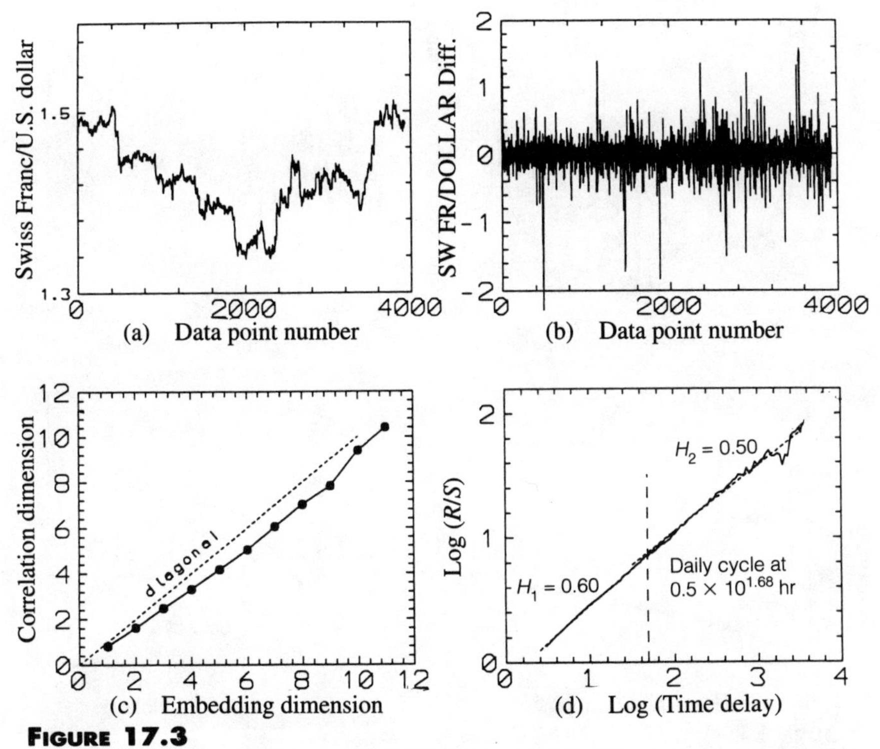

FIGURE 17.3

Results of chaos analysis and rescaled range analysis of 30-minute Swiss franc/U.S. dollar returns for September 1991 to February 1992. (a) Daily exchange rates. (b) Daily rate differences. (c) Correlation dimension. (d) Hurst analysis

of a biased random walk with memory effects where the trend is reinforced. A successful betting strategy for such a regime would be to speculate with the trend. For longer time lags the Hurst coefficient is 0.5, indicating that the process is essentially a random walk for which no profitable trading strategies can be developed. This does not, of course, exclude the possibility of developing successful forecasting methodologies based on models that consider cross-correlated data as well (such as inflation rates, interest rates, etc.). The crossover in the slopes of the Hurst plot occurs at a time delay of 48 half-hour intervals, suggesting average daily cycles in this exchange rate.

The 30-minute trade volume chaos analysis and Hurst coefficient are shown in Figure 17.4. The chaos analysis (Figure 17.4c) shows that the correlation dimension does not saturate with increasing

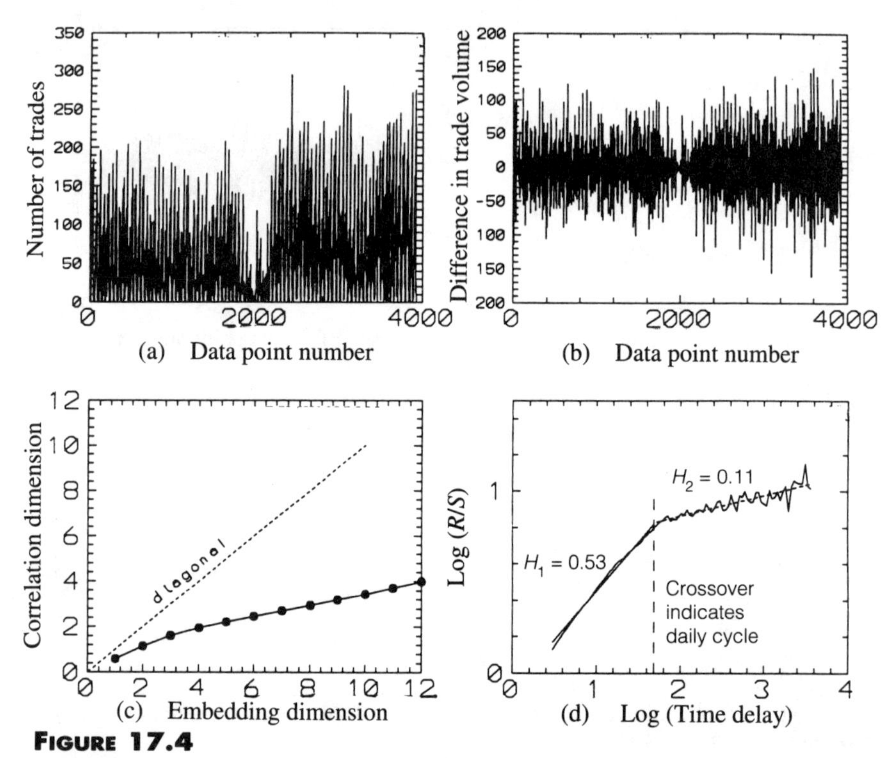

FIGURE 17.4

Results of chaos analysis and rescaled range analysis of 30-minute Swiss franc/U.S. dollar trade volume for September 1991 to February 1992. (a) Trade volume. (b) Trade volume differences. (c) Correlation dimension. (d) Hurst analysis

embeddings. The fact that the value of the correlation dimension remains well below the value of the embedding space indicates the presence of certain correlations that can be exploited for long-term predictions of the 30-minute trade volume. The Hurst analysis (Figure 17.4d) is consistent with this conclusion. While the trade volume follows essentially a random walk ($H = 0.5$) during short time lags (less than a day), longer time lags show a Hurst coefficient of 0.11, which is significantly less than 0.5 and indicates a strong antipersistent behavior. The slopes for $H = 0.5$ and $H = 0.11$ cross at a time delay corresponding to 48 half-hour intervals, indicating a daily cycle.

Figure 17.5 shows the results of chaos and rescaled range analyses of the 1-minute closing differences of the Swiss franc/U.S. dollar exchange rates. The chaos analysis (Figure 17.5c) excludes chaos and

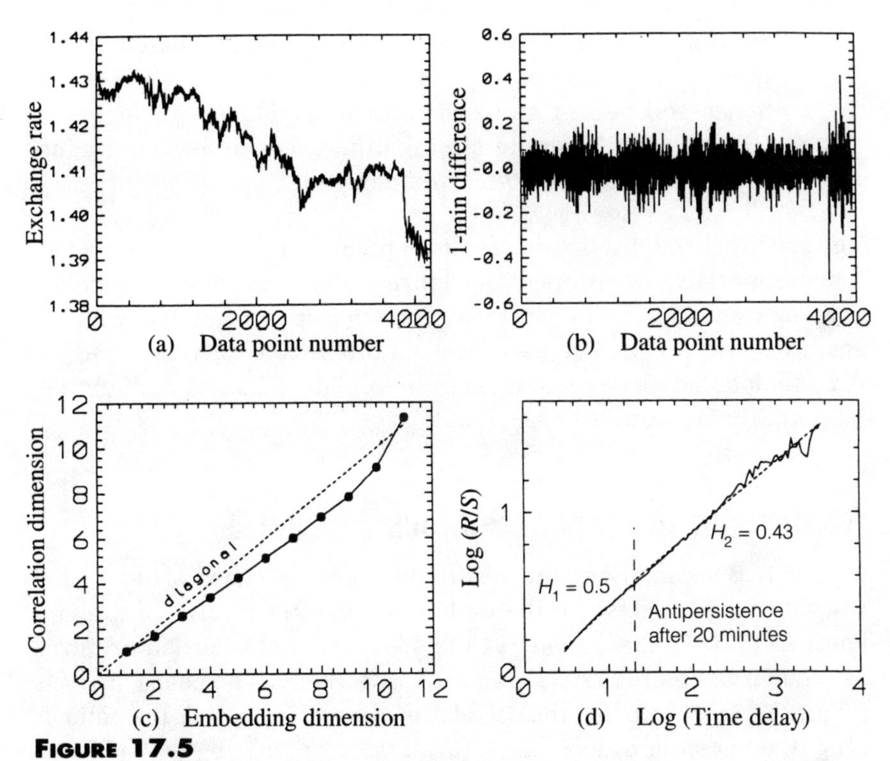

FIGURE 17.5

Results of chaos analysis and rescaled range analysis of 1-minute Swiss franc/U.S. dollar returns for 3–7 February 1992. (a) 1-minute exchange rate. (b) 1-minute difference. (c) Correlation dimension. (d) Hurst analysis

suggests that the 1-minute exchange rate differences follow essentially a random walk. The rescaled range analysis (Figure 17.5d) shows a short-term Hurst coefficient of $H = 0.5$ and a long-term Hurst coefficient of $H = 0.43$. The long-term slope crosses the short-term slope with a time delay that corresponds to 20 minutes, suggesting a 20-minute average cycle length. The antipersistent trend of the Hurst coefficient suggests that a profitable betting strategy could be formulated by speculating against the trend after a delay of at least 20 minutes.

These analyses allow the following general conclusions to be drawn regarding the Swiss franc/U.S. dollar exchange rates:

1. Long-term prediction of the trade volume is in principle feasible.

2. Successful betting strategies can be developed based on 30-minute closing difference.
3. Successful betting strategies can be developed while speculating on the 1-minute closing differences by betting against the trend based on 20-minute delays.

The seemingly contradictory results of points 1 and 2 confirm the fact that conservative investors with a longer-range view who extrapolate ongoing trends, as well as intraday traders (especially those who bet against the trend), can actually make a profit from currency trading. A more detailed discussion of chaos in foreign exchange markets can be found in DeGrauwe et al. (1993).

Analysis of the Japanese Stock Market

Figure 17.6 summarizes the results of applying chaos analysis and rescaled range analysis to the daily stock market returns of the Japanese TOPIX for the period 1949 to 1993. Daily closings and returns are shown in Figures 17.6(a) and (b), respectively. The chaos analysis (Figure 17.6c) shows that the correlation dimension equals the embedding dimension and does not saturate with increasing embeddings, suggesting that the TOPIX returns follow in first order a random walk.

The Hurst analysis (Figure 17.6d) shows two different values for the slope. For time delays of less than 10 years, the Hurst coefficient is 0.62, suggesting a positive bias or memory effect and indicating a persistent trend. For longer time delays the Hurst coefficient is 0.5, suggesting a random walk. The crossing of both slopes suggests average cycles of about 10 years. Note, however, that the crossover of both slopes is very shallow, leading to a large uncertainty in the exact value of the average cycle length. The relatively large value for the Hurst coefficient for shorter time lags shows that a successful betting strategy can be formulated by following the trend of the market.

Conclusions and Implications

The various examples of financial markets discussed in this chapter clearly indicate that chaos is the exception rather than the rule. We found high-order chaos for the U.S. Fed Funds (1975 to 1992), and low-order chaos for the yen/dollar returns during the 1970s (see Chap-

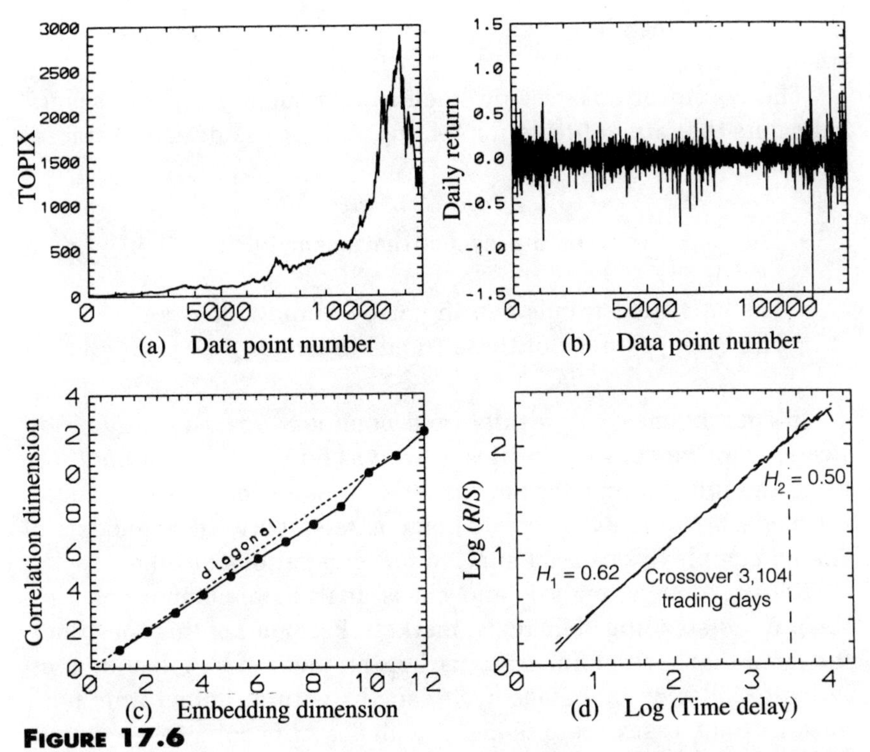

FIGURE 17.6

Results of chaos analysis and rescaled range analysis of daily Japanese TOPIX returns for 1949 to 1993. (a) Daily closings. (b) Daily returns. (c) Correlation dimension. (d) Hurst analysis

ter 15). Even when short-term prediction for a chaotic series is in principle possible, there usually are not enough data available to develop a reliable model for financial markets with high-order chaos. Chaos analysis often reveals that even when the data are not chaotic, there may be additional correlations present that can be exploited for long-term speculation (e.g., Fed Funds 1988–1992, 30-minute trade volume Swiss Franc/Dollar). Hurst analysis often reinforces these conclusions.

Rescaled range analysis shows that most financial markets follow a biased random walk. This bias indicates memory effects that can either enforce the trend or work against the trend. The R/S analysis further reveals the presence of average cycles in most financial markets. Rescaled range analysis allows successful betting strategies to be developed by indicating whether the trend should be followed or speculated against; it also shows over which time period these plays are appropriate.

The nonlinear analyses outlined in this chapter provide several indicators that are valuable for trading and for the design of trading systems, including:

- The periodicity of the cycles that seem to prevail in various securities
- The persistent trends within narrow time windows
- The disappearance of these trends in longer time windows

In consequence, the results from nonlinear analyses provide indications for modeling priorities as well as for determining the input space and time horizon for the desired output for the design of neural net models. Nonlinear analysis demonstrates why linear models of limited complexity can not survive for long periods of time.

Rescaled range analysis and chaos analysis are important for a better understanding of financial markets. Part Four of this book demonstrated that several financial markets are not random, having nonperiodic cycles with average cycle lengths within limited time horizons. In some cases there is clear evidence of chaotic regimes with "strange" attractors, which implies predictability of the market signal. In other cases, there are short- and/or long-term memory effects that allow trading with favorable odds.

Neural networks provide a reliable basis for modeling nonlinear, dynamic market signals. Nonlinear dynamics and chaos theory can provide important inputs for the design of trading systems using neural networks. These findings also have important implications for the design of benchmarks and risk management strategies. Traditional mean/variance optimizations between expected returns and absolute or relative risk measures are impractical in the case of fractal distributions of returns.

References

Brock, W., D. Hsieh, and B. LeBaron. 1991. *Nonlinear Dynamics, Chaos, and Instability: Statistical Theory and Economic Evidence*. Cambridge, MA: The MIT Press.

De Grauwe, P., H. DeWachter, and M. Embrechts. 1993. *Exchange Rate Theories: Chaotic Models of Foreign Exchange Markets*. Oxford: Blackwell.

Hannula, H. 1994. "Polarized Fractal Efficiency." *Technical Analysis of Stocks and Commodities* (Jan.).

Larrain, M. 1991. "Testing Chaos and Non-Linearities in T-Bill Rates." *Financial Analyst Journal* (Sept.-Oct.).

Peters, E. 1989. "Fractal Structure in the Capital Markets." *Financial Analyst Journal* (July-Aug.). See also *Chaos and Order in the Capital Markets: A New View of Cycles, Prices and Market Volatility*, New York: John Wiley & Sons (1991); "A Chaotic Attractor for the S&P 500," *Financial Analyst Journal* (Mar.-Apr. 1991).

PART FIVE

Risk Management and the Impact of Technology

shui di shi chuán:

Constant dripping wears away the stone; Little strokes fell great oaks. According to this Chinese idiom, even an infinitesimal force can accomplish a seemingly impossible feat with persistence.

CHAPTER 18

Risk Management Measures

Henry Green, Robert Mark, and Michael Pearson

You see, we thought, "We were invulnerable."

Former employee, Drexel Burnham Lambert
Fortune *magazine, May 1990*

The underlying structure of financial markets remains largely unknown, as evidenced by the many attempts to model these markets with deterministic, stochastic, and probabilistic methods— which have yielded limited success. Nonlinear dynamics and chaos theory, described in Part Four, still need to be effectively deployed for model building. Nevertheless, measuring and quantifying aggregate market risk is a necessity. This chapter presents a structured and robust methodology for measuring financial risk associated with trading, as well as an approach for effective risk management. The methodology outlined here is well established and currently applied in several financial institutions.

Financial Risk Defined

Trading operations are mainly concerned with price (or market) and credit risk. The primary determinants of price are movements in interest rate, foreign exchange rates, commodity risk, and equity risk. A concept known as risk measurement units (RMUs) aims to measure and quantify market risk within a consistent earnings-at-risk framework. The basic categories of risk for a financial institution are schematically diagrammed in Figure 18.1.

FIGURE 18.1
Risk categories.

The RMU approach for trading risk was first introduced at Manufacturers Hanover Bank in 1989. RMUs do not measure liquidity or operational risk, but they do measure gap risk and market-induced credit risk. The motivation behind utilizing the RMU framework to measure gap risk has been outlined by Mark (1991). Credit risk is sometimes cause for confusion, since the "credit risk" of certain financial assets is driven by market risk. Credit risk driven by market risk, such as a swap, is a fraction of the total credit risk and is sometimes referred to as a fractional exposure (FE). Green and Mark (1994) provide a series of examples that illustrate how the RMU method can be applied as a practical risk management methodology.

Risk Management Strategies

A well-designed, clearly articulated, strategic plan for any Treasury operation calls for the implementation of increasingly sophisticated systems and analytical tools. The plan should be designed to recognize that most of the major risks can be handled in the first 20% of the effort. Furthermore, the plan should be evolutionary (see Figure 18.2). If a fixed-income risk-control system is based solely on limiting the nominal amount traded, then the next higher level of risk control is based on a basis point value (BPV) approach (otherwise referred to as the .01 value). This follows because the natural counterpart to a cash amount is an interest rate point movement or, equivalently, the next level of sensitivity which is natural to risk assessment is arguably basis point movements. The resolution provided by BPV can facilitate the risk detection in certain traded instruments and, arguably, a portfolio of traded instruments, provided that a risk equivalence in BPV

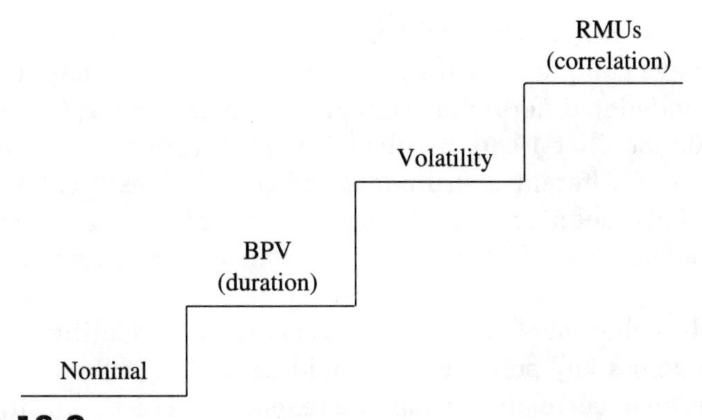

FIGURE 18.2
Evolution of risk measurement across cash markets

can be determined for the traded instruments. This approach calculates the change in price arising from a one basis point (bp) change in yield of some benchmark security and then compares other securities to the benchmark's change in price. The next higher level of sophistication combines BPV with volatility, which is achieved through using the RMU method.

An instrument-level risk measurement unit (IRMU) is approximated by multiplying the instrument's BPV by a multiple of its overnight volatility (set at two standard deviations) to arrive at a desired confidence level. For example, one RMU (set at, say, $1,000) is a basic measure of risk such that actual losses will exceed the number of RMUs on only one day in 40 (2.5% of the time). If a 4-year T-note has a BPV of $312 on a $1 million position and the change in yield volatility is 9 bp's, then 5.6 RMUs is an approximation of the overnight dollars-at-risk (DAR). Observe that $312 per $1 million times 9 bp's times 2 equals 5.6 RMUs per $1 million. Note that the selection of dollars is arbitrary; any currency can be denominated.

Measuring the price volatility of financial instruments is central to the RMU method. This is best approached by analyzing the underlying volatility of interest and exchange rates, because this allows a general risk management model to be built that can accommodate a broad variety of financial instruments. There is yet to be found a single adequate measure of volatility that can be applied to hedge complex portfolios. Determining the correct approach or best practice in the calculation of volatility remains an empirical process, and each instrument, market, and portfolio requires best practice in the calculation of volatility to be established by individual tuning. Numerous authors have compiled results of different volatility calculation techniques applied to different instruments and markets (e.g, Garman and Klass 1980; Schiller 1990). Because of the mixture that can result in a portfolio of different instruments and transactions involving multiple markets, defining a single measure of volatility for a portfolio remains a challenge. The standard covariance matrix approach is not enough.

A still higher level of sophistication combines volatility with correlations across key points on the yield curve to produce a portfolio RMU. Because correlation measurements as a statistic are independent of the underlying pricing or rate distributions, correlation measurements provide an invariant method for detecting and quantifying risk. This approach enables the DAR measure to capture nonparallel yield curve shifts. Similarly, a well-formulated approach for a foreign currency portfolio would incorporate volatility and correlation parameters into the DAR computation—for example, the risk associated with simultaneously buying and selling two negligibly correlated currencies (e.g., Deutsche marks and Australian dollars). The choice of currency pairs in this discussion is based on quantified measures, which we note may not hold indefinitely. Furthermore, options risk can be calculated utilizing RMU matrix techniques. For the case of options risk, RMUs have been developed with reference to option risk sensitivity parameters such as delta/gamma, vega, and theta risk.

The price risk of holding *two or more instruments* in a portfolio is not the sum of the individual price risks or volatilities. To the extent to which there is "netting off" of price risk, diversification reduces risk in a portfolio. In part, this is well known, as discussed by Markowitz (1952). Miller (1991) expands on the concept of market structures through the proposition of a model which is particular in its discussion of liquidity and how pricing, and thus price risk, can be rationalized in real world terms. The discussions in Miller are general and broad, but they do serve to suggest guidelines for understanding the interrelationships among market, liquidity, and credit risk. To measure the price risk of a portfolio, one therefore needs to measure the degree to which the prices of different instruments move together (i.e., are correlated). A portfolio RMU must take into account the covariances between the prices of different instruments. Hence, measuring the correlation between instruments is central to the RMU method. As an illustration, Table 18.1 shows the correlations between

TABLE 18.1 *Correlations between Volatilities of U.S. Treasury Securities (as of end of April 1993)*

	Fed Funds	2-year T-notes	5-year T-notes	10-year T-notes	30-year T-bonds	1 Std. dev.,%
Time horizon: 90 days						
Fed Funds	1.00					9.24
2-year T-notes	−0.025	1.00	—	—	—	1.54
5-year T-notes	−0.024	0.984	1.00	—	—	1.23
10-year T-notes	−0.017	0.964	0.989	1.00	—	0.93
30-year T-bonds	−0.005	0.929	0.964	0.991	1.00	0.93
Time horizon: 60 days						
Fed Funds	1.00					9.53
2-year T-notes	0.185	1.00	—	—	—	1.55
5-year T-notes	0.159	0.983	1.00	—	—	1.31
10-year T-notes	0.168	0.955	0.985	1.00	—	1.01
30-year T-bonds	0.179	0.922	0.961	0.991	1.00	0.65
Time horizon: 30 days						
Fed Funds	1.00					9.10
2-year T-notes	−0.235	1.00	—	—	—	1.56
5-year T-notes	−0.184	0.913	1.00	—	—	1.53
10-year T-notes	−0.083	0.773	0.954	1.00	—	1.23
30-year T-bonds	−0.01	0.590	0.843	0.956	1.00	0.76

the volatilities of the daily yield changes of U.S. Treasury securities. These correlations are provided for three time horizons—90, 60, and 30 days—measured at the end of April 1993. Time horizons refer to the historical averaging period used to determine the statistical mean and historical variance values. Volatility is interpreted as the square root of the variance.

Using the correlations of the volatilities of daily yield changes of U.S. Treasury securities over 90 days, Table 18.2 illustrates computations of the gross and net RMU position. Columns 1 and 2 in Table 18.2 show a portfolio of $10 million distributed among Fed Funds, 2-, 5-, and 10-year notes, and 30-year bonds. Column 3 contains the basis point value of each security; column 4 translates these values into 10-year note equivalents. Column 5 is the result of multiplying the nominal amounts for each security with the equivalent risk ratios. Column 6 shows one standard deviation of the daily volatilities of each security; column 7 shows two standard deviations. The RMU

TABLE 18.2 *Sample of Gross RMUs of a Portfolio of U.S. Treasury Securities*

Security	Nominal value, $million	bp value	Ratio of 10-year equivalents	Risk equivalent amount	1 Std. dev., %	2 Std. dev., %	RMU Position	RMU ^2
Fed Funds	6.00	0.27	0.00	0.00	9.24	18.48	0.30	0.09
2-year T-notes	2.50	177	0.27	0.68	1.537	3.07	13.60	185.03
5-year T-notes	1.00	400	0.61	0.61	1.231	2.46	9.85	96.98
10-year T-notes	0.40	654	1.00	0.40	0.929	1.86	4.86	23.62
30-year T-bonds	0.10	948	1.45	0.14	0.929	1.86	1.76	3.10
				1.83			30.37 (Gross RMUs)	

Sample of Net RMUs of a portfolio of U.S. Treasury Securities

	Portfolio including correlations		
RMU security correlations	RMU(*X*) × RMU(*Y*)	2 × correlation	2 × C × RMU(*X*) × RMU(*Y*)
2-5	134	1.97	263.63
2-10	66	1.93	127.47
2-30	24	1.86	44.52
5-10	48	1.98	94.68
5-30	17	1.93	33.44
10-30	9	1.98	16.97
FF-2	4	−0.05	−0.20
FF-5	3	−0.05	−0.14
FF-10	1	−0.03	−0.05
FF-30	1.0	−0.01	−0.01
			29.82 (Net RMUs)

positions in column 8 are the products of the amounts in columns 5 and 7. The gross RMU position of this $10 million portfolio is the sum of the RMUs for all securities in the portfolio. The net RMU position is the gross RMU plus the product of the correlation of two securities (combinations of Fed Funds, 2-, 5-, and 10-year notes, and 30-year bonds in Table 18.1, 90-day horizon) with the RMU of each security. As there are few negative correlations between the securities in this portfolio, the net RMU level is not very different from the gross RMU level (29.8 RMUs versus 30.4 RMUs). A portfolio without Fed Funds, with 40% in 2-year notes and equal amounts in the other securities (20% each), has 99.8 RMUs—three times higher than the portfolio described above.

A key property of an RMU-based risk system is that total risk can be aggregated across different business units to provide one overall DAR measure. Ultimately, a system can be designed where risk limits for lower-level organizational units can be dynamically shared within one overall limit. Furthermore, the same tools used to measure risk can be used to construct optimal hedges to reduce risk. Under an RMU-based system, risk can be added both horizontally and vertically across an organization to produce meaningful measures of total risk. That the RMU approach can be used to measure risk and make money is a proposition which has been well established in practice. Portfolios that are managed using full immunization techniques offer no upside return opportunities. Riskless strategies employing duration or financial futures hedging for interest rate risk management are too costly and represent attempts at overprotection due to uncertainty in the underlying risk measurements. The RMU method considers these shortcomings by being sensitive to the different volatilities of different instruments.

As discussed by Markowitz (1952), if assets are combined in a portfolio, the risk of the resulting portfolio should be less than the sum of the individual risks of the constituent assets. Simply stated, diversification reduces risk. One can argue that when the overall portfolio of an investment operation combined with hedging instruments is considered, a classical portfolio theory approach (such as that proposed by Markowitz) does not hold. The principal causes are due to taxes, transaction costs, variable market opportunities, and variable needs of the investment organization.

In the real world, financial risk management is a necessity. For the firm involved in numerous financial markets, involving numerous financial instruments subject to variable time horizons and to regulatory control, risk management must be expected to return a measure of its achievement through a positive contribution to the firm's net cash flow.

A detailed model representing the above concepts can be found in Green and Mark (1994). The premise for the model is that financial transactions originate as a conscious decision on behalf of an investor. The decision to invest (buy or sell) results in an action (either a trade or speculation), which in turn leads to the immediate exposure of the investor or firm executing the investment decision. The exposure can typically be foreign exchange and/or interest rate risk and can result in liquidity risk exposures; the potential consequences are clear. The processes that follow, including those of hedging, accounting, pricing, and evaluating market volatilities and all the other necessary activities, make up the overall process of financial risk management. Risk quantification (Figure 18.3) encompasses market risk, credit risk, liquidity risk, operational risk, regulatory risk, and valuation risk.

Position Management and Strategies

One thing to note straight away about trading is that no capital is required to trade in financial markets. All you need is someone to

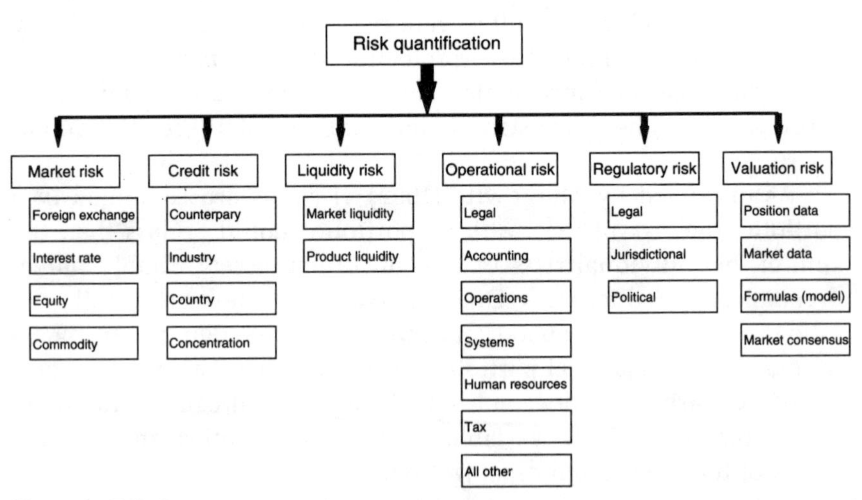

FIGURE 18.3
Risk quantification of different risk categories

give you sufficient credit limits to allow you to do the deals you want. Clearly, there is a direct relation between capital backing (or collateral) and lines granted, but there is no necessity for the counterparts actually to see the money. A pure trading operation never actually uses capital, save as a source of interest income to cover the cost of capital and as a demonstration of collateral to institutions.

Every instrument traded is a risk. It is a popular misconception that hedging and arbitrage trading are safe strategies to follow because one loss is countered by an offsetting gain. Except in a very few situations (which anyway would not produce interesting returns), this simply is not the case.

Table 18.3 illustrates a few examples of instruments, and combinations of instruments, and the risks that are taken in each case. The point is that a risk should only be taken (i.e., the instruments should only be traded) when a view has been taken on the particular risk involved.

With reference to neural networks, measures of financial risk need to include measures associated with the errors, uncertainties, and risks of the underlying forecasts and the subsequent trading decision models. The forecast model error is defined as the relative error between the actual market rate which results for the same forward date or period as forecast. The trading forecast cost can then be determined by taking the difference between the forecast rate and the actual market rate and multiplying this value by the amount of the capital used.

The output from any forecasting model must be treated in a very detailed manner for determining trading risk different from forecast

TABLE 18.3 *Traded Foreign Exchange (FX) Instruments and Corresponding Principal Risks*

Traded Instrument	Risk resulting from trade
Spot FX	Spot rates
Loans and deposits	Fixed interest rates
Swap FX	Relative fixed interest rates
Forward FX	Spot and relative fixed interest rates
IRS/FRAs	Forward/forward interest rates
Currency swaps/FXAs	Relative forward/forward interest rates
Bonds	Bond Prices
Options	Volatility
Futures	Underlying instruments
Cash/futures arbitrage	Basis

risk. The framework of RMU analysis allows the trading risk to be treated in a precise order as well as the position risk to be reported in terms of DAR relative to trading model or forecast error volatility.

Special mention should be made of the use of *options* and various options strategies. As with all other instruments, market makers will not give you something for nothing. If you are buying a hedge against a risk, you will most certainly pay for it. This raises the question: If you are removing a risk by hedging, why did you put the risk on in the first place? Remember that risks should be put on only in order to trade them actively. Also, assuming that the option costs a fair market price in terms of all relevant and readily ascertainable market conditions and levels, then you can only gain from the trade if you have a better measure of the future movement of the underlying risk, i.e., volatility. For example, a long straddle is an outright bet on increasing volatility. One should not take such a position without data and research to back it up.

Here is another example. An alternative to rolling foreign exchange (FX) spot positions would be to put all trades at a forward date as outright positions, probably by using a spot deal and a swap deal. While this should reduce transaction costs, there will be additional exposure to the risk of the relative fixed interest rates (interest rate differentials) of the two currencies involved.

One suggestion to overcome this would be to take all FX positions for value on International Money Market roll dates (the third Wednesdays of March, June, September, and December). The inherent interest rate risk could then be removed by taking a futures spread position (e.g., using Short Sterling and Euro-Dollar 3-month interest rate contracts). However, note that the earlier comments on nontraded instruments being nonprofitable still hold, but the hedge might be a reasonable trade-off.

This leads to the obvious conclusion that it is foolish to view any one instrument in isolation from all others. The data, research, and measurements required for investigating risk overlap considerably. Thus, one should be concerned with the relationships between all markets, instruments (e.g., cash and futures), maturities (e.g., spot and forward), and hedging strategies (e.g., futures spreads).

Improving Portfolio Risk Management

Managing portfolio risk requires clear policy and objectives. The key requirements for any financial risk management program are:

- Acquisition, measurement, and interpretation of up-to-date information, including assessments of uncertainty
- Insight and expert understanding of many diverse activities
- Clear distinction between investment decisions and processes that support the execution of investment
- A flexible infrastructure to manage outcomes

Maintaining the right level of risk and profitability requires setting policy and implementing an approach that can:

- Detect changes in risk
- Allow for managing fund movements without offsetting the covered positions
- Maintain liquidity
- Identify and approach opportunities that may provide profits beyond the risk-neutral position
- Provide the maintenance of losses within defined tolerances
- Effect radical strategy change for unexpected conditions
- Structure incomplete hedges to allow for special conditions
- Establish pricing relationships between all instruments traded and used for hedging (risk does not add up in a linear manner)
- Rationalize the time horizons for the different inputs (trades, hedge transactions) and outputs (profit measure, tax impact, settlement costs)
- Establish a track record to maintain confidence and facilitate retention of the experience and knowledge gained

Summary

This chapter provides an overview of the issues and requirements for establishing effective risk management strategies, policies, and systems. The four key primary objectives of any market risk program should be to:

- Control market risk
- Relate market risk measures to the level of capital consumed
- Provide a consistent methodology for aggregating market risk
- Introduce a standard way of incorporating risk into the credit risk (and, where possible, the liquidity risk) evaluation process

The evolution of market risk measurement normally proceeds from the lowest level, where risk is either controlled based on the nominal amount traded or on the basis of risk equivalence (e.g., duration-based risk management), to higher levels, where risk is controlled based on the RMU concept. The RMU concept takes into account the volatility of various instruments and the correlations between them. RMUs can be translated into dollars at risk.

Establishing an improved risk management system may require artificial intelligence tools and other advanced technologies discussed in this book. A recurring question about such technologies is, "What will happen if all the players in the market have or deploy the same tools?" Even when all players have the same tools, access to the same market information, the same pricing formulas, and the same risk management approach, there is ample proof to date that only a handful will profit. It is unlikely that human nature will suddenly become completely efficient because of more powerful tools. The next chapter elaborates on the influence of technology on modern finance theory and on the behavior of financial markets.

References

Garman, M. B., and M. J. Klass. 1980. "On the Estimation of Security Price Volatilities from Historical Data. *Journal of Business* 53(1).

Green, H. G. and R. Mark. 1994. "Managing a Portfolio of Positions." In *The Handbook of Interest Rate Risk Management*. Homewood, IL: Dow Jones Irwin.

Mark, R. 1991. "Risk Management." *International Derivative Review* (Mar.); "Units of Management." *Risk* (June); see also "Risk According to Garp" *Wall Street Computer Review* (Dec.).

Markowitz, H. M. 1952. "Portfolio Selection." *Journal of Finance* VII (Mar.).

Miller, M. H. 1991. *Financial Innovations and Market Volatility*. Cambridge, MA: Blackwell.

Schiller, R. J. 1990. *Market Volatility*. Cambridge, MA: The MIT Press.

CHAPTER 19

The Impact of Technology on Financial Markets

Guido J. Deboeck

Wall Street people . . . learn nothing . . . and forget everything.

B. Graham
Interview, 1976

Thus far we have provided an overview of advanced technology concepts and techniques and have demonstrated their application for trading, risk, and portfolio management. Several systems described in this book have been in operation for a number of years. These techniques are no longer the exclusive territory of researchers, rocket scientists, and academics, but are deployed in many financial institutions around the world. Some institutions have developed new financial products using them (e.g., fuzzy unit trusts launched by Yamaichi in Japan, disciplined equity funds launched by Fidelity). Others use them to promote their business to potential clients (e.g., neurocomputing advertisements by Dai-ichi Kanyo Bank). This chapter steps back from the specifics and assesses the potential impact of technology on financial markets.

A Historical Perspective

The most famous insights in the history of modern finance and investment appeared in a short paper published in March 1952 in the *Journal of Finance*. The paper, "Portfolio Selection," was by Harry

Markowitz, a 25-year-old graduate student at the University of Chicago. According to Markowitz, there is a rule that "the investor should diversify and that he/she should maximize expected return" simultaneously. The rule states that the investor should diversify funds among those securities that give maximum expected return. In setting forth this paradigm, Markowitz "commends" to investors an *efficient portfolio*—one that offers the highest expected return for any given degree of risk, or the lowest degree of risk for any given expected return.

Markowitz's contribution was to distinguish between the riskiness of an individual stock and the riskiness of an entire portfolio. The riskiness of a portfolio depends on the covariance of its holdings, not on the average riskiness of the separate investments. *Covariance* is the correlation between the variance of two investments. *Variance* means the distance from an average. Professor Albert Coppe, who in the sixties taught statistics at the Catholic University of Leuven, used to say that a case of high variance is a man who has his feet in the oven and his head in the refrigerator, but on average feels pretty well.

A combination of very risky holdings may constitute a low-risk portfolio as long as the holdings do not move in lockstep with one another, that is, as long as they have low covariance. In his 1987 book, *Mean-Variance Analysis in Portfolio Choice and Capital Markets*, Markowitz develops a theory of investing that includes risk as well as return. To follow Markowitz's prescriptions, investors must analyze all possible combinations of assets, searching for efficient portfolios among them. this procedure is complicated and time consuming. Investors not only must make reliable estimates of variability for each individual security, but they also must estimate the expected return, a challenging task under any circumstances. Markowitz suggested the use of linear programming to find the optimum combination.

Linear programming is simple if one has access to a computer. Easy access to computers, however, is a relatively recent phenomenon. When William Sharpe, a student of Markowitz, used linear programming in 1961, he reported that it required 33 minutes to solve a 100-security problem on the best commercially available IBM computer at that time.

Sharpe simplified Markowitz's approach by eliminating the calculation of covariances between each pair of securities. He suggested that only the relationship of each of the securities to the dominant factor need be calculated. The most important single influence thus became an index expressing the rate of change in the stock market as a whole. Sharpe's simplified model was a giant step forward in bringing Markowitz's ideas on portfolio selection closer to real world applications.

The big attraction of Sharpe's single-index model was that computing time required to solve a 100-security problem on a state-of-the art mainframe computer was reduced from 33 minutes to 30 seconds. If the Markowitz model were run today on a 386 machine with a math coprocessor it would also take less than a minute. The simplified Sharpe model, on the other hand, would be even faster, executing in an instant.

With the introduction of the IBM PC in the early 1980s, Sharpe immediately recognized the value of the PC in solving portfolio management problems. The software he produced made him a driving force in the rapid proliferation of desktop computing among professional investors. Sharpe's major breakthrough came with the capital asset pricing model (CAPM). Sharpe argued that an optimal investment strategy is one which buys and holds as widely diversified a basket of stocks as possible.

Barr Rosenberg repackaged Markowitz's and Sharpe's ideas in an enriched and more applicable form. He introduced the concept of *extra market covariance*, which means that many stocks move together independently of what the market does as a whole. For example, stocks of companies in the same industry, stocks that are small in size, or stocks that are exposed to fluctuations of the value of the dollar may move independently of the market.

Rosenberg introduced a new optimizer. This program identifies efficient portfolios in a universe of securities, each with its own expected return and risk. In 1985, Rosenberg established Rosenberg Institutional Equity Management (RIEM). He and his associates created a database covering 3,500 U.S.-based and 1,800 Japanese-based companies, broken down into 150 categories. Balance sheets, earnings, and sales for each company were compared with data for similar companies and were adjusted for size and variations in capital structure. His optimizer program continuously reviewed these companies' securities and selected efficient portfolios based on the prices continuously fed into the computer from the ticker tape. According to Rosenberg, RIEM was the only investment organization that had a real-time optimizer at that time. The computer made all the investment decisions, and those decisions could not be humanly overridden! The only way people could influence the outcome was by changing the programs or the data. Peter Bernstein called this "pure Rosenbergiana"; the nitty-gritty was essential to Rosenberg's success. Trading systems became the state of the art.

In 1989 the Dow Jones published *Guide to Trading Systems,* in which Bruce Babcock outlined numerous trading systems based on technical analysis indicators. He explained the design and implementation of trading system developments and showed how to create simple and advanced systems, including approaches to money management. Prior to Babcock's book, the design of trading systems was, according to Larry Williams, in the hands of less than a hundred people in the United States. Williams wrote in the foreword, "Thanks to Bruce Babcock, now everyone can understand how systems are put together."

Babcock's book shows that the main reason why traders fail is because of lack of discipline. For all traders there is a continuum between 0% mechanical and 100% automated. Successful traders have a relatively mechanical approach; it is the only way to avoid the destructive emotionalism that permeates trading. Babcock describes a mechanical approach as one where:

- A predetermined group of securities or markets is followed.
- Mathematical formulas are applied to prices that tell when to buy and when to sell.
- Entry rules exist, along with exit rules for losing trades.
- Rules exist for when to start trading and stop trading.

A user of a trading system makes only the initial choice of system and markets, and applies the system rules to market price action. If the system is computerized, he or she has to provide data to the computer, run the system, and place the orders the system dictates. The key to creating a successful mechanical system, as discussed in the Introduction to this book, is to avoid curve-fitting.

Trading systems became state of the art in the eighties. In the early nineties they moved from mainframe to personal computers. In 1990, Futures Truth Inc. in North Carolina released Excalibur™, a software package to test trading systems. Excalibur allows testing of a user-defined rule or set of rules. Extensive experimentation with Excalibur, which is written in FORTRAN, shows that few mechanical trading systems are profitable. The limitations of mechanical systems motivated research on machine-learning techniques. It was found that properly designed neural-based trading systems can outperform mechanical systems. As demonstrated in Chapters 3 to 7, the performance of neural-based systems exceeds the performance of the majority of mechanical systems. Furthermore, genetic algorithms permit optimization of trading strategies and configuration of portfolios of models that achieve multiple objectives under given risk constraints.

Nonlinear dynamics and chaos theory, discussed in Chapters 15 to 18, explain why mechanical systems are less effective than models based on machine learning. The coexistence of local randomness and global determinism and of periodic and nonperiodic cycles in financial markets requires nonlinear, adaptive technologies for capturing trends as well as structural changes in financial markets.

In fact, there appears to be a direct correlation between access to more powerful workstations, more sophisticated analytics, modeling software and the ability to diagnose and understand the complexity of financial markets. Until powerful workstations became available, testing of earlier theories of modern portfolio management was limited, expensive, and time consuming. The earlier theories were constrained by their inability to test and verify market behavior. Today, vast quantities of data can be accessed and processed at high speeds, and advanced analytical tools and models can be used to produce empirical evidence of the increasing complexity of financial markets. Without fast computers and sophisticated software, the fractal nature of financial markets is impossible to figure out!

In sum, the more distance we gain from daily market news and the more sophisticated the analytical tools we deploy, the better we can gain an understanding of market behavior. As demonstrated by several studies in this book, profitable trading systems have been designed that require infrequent retraining and produce profitability at lower risk than achieved by many human traders. As a consequence, traders watching Reuters, Telerate, or Bloomberg screens are placed at a disadvantage.

A Taxonomy of Market Hypotheses

Must we repeat that disciplined trading is imperative? This section provides yet another perspective: a comparison of current practices of traders and their underlying assumptions, with new market hypotheses and their underlying assumptions. The evolution in thinking about trading may well be attributable to the rise of nonlinear dynamics.

Table 19.1 provides an overview of new insights into the behavior of financial markets. This table is based on Peters' book *Chaos and Order in Financial Markets*, his most recent work entitled *Fractal Market Analysis*, Tonis Vaga's coherent market hypothesis, and the work of many who over the past two or three years have demonstrated the nonlinear properties of financial markets. Table 19.1 categorizes market hypotheses based on assumptions concerning market signals (stochastic or deterministic) and approaches to modeling (linear or nonlinear).

If one assumes that market signals are damped harmonic oscillators with random noise or Gaussian distributions of returns, then

TABLE 19.1 *Market Hypotheses, Trading Strategies, and Model Applicability*

	Linear models	Nonlinear models
Stochastic market signals		
Market signal	I. Damped harmonic oscillator + random forces (Gaussian distribution of returns)	IV. Stochastic chaos: damped anharmonic oscillator + random forces
Market hypothesis	Efficient market hypothesis	Coherent market hypothesis
Trading strategies	Indexing and noise trading	Adaptive neural models Trading based on fractals and Lyapunov exponents
Applicability	Random walk markets	Bear/bull markets Trading markets Random walk markets
Deterministic market signals		
Market signal	II. Damped harmonic oscillator + random forces (Gaussian distribution of returns)	III. Deterministic chaos: damped anharmonic oscillator + random forces
Market hypothesis	Trending market hypothesis	Fractal market hypothesis
Trading strategies	Fundamental and technical Trend-following Mechanical models	Static neural net models Nonlinear models
Applicability	Bull/bear markets only	Bull/bear markets Trading markets

markets are efficient. In this case, predicting market prices is difficult or impossible and an index or benchmark approach to trading is the only way to produce returns equivalent to market returns.

If one assumes the existence of temporary market inefficiencies—that is, exceptions to the market efficiency—then it is possible to accept a *trending market hypothesis*. In this case, historical analyses can be used to define "clever" benchmarks that can outperform a market index by benefiting from temporary inefficiencies. As temporary inefficiencies disappear, benchmark compositions need to be updated. Mechanical trading systems and linear models of limited complexity can add value by capturing ongoing trends for short periods of time.

If market signals are damped anharmonic oscillators with random noise that is approximately Gaussian, a different market hypothesis (as demonstrated by several authors in Part Four) must be considered. Peters and others have suggested a *fractal market hypothesis*.

The fractal market hypothesis proposes that markets are nonlinear, dynamic systems with noninteger fractal dimensions, the predictability of which can be estimated. This hypothesis permits the dimension of the probability distribution of price or yield changes to range between 1 and 2 and to be either a fraction or an integer value. The efficient market hypothesis assumes that the probability density function of the price or yield changes equals exactly 2. In sum, the fractal market hypothesis provides a more general paradigm of the behavior of financial markets.

A refinement of the fractal market hypothesis is the *coherent market hypothesis* (Vaga 1990), which suggests that there are at least four different markets: coherent bull markets, coherent bear markets, transition markets, and random walk markets. Each requires a different investment strategy and risk management approach. For example, in coherent bull markets, a buy/hold strategy may have a better reward/risk ratio than any other strategy.

The efficient market hypothesis has been the working hypothesis for the last 40 years. The work of Markowitz, Sharpe, Rosenberg, and many others introduced investors to index-based investment strategies that essentially produce a rate of return equivalent to the rate of return of the overall market. Passive trading strategies led many to adopt mechanical trading models that either follow trends or are directed to exploit the volatility in markets. Tactical asset allocation strategies are then focused on the timing of switches between one approach versus another.

The newer fractal market hypothesis assumes that market signals are nonlinear dynamic systems, which may have persistent or antipersistent trends. This hypothesis involves fractal dimensions and/or attributes of chaotic series. It has far-reaching implications for the design of automated trading systems. As illustrated in Part Four, the identification of persistent or antipersistent trends within particular time horizons, estimation of the correlational dimensions, and evidence of chaos are invaluable criteria for the design of nonlinear dynamic systems—including neural networks and fuzzy trading systems. As more empirical evidence of nonlinearities in financial markets becomes public, it is not inconceivable that the fundamental paradigms of trading, risk, and portfolio management will be affected. Even more challenging is the potential impact of the coherent market hypothesis and the underlying theory of social imitation and the behavior of crowds.

In fact, the two-dimensional representation of market hypotheses and investment models in Table 19.1 may describe a limit cycle or chaotic process in the evolution of modern financial theory. It is interesting to observe that thinking about financial markets "circles" from quadrant I, the efficient market hypothesis; to quadrant II, the trending market hypothesis (currently practiced by many institutions); to quadrant III, the fractal market hypothesis (recognized by more and more financial professionals); to quadrant IV, the coherent market hypothesis (still largely unappreciated in the financial community).

New Market Paradigms Based on Rational Beliefs

Until recently no solid alternative theoretical framework existed for the behavior of financial markets. A recent paper by Mordecai Kurz of Stanford University, however, proposes a new theory of asset pricing based on the *rational belief equilibrium* (RBE). H. Brock of Strategic Economic Decisions Inc. (Menlo Park, California), who introduced us to Professor Kurz's important work, describes it as follows:

> Professor Kurz systematically [generalized] the existing theory whereby the efficient market hypothesis and the CAPM are simply subsumed as a special case that would only be valid in very special and exceptional sets of circumstances. In other words, the two most important theories of modern finance become special cases in the same way that Einstein's relativity theory generalized classical New-

tonian physics which remains within relatively theory as a limiting special case.

RBE deliberately contrasts with the classical *rational expectations equilibrium* (REE). The difference between both concepts lies in the kinds of information and beliefs investors are assumed to hold. In Kurz's model, different market participants are assumed to hold *different beliefs* about future events (including future market prices) even though they all share the *same information or data*. The difference in their beliefs results from the fact that, even though they share the same data, they possess widely differing "mental models" of the world. Accordingly, they draw different *inferences* from the same generally available factual information and have different beliefs about the future–individual forecasts, as it were. At an analytical level, these beliefs can be represented by probabilistic forecasts of future events. In Peter's *Fractal Market Analysis* the emphasis is on different investment horizons; in Kurz's work this is broadened to different inferences based on different beliefs.

The distinction between different investment horizons and different inferences (based on the same information) is not made in classical efficient market theory. Peters demonstrated how markets become unstable when all market participants adopt the same investment horizon. This is reinforced by Kurz, who demonstrates that while beliefs will generally differ, each of these divergent beliefs (forecasts) can be deemed rational to the extent that it cannot be refuted by the existing data utilizing generally accepted modes of statistical inferences. Kurz shows that any given database will generate a multiplicity of different yet rational beliefs. While each investor's beliefs may be rational in Kurz's sense of nonrefutability-by-the-data, the beliefs of most, if not all, investors will not reflect the true probability distribution over future events–including the true future prices of financial assets and yields. The latter is explained by the fact that existing data alone cannot produce true distribution, a principle proved by Kurz. This is caused by ongoing structural changes in the environment that cannot be identified solely by analyzing available times-series data drawn from history. The important role of structural changes invalidates traditional methods of drawing inferences purely on a statistical basis. Figure 19.1 graphically illustrates Kurz's principal results and contrasts them with those of classical efficient market theory.

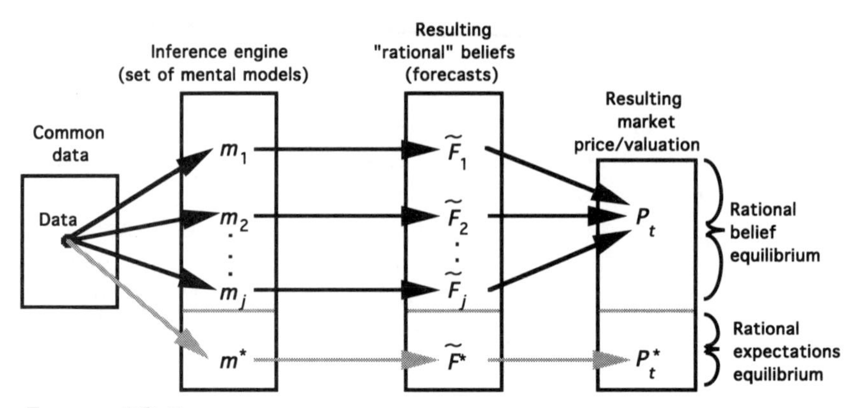

FIGURE 19.1

A rational belief equilibrium (Source: Forecast and Risk Assessment, Strategic Economic Decisions Inc., February 1993)

Professor Kurz's concept of rational beliefs has some counterintuitive implications for the behavior of financial markets. The central concept of RBE is a sequence of market prices generated over time by market participants, aiming to maximize profit and armed with his or her own beliefs about the future. Three important implications are:

1. The market will generally misprice assets: this mispricing reflects not only the existence of a subset of wrong beliefs, but also the counterintuitive role played by the distribution of beliefs throughout the market. The market as a whole misprices assets even though the beliefs of all market participants are assumed to be rational.
2. Mispricing is not attributable to any "inefficiencies" in the processing of information. Professor Kurz writes:

   > The theory of rational beliefs admits rational agents who make incorrect assessments of economic situations even though they have efficiently processed all information. Consequently, the theory admits the possibility of incorrect assessments by markets as a whole. Hence, the central difference between a RBE and "efficient market theory" has to do with the evaluation of information rather than the efficiency of its processing.

3. Market volatility: in classical theory price volatility reflects the revision of individual expectations about the future as

derived from news announcements. Kurz shows that the process by which expectations are changed when agents hold different yet rational beliefs is far more complex than hitherto imagined—and generates more volatility than would result were everyone to share the same beliefs.

A corollary of this analysis of volatility is a new theory of financial market "bubbles" and "crashes"—dramatic price movements not attributable to "news." Kurz's theory shows that market crashes can result from rational behavior on the part of market participants equipped with completely rational beliefs. Kurz's theory can, in principle, explain the magnitude of a given price bubble in terms of both price and the volume of trading at each point in time.

Impact of Neural, Genetic, and Fuzzy Modeling

Are the technologies described in this book the frontier in technology innovations and applications in finance? What impact will they have on financial markets?

Neural network concepts have been around since 1943—more than 50 years! In 1972 Paul Werbos published a dissertation entitled "Beyond Regression Analysis," which outlined the principles of back-propagation. David Rummelhart and James McClelland popularized back-propagation in 1988 with their work, *Explorations in Parallel Distributed Processing: A Handbook of Models, Programs and Exercises*. Since then, many textbooks and collections of articles have been published.

As discussed in Part One of this book, neural nets are very good at finding relationships in a huge input/output product space even if the patterns are ill defined. The performance of a neural net trading system will depend, however, on the information that flows into the system, on how the input information is preprocessed, on the desired output, and on the quality of the neural net design. The neural net design in turn depends on the appropriateness of the neural net architecture, the learning algorithm, the adequacy of neural net training, and the relevance of the training samples. In consequence, no matter how sophisticated the neural net technology, the design of a neural trading system remains an art.

This art, especially in terms of training and configuring neural nets for trading, can be simplified through the use of genetic algorithms (GAs). As demonstrated in Part Two, GAs are powerful tools for finding optimal parameters for complex problems with given objective functions and one or more constraints. GAs can be used for selecting inputs, configuring, as well as finding optimal weights or the best neural nets for a particular problem. The main drawback of using GAs to optimize neural nets for trading is that they may produce results that are undesirable or that do not generalize properly. The end product from a GA process will largely depend on the specification of the objective function.

The drawbacks of machine-learning techniques for trading are (1) lack of explanatory capability, (2) difficulty of including structured knowledge; and (3) bias toward quantitative data. In sum, the lack of confidence intervals and the inability to trace back outputs or even to include structured knowledge have prevented rapid diffusion and easy acceptance of neural net technology.

Fuzzy logic and mathematical techniques can improve the effectiveness of neural networks. They are able to embed structured knowledge about financial markets, have more explanatory capability, and are inherently more stable. Several examples of such applications were presented in Part Three.

What impact will advanced technology have in the future?

Neural networks are used for credit card fraud tracking, bankruptcy predictions, and bond ratings and pricing. Several banks use neural nets for trading foreign exchange, commodity, stock, and/or fixed-income markets. Fuzzy systems are used for portfolio management, trading, and commercial applications. Genetic algorithms are used for stock selection, asset allocation, and portfolio composition. GAs are also used in design applications, for work scheduling, and for planning.

Empirical findings from nonlinear dynamics and chaos theory have made the efficient market hypothesis (EMH) and capital asset pricing model (CAPM) obsolete. Edgar Peters and others have explained why these theories no longer hold. The theory of rational beliefs proposed by Professor M. Kurz, and reviewed in the previous section, is the first systematic generalization of the efficient market hypothesis. EMH and CAPM are simply subsumed as a special case that is valid only in very special and exceptional sets of circumstances. Thus, the two most important theories of modern finance become special cases in the same way that Einstein's relativity theory generalized classical Newtonian physics.

The extent of the impact of these advanced technologies over the next few years will depend on how they are integrated and made accessible. Ease of use has always had a dramatic impact on the adoption of technology.

Tools for the design of neural networks, genetic algorithms, and fuzzy systems are at present still too difficult to use. Some are prototypes, stand-alone C routines, or products with a poor or nonstandard human interface. Most software products are single-function tools aimed at similar operations (e.g., training neural nets). Multipurpose software tools that provide easy access to neural net capabilities, genetic algorithms, and fuzzy logic, including automated extraction of fuzzy rule bases, have yet to be commercialized. Until this happens, many applications will continue to be the result of painstaking trial and error, rather than engineering.

Finally, most commercial products for applying advanced technology are too expensive to be widely adopted. Site licenses of these software tools for large organizations are necessary. Thus, advanced technology is not yet as popular as Excel™, WingZ™, or Lotus 1-2-3™, but the development of spreadsheet extensions using these algorithms surely points to the future.

Arthur C. Clarke said, "Any sufficiently advanced technology is indistinguishable from magic." However, this does not mean that the use of advanced technology automatically produces magical results. The technologies described in this book should be applied judiciously, in conjunction with hard-won knowledge and analysis, to produce a trading edge.

Conclusions

In *A Short History of Financial Euphoria: Financial Genius Is Before the Fall*, John Kenneth Galbraith suggested an unfailing rule by which the individual investor and, needless to say, the pension and other institutional-fund manager should be guided:

> . . .when a mood of excitement pervades a market or surrounds an investment prospect . . . all sensible people should circle the wagons; it is the time for caution. Perhaps, indeed, there is opportunity. Maybe there is that treasure on the floor of the Red Sea. A rich history provides proof, however, that, as often or more often, there is only delusion and self-delusion.

Unless many of those cited in this book or in a recent survey of new technologies in finance, published in *The Economist* (October 1993), are suffering from mass delusion, there are definitely patterns in financial markets that can provide reliable guidance for the future, and cleverly programmed computers with a lot of horsepower can uncover these patterns. The evidence provided in this book on financial applications of neural networks, genetic algorithms, fuzzy logic, and recent findings from nonlinear dynamics and chaos theory points to new levels of understanding of the behavior of financial markets. The hypotheses advanced by Peters and Vaga and the theory of rational beliefs proposed by Professor Kurz are new milestones in the *chaotic* evolution of modern finance theory. The influence and impact of technology are undeniable!

The spread of machine-learning techniques and model-based trading in financial markets will profoundly affect the future design of financial instruments. New unit trusts and mutual funds based on disciplined trading are already sprouting; some are well ahead both in attracting capital and outperforming standard index-based funds.

The ability to forecast financial markets within limited time horizons will reduce risk without increasing the price of safety. Even limited predictability will bring more traders to the market, which will increase the market's liquidity. Increased liquidity will decrease the bid-ask spread and lower transaction costs. This will in turn create new opportunities for more investors to participate and to create greater wealth. In the same way that decreasing the cost of cars increases the demand for them and the profit to be made from building them, automation of trading will prove to be an engine of economic growth.

References

Bernstein, P. 1992. *Capital Ideas: The Improbable Origins of Modern Wall Street.* New York: Free Press.

Brock, H. W. 1993. "Forecast and Risk Assessment." Technical report. Strategic Economic Decisions Inc.

Brock, H. W. 1993. "Inference from a Decade in the Business." Technical report. Strategic Economic Decisions Inc.

Galbraith, J. K. 1990. *A Short History of Financial Euphoria: Financial Genius Is Before the Fall.* Whittle Direct Books.

Kurz, M. 1992. "Asset Prices with Rational Beliefs." Unpublished paper. Stanford University.

Peters, E. 1989. "Fractal Structure in the Capital Markets." *Financial Analyst Journal* (July-Aug.).
Peters, E. 1991. *Chaos and Order in the Capital Markets: A New View of Cycles, Prices and Market Volatility*. New York: John Wiley & Sons.
Peters, E. 1991. "A Chaotic Attractor for the S&P 500." *Financial Analyst Journal* (Mar.-Apr.).
Peters, E. 1994. *Fractal Market Analysis*. New York: John Wiley & Sons.
Vaga, T. 1990. "The Coherent Market Hypothesis." *Financial Analyst Journal* (Nov.-Dec.).

CHAPTER 20

The Cutting Edge of Trading Technology

Guido J. Deboeck

God pity a one-dream man.

Robert H. Goddard
(U.S. rocket scientist and pioneer), 1882–1945
Notebooks, Goddard Memorial Library,
Clark University

Neural networks, genetic algorithms, fuzzy logic, and non-linear dynamics are by no means new concepts. What is relatively new is the use of these advanced technologies for trading, risk, and portfolio management. Although such financial applications continue to be developed and to spread rapidly around the globe, they no longer represent the frontier of innovation. This chapter discusses technological developments at the cutting edge, including virtual reality and robotic trading.

Virtual Reality

Virtual reality (VR) is the human experience of perceiving and interacting through sensors and effectors with a synthetic simulated environment containing simulated objects, as if they were real. The basic principle of VR is complete immergence in a computer model or problem simulation.

How can this be applied to financial markets? Consider a fund manager responsible for tracking the performance of thousands of companies. VR systems can display market information dynamically in a spatially arranged world. Imagine circling Paris, London, Tokyo, or New York in a helicopter and being able to swoop down for a closer view of individual skyscrapers. The buildings are actually 3-D bar graphs that sport corporate logos. The bars shrink and grow in relation to the corporations' stock values, and particular stocks can be clustered in neighborhoods by industry or market. Blinkers and spinners can be added to the tops of the bars to attract attention if, for example, the value of a stock starts to fall.

One such VR system for financial markets has been developed by Maxus Systems International, which specializes in developing international portfolio management systems that include sophisticated analytic support. Maxus integrates data input from existing data sources with proprietary or in-house databases, including links to existing back office systems. The original Maxus system, Capri™, is a Windows-based portfolio management system with real-time access to Reuters and Knight Ridder data feeds. The virtual reality extension of Capri provides ten embedded dimensions of market data on a huge checkerboard (see Figure 20.1). One axis represents different markets,

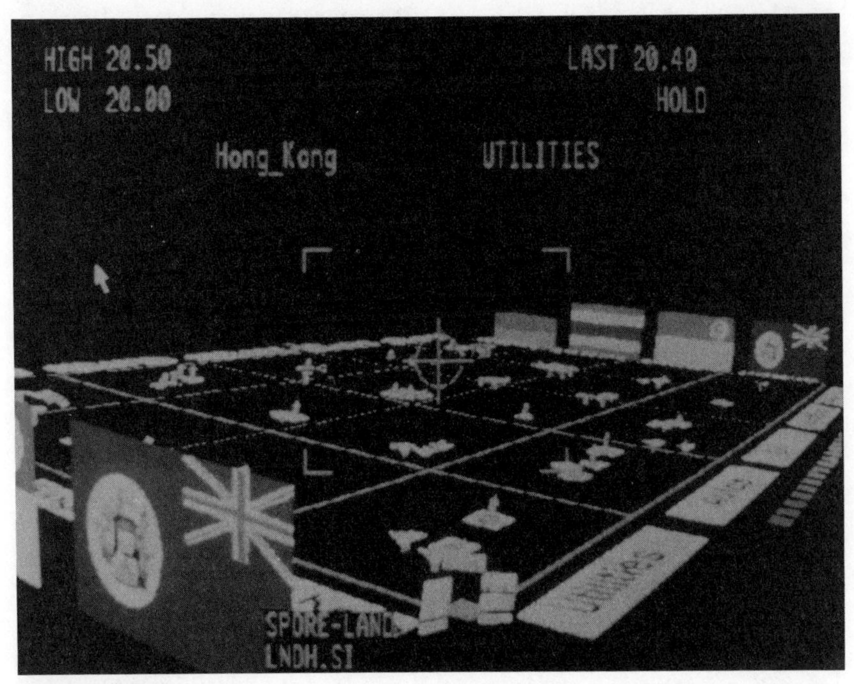

FIGURE 20.1
Virtual reality "checkerboard" of market data, divided by country and by sector

such as the U.S., Japan, Hong Kong, Germany, and France. Another axis represents industry groups, and each grid on the board has chips with different company logos and of different shapes and colors. These chips move around and attract attention to particular developments—for example, large jumps in market prices.

Using a joystick, one can fly into this cyberspace. Say we fly over the checkerboard and zoom in on German stocks. To obtain more detail we descend to ground zero, where detailed information on percentage changes, price-earnings ratios, and price-book values is recorded. After looking over the scene, we ascend, fly over Hong Kong, and check out the Japanese market. The variations in shape, color, and movement on the checkerboard between the German and the Japanese stock markets provide an instant impression as to which is currently more attractive. Comparisons between sectors are easy to make; furthermore, selecting the best companies in each sector around the globe becomes a snap. Should we become "lost in space" by navigating into the nongraphics void, a double tetrahedron icon named LIA (for Limited Intelligent Agent) can be called upon to vocally lead us back to populated cyberspace.

The tremendous advantage of VR systems is that a portfolio manager does not have to wade through tons of different reports to get the information, and is able to better see the confluences between different markets, industries, and companies belonging to the same sector. Virtual reality allows *complete immergence in financial markets* and thus lessens distraction from outside influences—such as the yelling traders, ringing phones, and annoying loudspeakers that make many trading floors a real zoo!

In a presentation in Washington in June 1993 Dr. Joel Orr, president of Orr and Associates, spoke about *pezonomics. Pezo* is derived from the Greek and means *play*. Since in play humans excel in innovation and productivity, Orr suggested that it may be worthwhile to study how we play and how play can enrich us. VR systems create spatial and temporal immediacy, allowing us to "wade" in the data. The world has become so complex that standard synthesis techniques are no longer adequate. Virtual reality is not merely a better interface between humans and machines, it is an essential tool for learning the laws of play—a tool that can extend human capabilities.

The main ideas behind the VR system developed by Maxus Systems International came from playing Nintendo™ video games. As a generation of Nintendo players grows up, their impact on trading technology should not be underestimated.

Robotic Trading

Better visualization and possibly sonification of data can provide a greater overview of and deeper insights into the interrelationships and dynamics of financial markets. Virtual reality can provide more advanced syntheses of financial data by allowing humans to be completely immerged in markets.

But what if humans were completely *removed* from markets? Can trading be automated to the point where it could be taken over by robots? Previous discussions of neural networks, genetic algorithms, and fuzzy systems in this book have showed the possibility of better pattern recognition, forecasting, optimization, and model design. The question is, can these be integrated to the point where robotic trading becomes feasible?

Figure 20.2 provides a robotic view of the applications of artificial intelligence (AI) for trading. This view includes *sensors* that collect data from market data feeds; *control mechanisms* that derive from rule-based systems, neural networks, or other model-based systems; and *effectors* that provide orders, cancellations, and so on.

In the September 13, 1991, issue of *Science* magazine, Rodney Brooks of MIT outlined new approaches to robotics. Early mobile robots, such as SRI's "Shakey," required huge computational processes to build up a detailed model of the environment, to plan actions, and to execute them, usually slowly. Professor Brooks observed that very simple "512 neuron" animals can run circles around these cum-

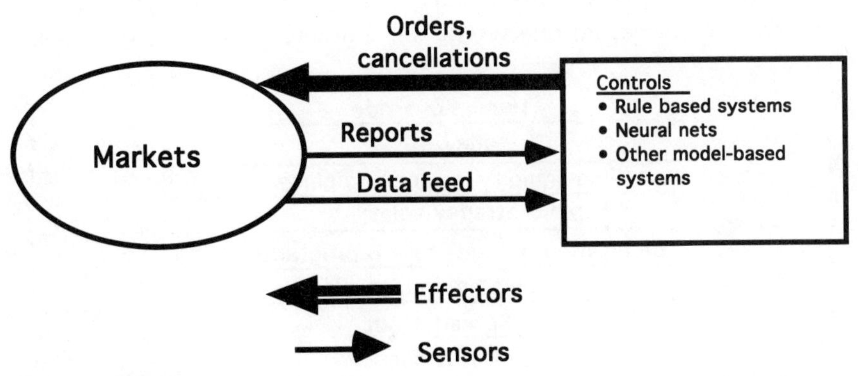

FIGURE 20.2
A robotic view of AI applied to trading, using sensors, controls, and effectors. (Source: D. Leinweber, "AI and Automated Trading," AI Applications in Wall Street, *New York, 20 April 1993*)

bersome robots. He designed a multilevel control strategy called the *subsumption architecture*. The basic idea is to build intelligent behaviors level by level, starting with commands such as "don't fall over," "move forward," "avoid obstacles," "explore your environment," "make maps as you go along," and so forth. In training robots for trading, the same basic ideas may apply: "don't make big losses," "follow the trend," "avoid being whipped," "keep track of your transactions," "follow your rules."

Traditional decomposition for intelligent control systems within AI uses a chain of information-processing modules proceeding from sensing to action. In Brooks' new approach, the decomposition is in terms of behavior-generating modules, each of which connects sensing to action. Layers are added incrementally, and newer layers depend on the successful operation of earlier layers. Figure 20.3 shows the application of the subsumption architecture to trading. The link between sensors and effectors shows a number of behaviors, starting at the lowest level with "get it done" (or execute the trade) and moving upward to "don't be predictable" (or don't let others know what you are doing), to "exploit intraday volatility," to "use human override when all else fails."

Robots based on these ideas are simple, fast, and effective. David Leinweber, director of research at First Quadrant Corporation in Pasadena, California, has suggested that the subsumption architecture of Brooks can be applied to trading to provide *intelligent behavior from the bottom up*! At the most elementary level of trading is the mindless, monolithic, big market order that, unless there was a huge quote at an attractive price, no one would send down to the market (except,

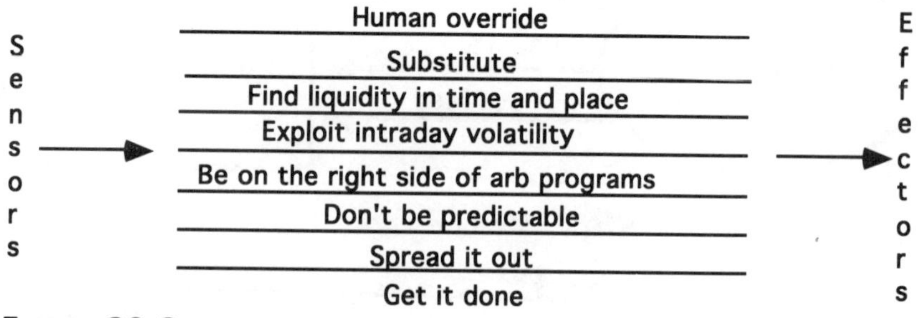

FIGURE 20.3

A subsumption architecture for precommitted trading. (Source: D. Leinweber, "AI and Automated Trading," AI Applications in Wall Street, *New York, 20 April 1993)*

perhaps, in the case of a mutual fund manager obliged to liquidate because of extraordinary demands from customers who want to get out). The next level consists of equally spaced, equally sized orders made in an easily predictable, nonstochastic manner. Next is to make seemingly random variations in order size and interorder timing. This begins to look a bit like the type of data present in real market transactions, but it is still pretty mindless.

Moving up to the next level, the old adage of using the "tide" to advantage comes into play. It is indeed conceivable to use sensor data to trade to your advantage. Many traders with a sufficient number of computer screens and a sufficient amount of experience can become skilled at *guessing* the direction of the market. Experienced traders are said to correctly predict the market slightly more than half the time. Once they achieve 55% winning trades, their behavior may be considered to be vaguely "intelligent." Note that we are still describing deals made right at the flash of the moment, over the phone, usually duly recorded to protect all parties involved. What about all the other possibilities, such as computer trading programs, feedback of execution costs, and the like?

Increasingly sophisticated control strategies that *decide* when to buy and sell and that provide specific instructions regarding transaction size, order types, and price can be effective alternatives to skilled human guesswork. There is indeed evidence that mechanical, neural, and fuzzy trading systems manage to achieve 60% to 65% winning traders and produce profit/risk ratios that exceed those of experienced traders. This raises significant questions about the value of human-based trading. More importantly, it raises questions about the appropriate mix of human- and computer-based trading.

To demonstrate the seriousness of this challenge, the Santa Fe Institute and the Economic Science Laboratory of the University of Arizona are cosponsoring a study to compare the performance of human and program traders. The main objective of this study, launched in January 1993, is to find out whether humans can learn to exploit the limitations and idiosyncracies of computers in repeated interactions. The cosponsors have created a computerized market, the Arizona Token Exchange (AZTE), in which a fictional commodity, "tokens," are traded. The market is a simplified version of commodity exchanges, such as the Chicago Board of Trade, where buyers and sellers are able to call out bids and ask to buy or sell units of the commodity. In each trading session traders are assigned the role of buyer or seller and are given an allocation of tokens. A seller's objective is to sell tokens for as much as possible above the token cost, and a buyer's objective is to buy tokens as cheaply as possible below their redemption value. By ranking the token costs and redemption values, well-defined supply and demand curves can be constructed. The intersection of these curves defines the so-called competitive equilibrium price and quantity, at which neoclassical economic theory predicts all trading will occur. The complication is that in the AZTE, each trader's token costs and redemption values are private information and differ from trader to trader. Thus, AZTE traders face a complex sequential decision problem: how much should they bid or ask for their own tokens, how soon should they place a bid or ask, and under what circumstances should they accept an outstanding bid or ask of some other trader? An additional complication is that each trading session runs for a fixed amount of time. This creates a difficult trade-off, for if traders spend too much time looking for a good deal, they may find themselves locked out of the market without trading anything.

Unlike real commodities markets where most traders are human, in the AZTE all opponents are computer programs. The opponent programs are selected from a field of more than 30 different trading strategies, including winners of the Santa Fe Institute's Double Auction Tournament. The program traders range from simple rules of thumb (such as "zero-intelligence" strategy) to sophisticated optimizing/learning algorithms (such as neural nets and genetic algorithms).

To trade on the AZTE, you need a Unix or PC-compatible computer linked to the Internet computer network. The trading interface software, which allows you to log on and trade at any time, for as long as you like, can be obtained by calling or writing the University of Arizona.[1] To qualify for an AZTE trading account, you need to file an application and a release form stating whether or not you wish to remain anonymous in published analyses of the outcome of this experiment. Upon receipt of your application, a trading account will be set up. Your dollar earnings will accumulate in your account until you decide to withdraw, at which time AZTE will close your account and mail a check for the total amount of the earnings.

The software and ASCII trader's manual (including the application form) is available via anonymous ftp on "fido.econ.arizona.edu", in the azte subdirectory. The manual (azte.man) explains how the software works and what is required to use it. You should first read through the manual, and then obtain the appropriate trading interface for your system. The DOS interface requires VGA graphics resolution and the use of Clarkson packet drivers for the network interface card. The Clarkson drivers are also available via ftp on "fido.econ.arizona.edu". If you don't have access to anonymous ftp, AZTE will mail you a diskette containing the software and trader's manual.

When intelligent computer-based trading strategies are deployed, an important issue remains: what is the role of the person who manages the robots and models? He or she can take the role of active gatekeeper, approving each step, or can remain passive, letting robots and model recommendations go through unless explicitly disapproved. If a person sits on top and uses the outputs for decision support, sometimes endorsing and sometimes rejecting the recommendations, then, according to Jim Hall, the system acquires "an infinite number of degrees of freedom." In this case acceptance of the results of the model will often depend on whether the recommendations confirm the "gut feelings" of the trader! As Klimasauskas pointed out in Chapter 1, *never override your model*!

Conclusions

Virtual reality and robotic trading provide new challenges and opportunities. As learning is most productive in a play environment and humans excel in innovation and productivity in play, the use of such cutting-edge technology may further enhance our capability to learn about the behavior of financial markets and profit from those markets. While there is no magic, no free lunch, in all of these techniques, the exploration of advanced trading technology described in this book provides a trading edge to all those who have the courage to stop guessing the next move of the financial markets.

Note

1. Contact Shawn LaMaster, Manager, Economic Science Systems, Development Economic Science Laboratory, McClelland Hall, Room 116, University of Arizona, Tucson, AZ 85719; (602) 621–6218; Internet: lamaster @ziggy.econ.arizona.edu

Glossary

Adaptability. The ability of computing systems (neural or otherwise) to adapt themselves to data. Such an ability is synonymous with learning.

Adaptive fuzzy system. A fuzzy system that forms its rules from data, rather than being given rules by a human expert. Fuzzy rules are extracted from data feeds. An adaptive fuzzy system acts as a human expert in real time.

Adaptive resonance theory. An unsupervised neural network paradigm that addresses issues of self-organization or clustering of patterns.

Antipersistence. A reversal of a time series, occurring more often than reversal would occur in a random series. If the system has been up in the previous period, it is likely to be down in the next period, and vice versa.

Artificial intelligence (AI). AI views the mind as a type of logical symbol processor that works with strings of text or symbols much as a computer works with strings of 0s and 1s. In practice, AI means expert systems or decision trees.

Attractor. In a nonlinear dynamic series, a point of stability. See also *Limit cycle; Point attractor; Strange attractor.*

Autoassociative memory. A memory designed to transform an input pattern to itself. If the input pattern is noisy, degraded, or incomplete, the memory will still recall the original or learned pattern.

Backpropagation. A learning scheme by which a multilayer network is organized for pattern recognition or classification utilizing an external teacher and error feedback (or propagation).

BDS statistics. Brock-Dechert-Scheinkman test for serial independence in a time series.

Beta. The standardized covariance of an instrument with its complete class of similar instruments. That is, the overall movement relative to its market.

Capital asset pricing model (CAPM). An equilibrium-based asset pricing model developed independently by Sharpe, Linter, and Mossin. The simplest version states that assets are priced according to their relationship to the market portfolio of all risky assets, as determined by the securities' beta.

Chaos. A deterministic, nonlinear dynamic system that can produce random-looking results; an aperiodic equilibrium state of a dynamic system. A system in a chaotic equilibrium seems to wander "at random" through states. Yet the behavior is deterministic: a mathematical equation describes it exactly. If the equation is known, any point of the chaotic path or trajectory can be predicted. Any two starting points of a chaotic system, no matter how close, will give rise to two paths that will diverge in time. Most dynamic systems have chaotic equilibria. These range from the interaction of subatomic particles to the bubbles in a hot tub and from the swirl of clouds in the sky to the distribution of galaxies in space. A chaotic system must have a fractal dimension and must exhibit sensitive dependence on initial conditions. See also *Fractal; Lyapunov exponent; Strange attractor.*

Coherent market hypothesis. A theory stating that the market distribution can be determined by a combination of group sentiment and fundamental bias. Depending on combinations of these two factors, the market can be in one of four states: random walk, unstable transition, chaos, or coherence.

Competitive learning. A learning rule where processing elements compete to respond to a given input stimulus. The winner then adapts to make itself more like the input.

Control strategy. Specifies the order in which network weights are updated for digital implementations of learning functions.

Correlation dimension. An estimate of the fractal dimension that (1) measures the probability that two points chosen at random will be within a certain distance of each other and (2) examines how this probability changes as the distance is increased. A dependent system will be held together by its correlations and will retain its dimension in whatever embedding dimension it is placed, as long as the embedding dimension is greater than its fractal dimension. White noise will fill its space because its components are uncorrelated, and its correlation dimension is equal to whatever dimension in which it is placed.

Correlation integral. The probability that two points are within a certain distance from one another is given by the sum or integral over fixed-size boxes in state space; used in the calculation of the correlation dimension.

Crossover. A procedure or operator in genetic algorithms used to introduce diversity during reproduction. One-point crossover takes two child chromosomes, selects at random one point on those children, and swaps the genetic material on the children at that point. Two-point crossover works the same, but selects two points at random. The use of crossover is a very important feature of the genetic algorithm, and may be critical to its success.

Delta rule learning. A type of learning where weights are modified to reduce the difference between the desired output and the actual output of a processing element. Synonymous with Widrow-Hoff learning. Realizes the minimization of mean squared error.

Divergence. A situation where two indicators are not confirming each other. For example, in oscillator analysis, prices trend higher while an oscillator starts to drop. Divergence usually warns of a trend reversal.

Dynamic system. A system that changes with time. A simple version of a dynamic system is a set of linear simultaneous equations. Nonlinear simultaneous equations are nonlinear dynamic systems. In math, a system described by a first-order differential or difference equation; a system whose change is some function of time or of system parameters. In a broad sense, everything is a dynamic system, including the universe and all its components. The starting point of a dynamic system is the initial condition. The final point or points is the equilibrium state. In between lie the transient states. A dynamic system can have two types of equilibrium states: periodic and aperiodic. Aperiodic equilibria are *chaotic* attractors. Once the system falls in one of these regions it moves around forever, or until something bumps it into a new state, with no apparent structure or periodicity to the movement. The simplest periodic equilibrium is the fixed point attractor. See also *Limit cycle; Point attractor; Strange attractor.*

Efficient frontier. In mean/variance analysis, the curve formed by the set of efficient portfolios—that is, those portfolios of risky assets that have the highest level of expected return for their level of risk. See also *Modern portfolio theory.*

Efficient market hypothesis (EMH). A theory that states, in its semistrong form, that because current prices reflect all public information it is impossible for one market participant to have an advantage over another and reap excess profits.

Euclidean geometry. Plane or "high school" geometry, based on a few ideal, smooth, symmetrical shapes.

Expert system. A computer system that tries to simulate a human expert; a search tree and method of traversal in artificial intelligence. The expert provides knowledge as if-then rules and a programmer codes these in software. Expert systems define a large logic tree or several small trees. An expert system has two components: the knowledge base and the inference engine. The knowledge base is the tree or trees of bivalent rules. The inference engine is some scheme for reasoning or "chaining" the rules. Fuzzy systems are a type of expert system since they too store knowledge as rules—but as fuzzy rules or fuzzy patches. Expert systems work with black/white logic and symbols. Fuzzy systems work with fuzzy sets and have a numerical or mathematical basis that permits both math analysis and simple chip design.

Feed-forward network. A network in which information flows in one direction. In such networks there are no feedback loops from a processing element to a previous one.

Fractal. An object in which the parts are in some way related to the whole; that is, the individual components are "self-similar at all magnifications." An example is the branching network in a tree. Each branch and each successively smaller branch are different, but all are qualitatively similar to the structure of the whole tree.

Fractal dimension. A dimension of a fractal structure. A classical way of describing a fractal dimension relates to measuring the length of a wiggly line using a measuring stick. For example, it would seem that one way to measure the length of a coastline would be to count how many times one can fit a measuring stick of a known length along the coastline and then multiply this number with the length of the measuring stick. While this seems obvious, it appears that the length of a wiggly line actually depends on the length of the measuring stick.

Fractal distribution. A probability density function that is statistically self-similar. That is, in different increments of time, the statistical characteristics remain the same.

Fractal market analysis. A method to measure the memory, the fractal dimension, and/or the correlation dimension in a time series. See also *Correlation dimension; Fractal dimension; Rescaled range (R/S) analysis.*

Fractional Brownian Motion. A biased random walk; comparable to shooting craps with loaded dice. Unlike standard Brownian motion, the odds are biased in one direction or the other.

Fuzzy logic. (1) Multivalued or "vague" logic where everything is a matter of degree, including truth and set membership. (2) Reasoning with fuzzy sets or with sets of fuzzy rules. This term dates back to the first work on fuzzy sets in the 1960s and 1970s by Lotfi Zadeh at the University of California at Berkeley. Zadeh chose the adjective "fuzzy" over the traditional adjective "vague" in his 1965 paper "Fuzzy Sets,' and the name has stuck. Other synonyms: ray logic, cloudy logic, continuous logic.

Fuzzy rule. A conditional of the form "if X is A, then Y is B," where A and B are fuzzy sets. In math terms, a rule is a relation between fuzzy sets. Each rule defines a fuzzy patch (the product $A \times B$) in the system "state space." The wider the fuzzy sets A and B, the wider and more uncertain the fuzzy patch. Fuzzy rules are the knowledge building blocks in a fuzzy system. In math terms, each fuzzy rule acts as an associative memory that associates the fuzzy response B with the fuzzy stimulus A.

Fuzzy set. A set whose members belong to it to some degree. In contrast, a standard or nonfuzzy set contains its members entirely or not at all. The set of even numbers has no fuzzy members.

Fuzzy system. A set of fuzzy rules that converts inputs to outputs. In the simplest case, an expert states the rules using words or symbols. In more complex cases, a neural system learns the rules from data or by watching the behavior of human experts. Each input to the fuzzy system fires all the rules to some degree, as in a massive associative memory. The closer the input matches the "if" part of a fuzzy rule, the more the "then" part fires. The fuzzy system adds up all these output or "then" sets and takes their average or centroid value. The centroid is the output of the fuzzy system. Fuzzy chips perform this associative mapping from input to output thousands or millions of times per second. Each map from input to output defines one FLIPS—or fuzzy logical inference per second. The fuzzy approximation theorem (FAT) shows that a fuzzy system can model any continuous system.

Gaussian system. A system whose probabilities are well described by a normal distribution, or a bell-shaped curve.

Generalization. The ability of a neural computing system to generalize from the input/output examples it was trained on to produce a sensible output to a previously unseen input. Compromise of the variance-bias dilemma.

Genetic algorithm (GA). A problem-solving technique that evolves solutions as nature does, rather than looking for solutions in a more principled way. Genetic algorithms, sometimes hybridized with other optimization algorithms, are the best optimization algorithm available across a wide range of problem types.

Hausdorff dimension. A different type of topological dimension which has the interesting characteristic that it can also take noninteger values. The Hausdorff dimension can be approximated by the fractal dimension or correlation dimension.

Hidden neuron. A usually nonlinear (sometimes linear) processing element with no direct connections to either inputs or outputs. It often provides the learning capacity of a neural network.

Hurst exponent (*H*). A measure of bias in fractional Brownian motion. $H = 0.50$ for Brownian motion; $0.5 < H < 1.0$ for persistent or trend-reinforcing series; $0 < H < 0.5$ for an antipersistent or mean-reverting system. The inverse of the Hurst exponent is equal to alpha, the characteristic exponent for fractal or Pareto distributions.

Layer. The main architectural component of a neural network, consisting of a number of processing elements of equal functionality and occupying a position in the network corresponding to a particular stage of processing.

Learning schedule. A schedule that specifies how parameters associated with learning change over the course of training a network.

Learning. Self-adaptation at the processing element level. Weighted connections between processing elements are adjusted to achieve specific results, eliminating the need to write a specific algorithm for each problem.

Leptokurtosis. The condition that a probability density curve has fatter tails and a higher peak at the mean than the normal distribution.

Limit cycle. An attractor (for nonlinear dynamic systems) that has periodic cycles or orbits in phase space. An example is an undamped pendulum, which will have a closed-circle orbit equal to the amplitude of the pendulum's swing. See also *Attractor*; *Phase space*.

Lyapunov exponent. A measure of the dynamics of an attractor. Each dimension has a Lyapunov exponent. A positive exponent measures sensitive dependence on initial conditions, or how much a forecast can diverge, based on different estimates of starting conditions. In another view, a Lyapunov exponent is the loss of predictive ability as one looks forward in time. Strange attractors are characterized by exhibiting at least one positive exponent. A negative exponent measures how points converge toward one another. Point attractors are characterized by all negative variables. See also *Attractor; Limit cycle; Point attractor; Strange attractor.*

Mechanical trading system. A mechanical trading approach that has (1) a predetermined group of securities or markets that are followed, (2) mathematical formulas applied to prices that tell when to buy and when to sell, (3) entry rules and exit rules for profitable and losing trades, and (4) rules for when to start trading and stop trading. The user of a mechanical trading system chooses markets to trade and applies the system rules to market price action. If the system is computerized, data must be provided to the computer to run the system and place the orders the system dictates. The key to creating a successful mechanical system is to avoid curve-fitting the system. A mechanical trading approach circumvents the destructive emotionalism that permeates discretionary trading.

Membership function. A function that represents the possibility of belonging to a set of crisp values. The membership functions are convex functions that range between 0 and 1.

Modern portfolio theory (MPT). Analysis of a portfolio of assets based on the expected return (or mean expected value) compared with the risk (or standard deviation) of the return of the securities in the portfolio. According to MPT, investors require a portfolio with the highest expected return for a given level of risk. See also *Efficient frontier.*

Moving average. A trend-following indicator that works best in a trending environment. Moving averages smooth out price action but operate with a time lag. Any number of moving averages can be employed, with different time spans, to generate buy and sell signals. When only one average is employed, a buy signal is given when the price closes above the average. When two averages are employed, a buy signal is given when the shorter average crosses above the longer average. Technicians use three types of averages: simple, weighted, and exponentially smoothed.

Mutation. An operator in genetic algorithm that introduces diversity during reproduction. At a very low level of probability, binary mutation replaces bits on a chromosome with randomly generated bits.

Network. A mathematical model of a computing system.

Neural computing. A fast-growing field of computing technology inspired by studies of the brain. Neural computing is ideally suited for pattern matching, pattern recognition, and control function synthesis.

Neural network. A nonlinear dynamic system that maps inputs to outputs; also called a neurocomputer. Its equilibrium states can recall or recognize a stored pattern or can solve a mathematical or computational problem.

Neural net trading system. An automated way of trading a financial security or financial market based on neural network technology. A neural net trading system can be used as a decision-support or decision-making system.

Neuron. A nerve cell in the physiological nervous system.

Nonlinearity. Refers to a mapping that is nonlinear, or where the input is not a multiple of the output. A nonlinear network can be achieved by using nonlinear transfer functions, by competition among neurons, or by normalization.

Order parameter. In a nonlinear dynamic system, a variable that summarizes the individual variables that can affect the system. For example, the Dow Jones is an order parameter, because it summarizes the changes of some 30 stocks.

Oscillator. Technical indicator for determining when a market is in an overbought or oversold condition. Oscillators are plotted at the bottom of a price chart. When the oscillator reaches an upper extreme, the market is oversold. Two types of oscillators use momentum and rates of change.

Parallel processing. A form of computing in which many computations are processed simultaneously. One of the unique features of neural computing is that it provides an inherently clean and simple mechanism for dividing the computational task into subunits. This inherent parallelism makes it an ideal candidate for highly parallel architectures.

Pattern recognition. The categorization of patterns in some domain into meaningful classes. A pattern usually has the form of a vector of measurement values.

Persistence. A tendency of a series to follow trends. If the system has increased in the previous period, chances are that it will continue to increase in the next period. Persistent time series have a long memory; long-term correlation exists between current events and future events. See also *Antipersistence; Hurst exponent; Rescaled range (R/S) analysis.*

Phase space. A graph that shows all possible states of a system. In phase space, the value of a variable is plotted against possible values of the other variables at the same time. If a system has three descriptive variables, the phase space is plotted in three dimensions, with each variable taking one dimension.

Point attractor. In nonlinear dynamics, an attractor where all orbits in phase space are drawn to one point or value. Essentially, any system that tends to a stable, single-valued equilibrium will have a point attractor. A pendulum damped by friction will always stop. Its phase space will always be drawn to the point where velocity and position are equal to zero. See also *Attractor; Phase space.*

Price patterns. Patterns that appear on price charts that have predictive value. Patterns are divided into reversal patterns and continuation patterns.

Processing element. The neuronlike unit that, together with many other processing elements, forms a neural computing network. Computational abstraction of a neuron.

Random walk. Brownian motion, where the previous change in the value of a variable is unrelated to future or past changes. See also *White noise.*

Rate of change. A technique used to construct an overbought/oversold oscillator. Rate of change employs a price ratio over a selected span of time. To construct a 10-day rate-of-change oscillator, the last closing price is divided by the closing price 10 days earlier. The resulting value is plotted above or below a value of 100.

Ratio analysis. The use of a ratio to compare the relative strength between two entities. Dividing an individual stock or industry group by the S&P 500 index can determine whether that stock or industry group is outperforming or underperforming the stock market as a whole. Ratio analysis can be used to compare any two entities. A rising ratio indicates that the numerator in the ratio is outperforming the denominator. Ratio analysis can also be used to compare market sectors, such as the bond market to the stock market or commodities to bonds. Technical analysis can be applied to the ratio line itself to determine important turning points.

Relative strength index (RSI). A popular oscillator developed by Welles Wilder, Jr., and described in his 1978 book, *New Concepts in Technical Trading Systems.* RSI is plotted on a vertical from 0 to 100. Values above 75 are considered to be overbought and values below oversold. When prices are above 75 or below 25 and diverge from price action, a warning is given of a possible trend reversal. RSI usually employs time spans of 9 or 14 days.

Reproduction. A procedure applied in genetic algorithm whereby problem solutions, or parents, are randomly selected from a population of problem solutions. Each parent's selection chances are biased so that parents with the highest evaluations are most likely to reproduce. Children are made by copying the parents, and the parents are returned to the population.

Rescaled range (*R/S*) analysis. Developed by H. E. Hurst to determine long-memory effects and fractional Brownian motion; a measurement of how the distance covered by a particle increases over longer and longer time scales. For Brownian motion, the distance covered increases with the square root of time. A series that increases at a different rate is not random. See also *Antipersistence; Fractional Brownian motion; Hurst exponent, Persistence.*

Resistance. The opposite of support. Resistance is marked by a previous price peak and provides enough of a barrier above the market to halt a price advance.

Roulette wheel parent selection. A technique to determine which problem solutions or population members are chosen for reproduction. Using this technique, each chromosome's evaluation is proportional to the size of its slice on a roulette wheel. Selection of parents to reproduce is carried out through successive spins of the roulette wheel spinner. See also *Genetic algorithm.*

Self-organization. The adaptation of the weights in a neural network in response to a learning stimulus. Usually this is unsupervised.

Set theory. The study of sets or classes of objects. The set is the basic unit in mathematics, just as the symbol is the basic unit in logic. Logic and set theory make up the foundations of math. In theory, all the symbols of advanced calculus and nuclear physics are just shorthand for the longhand of sets and logic. Classical set theory does not acknowledge the fuzzy or multivalued set whose members belong to some degree. Classical set theory is bivalent; each set contains its members entirely or not at all.

Simple average. A moving average that gives equal weight to each day's price data.

Simulated annealing. A technique used to search for global minima in an energy surface in which states are updated based on a statistical rule rather than deterministically; this update rule changes to become more deterministic as the search progresses.

Speech recognition. The decoding of a sound pattern into phonemes or words.

Stochastics. An overbought/oversold oscillator that is based on the principle that as prices advance, the closing price moves to the upper end of its range. In a downtrend, closing prices usually appear near the bottom of their recent range. Time periods of 9 and 14 days are usually employed in its construction. Stochastics uses two lines: %K and its two-day average, %D. These two lines fluctuate in a vertical range between 0 and 100. Readings above 80 are overbought, while readings below 20 are oversold. When the faster %K crosses below the %D line and the lines are above 80, a sell signal is given. There are two stochastics versions: fast stochastics and slow stochastics. Most traders use the slower version because of its smoother look and more reliable signals.

Strange attractor. An attractor in phase space, where the points never repeat themselves and the orbits never intersect, but both the points and the orbits stay within the same region of phase space. Unlike limit cycles or point attractors, strange attractors are nonperiodic and generally have a fractal dimension. They are a configuration of a nonlinear chaotic system. See also *Attractor; Chaos; Limit cycle; Point attractor.*

Subsethood. The degree to which one set contains another set. In classical set theory, a set contains subsets entirely or not at all. In fuzzy logic, this is a matter of degree; the subsethood or containment value can take any value between 0 and 100%. The measure of subsethood derives from the subsethood theorem, which provides a new way to view the probability of an event. According to this theorem, probability equals the whole in the part; in other words, the probability of the part, or event, is the degree to which the whole or the "space" of all events is contained in the part. This relation cannot hold if subsethood is not fuzzy and can take on only the extreme black/white values of 0 and 100%.

Summation function. The part of a processing element that adds the signals which enter the element.

Supervised learning. Learning in which a system is trained by using a teacher to show the system the desired response to an input stimulus, usually in the form of a desired output.

Threshold. A constant which is used as a comparison level by a variable. If the variable has a value above the threshold, some action is taken (for example, a neuron fires); if its value is below the threshold, no action is taken.

Trading system. An automated way of trading a financial security or financial market. Such a system can be designed on the basis of simple if-then rules embedded in a mechanical trading system or on the basis of neural networks, genetic algorithms, fuzzy logic, or combinations of these technologies. A trading system can be used for decision support or decision making.

Training. Exposing a neural computing system to a set of example stimuli to achieve a particular user-defined goal.

Transfer function. The component of a processing element through which the sum is passed (transformed) to create net output. It is usually nonlinear.

Trend. Refers to the direction of prices. Rising peaks and troughs constitute an uptrend; falling peaks and troughs constitute a downtrend. A trading range is characterized by horizontal peaks and troughs. Trends are generally classified as major (longer than six months), intermediate (one to six months), or minor (less than a month).

Trendlines. Straight lines drawn on a chart below reaction lows in an uptrend, or above rally peaks in a downtrend, that determine the steepness of the current trend. The breaking of a trendline usually signals a trend change.

Universe of discourse. The range over which a variable is defined.

Unsupervised learning. Learning in which no teacher is used to show the correct response to a given input stimulus; the system must organize itself purely on the basis of the input stimuli it receives.

Weighted average. A moving average that uses a selected time span but gives greater weight to more recent price data.

Weighted connections. The channels through which information enters processing elements in a neural computing system, throughout which memory is distributed. Also called interconnects.

White noise. The audio equivalent of Brownian motion; sounds that are unrelated and sound like a hiss. The video equivalent of white noise is "snow" on a television receiver screen.

Bibliography

Abarbanel, H. D. I. "Signal Processing on Strange Attractors." *Seminar on Nonlinear Signal Processing*. Randle/SAIC, March 1990.

Abarbanel, H. D. I., R. Brown, and M. Kennel. "Variation of Lyapunov Exponents on a Strange Attractor." *Journal of Nonlinear Science* 1, 1991.

Abraham, R. H., and C. D. Shaw. *Dynamics: The Geometry of Behavior. Part One—Periodic Behavior. Part Two—Global Behavior*. Santa Cruz, CA: Aerial Press, 1982.

Ammeraal, L. *Programs and Data Structures in C*. New York: John Wiley & Sons, 1987.

Araki, S., et al. "A Self-Generating Method of Fuzzy Inference Rules." *Fuzzy Engineering toward Human Friendly Systems*, vol. 2. International Fuzzy Systems Engineering, 1991.

Asakwa, T. K. "Stock Market Prediction System with Modular Neural Networks." In *Proceedings of the International Joint Conference on Neural Networks* (San Diego), 1990.

Babcock, B. *Trading Systems*. Homewood, IL: Dow Jones Irwin, 1989.

Baker, G. L., and J. P. Gollub. *Chaotic Dynamics: An Introduction*. Cambridge: Cambridge University Press, 1990.

Benachenou, D., M. Cader, and G. Deboeck. "Implementation of Neural Trading System." In *Proceedings of the International Joint Conference on Neural Networks* (Beijing), 1992.

Bergerson, K., and D. Wunsch. "A Commodity Trading Model Based on a Neural Network-Expert System Hybrid." In *Proceedings of the International Joint Conference on Neural Networks* (Seattle), 1991.

Bernstein, J. *Timing Signals in the Futures Market*. Chicago: Probus, 1992.

Bernstein, P. *Capital Ideas: The Improbable Origins of Modern Wall Street*. New York: Free Press, 1992.

Bernstein, P. L. "Flows of Funds and Flows of Expectations." *Journal of Portfolia Management* 16(4), 1990.

Bezdek, J. C. *Pattern Recognition with Fuzzy Objective Function Algorithm*. New York: Plenum Press, 1981.

Bochereau, L., and Bourgine. "Extraction of Semantic Features and Logical Rules from a Multilayer Neural Network." In *Proceedings of the International Joint Conference on Neural Networks* (San Diego), 1990.

Bornholdt, S. "General Asymmetric Neural Networks and Structure Design by Genetic Algorithms." *Neural Networks* 5, 1992.

Brock, H. W. *Forecast and Risk Assessment.* Strategic Economic Decisions Inc., Feb. 1993.

Brock, H. W. *Inference from a Decade in the Business.* Strategic Economic Decisions Inc., 1993.

Brock, W. A. "Distinguishing Random and Deterministic Systems: Abridged Version." *Journal of Economic Theory* 40, 1986.

Brock, W., D. Hsieh, and B. LeBaron. *Nonlinear Dynamics, Chaos, and Instability: Statistical Theory and Economic Evidence.* Cambridge, MA: The MIT Press, 1991.

Casdagli, M. and S. Eubank, eds. *Nonlinear Modeling and Forecasting, Papers.* Workshop sponsored by the Santa Fe Institute and NATO. Reading, MA: Addison-Wesley, 1992.

Colby, R. W., and T. A. Meyers. *The Encyclopedia of Technical Market Indicators.* Homewood, IL: Business One Irwin, 1988.

Colin, A. "Neural Networks and Genetic Algorithms for Exchange Rate Forecasting." In *Proceedings of the International Joint Conference on Neural Networks* (Beijing), 1992.

Cox, E. "Fuzzy Fundamentals." *IIEE Spectrum*, Oct. 1992.

Cox, E. "Adaptive Fuzzy Systems." *IIEE Spectrum*, Feb. 1993.

Danon, Y., and M. J. Embrechts. "Least Squares Fitting Using Artificial Neural Networks." Troy, NY: Rensselaer Polytechnic Institute, June 1992.

Davis, L. *Handbook of Genetic Algorithms.* New York: Van Nostrand Reinhold, 1991.

Deboeck, G. "Pre-processing and Evaluation of Neural Nets for Trading Stocks." *Advanced Technology for Developers*, Aug. 1992.

Deboeck, G. "Nonlinear Dynamic Analysis Techniques for Preprocessing of Data for Neural Nets." *Advanced Technology for Developers*, Sept. 1992.

Deboeck, G. "Basic Techniques for Fuzzy Model Design." *Advanced Technology for Developers*, Nov. 1992.

Deboeck, G. "Neural, Genetic and Fuzzy System Applications in Wall Street." *International Journal on Neural and Mass-Parallel Computing and Information Systems, Special Issue on PASE '93*, No. 6, 1993.

Deboeck, G. "Neural, Genetic, Fuzzy Approaches to Design." *AI in Finance* (special report of *AI Expert* magazine), Dec. 1993.

Deboeck, G., H. Green, M. Yoda, and G. S. Jang. "Design Principles for Neural and Fuzzy Trading Systems." In *Proceedings of the International Joint Conference on Neural Networks* (Beijing), 1992.

Deboeck, T., and G. Deboeck. "GenNet: Genetic Optimization of a Neural Net for Trading." *Advanced Technology for Developers*, Oct. 1992.

De Grauwe, P., H. Dewachter, and M. Embrechts. *Exchange Rate Theory: Chaotic Models of Foreign Exchange Markets.* Oxford: Blackwell, 1993.

Dorf, R. C. *Modern Control Systems.* Reading, MA: Addison-Wesley, 1967.

Douglas, M. *The Disciplined Trader: Developing Winning Attitudes.* New York: Institute of Finance, 1990.

Dutta, S., and S. Shekhar. "Bond Rating: A Non-Conservative Application of Neural Networks." In *Proceedings of the IIEE International Conference on Neural Networks*, vol. 2, 1988.

Eckmann, J. P., and D. Ruelle. "Ergodic Theory of Chaotic and Strange Attractors." *Review Modern Physics* 57(3), 1985.

Elder, J. F. IV, and M. T. Finn. "Creating 'Optimally Complex' Models for Forecasting." *Financial Analysts Journal*, Jan.-Feb. 1991.

Fahlman, S. E. "An Empirical Study of Learning Speed in Back-propagation Networks." In *CMU*. Pittsburgh: Carnegie-Mellon University, Computer Science Dept., 1988.

Fahlman, S. E., and C. Lebiere. "The Cascade-Correlation Learning Algorithm." In *Advances in Neural Information Processing Systems*, vol. 2. D. S. Touretzky, ed. Palo Alto, CA: Morgan Kaufman, 1990.

Feder, J. *Fractals*. New York: Plenum Press, 1988.

Feigenbaum, M. J. "Universal Behaviour in Nonlinear Systems." *Los Alamos Science*, 1980.

Felsen, J., "Learning Pattern Recognition Techniques Applied to Stock Market Forecasting." IEEE Transactions on Systems, Man and Cybernetics SMC-5(6), 1975.

Fortin, C., R. Kumaresan, W. Ohley, and S. Hoefer. "Fractal Dimension in the Analysis of Medical Images." *IEEE Engineering in Medicine and Biology* 11(2), 1992.

Frank, G. W., T. Lookman, M. A. H. Nerenberg, C. Essex, J. Lemieux, and W. Blume. "Chaotic Time Series Analyses of Epileptic Seizures." *Physica D* 46, 1990.

Fraser, A. M. "Information and Entropy in Strange Attractors." *IEEE Transactions on Information Theory* 35(2) 1989.

Fraser, A. M., and H. L. Swinney. "Independent Coordinates for Strange Attractors from Mutual Information." *Physical Review* 33, Feb. 1986.

Frison, T. "Contemporary Signal Processing." Technical Report. Great Falls, VA: Randle Inc., 1991.

Galbraith, J. K. *A Short History of Financial Euphoria: Financial Genius Is Before the Fall*. Whittle Direct Books, 1990.

Garman, M. B., and M. J. Klass. "On the Estimation of Security Price Volatilities from Historical Data." *Journal of Business* 53(1), 1980.

Gershenfeld, N. "An Experimentalist's Introduction to the Observation of Dynamical Systems." In *Directions in Chaos*, vol. 2. H. B. Lin, ed. World Scientific, 1988.

Glass, L., and M. C. Mackey. *From Clocks to Chaos*. Princeton, NJ: Princeton University Press, 1988.

Gleick, J. *Chaos: The Making of a New Science*. New York: Viking Press, 1987.

Goldberg, D. E. *Genetic Algorithms in Search, Optimization and Machine Learning*. Reading, MA: Addison-Wesley, 1989.

Grassberger, P. "Do Climatic Attractors Exist?" *Nature* 323, Oct. 1986.

Grassberger, P. "An Optimized Box-assisted Algorithm for Fractal Dimensions." *Physics Letters A* 148, Aug. 1990.

Grasserger, P., and I. Procaccia. "Measuring the Strangeness of Strange Attractors." *Physica D* 189, 1983; see also "An Optimized Box-assisted Algorithm for Fractal Dimensions." *Physics Letters A* 148 Aug. 1990.

Green, H. "Neural Networks in Financial Services: Survey." In *Proceedings of Expert Systems and Neural Networks in Trading* (New York), 1991.

Green, H. G. *Artificial Intelligence Tools in Financial Markets: An Updated Survey*. London: Centre for Cognitive Systems, Imperial College of Science, Technology and Medicine, 1993.

Green, H. G. and R. Mark. "Managing a Portfolio of Positions Unit." In *The Handbook of Interest Rate Risk Management*. J. Francis and A. Wolf, eds. Homewood, IL: Dow Jones Irwin, 1994.

Hammel, S., and M. Hammel. "A Noise Reduction Method for Chaotic Systems." *Physics Letters A* 148(8, 9), 1990.

Harp, S., and T. Samad. "Genetic Synthesis of Neural Network Architecture." In *The Handbook of Genetic Algorithms*. L. Davis, Ed. New York: Van Nostrand Reinhold, 1991.

Hausdorff, F. "Dimension and Ausseres Mass." *Math Annalen* 79, 1918.

Hawley, D. D., J. D. Johnson, and D. Raina. "Artificial Neural Systems: A New Tool for Financial Decision-making." *Financial Analysts Journal*, Nov.-Dec. 1990.

Hirose, Y., K. Yamashita, and S. Hijiya. "Back-Propagation Algorithm Which Varies the Number of Hidden Units." *Neural Networks* 4, 1991.

Holland, J. *Adaptation in Natural and Artificial* Systems. Originally published by the University of Michigan Press (Ann Arbor) in 1975. Reissued by The MIT Press (Cambridge, MA) in 1992.

Humpert, B. "Neurocomputing in Financial Services." *Journal of Expert Systems for Information Management*. 2(3), 1989.

Hurst, H. E. "Long-term Storage of Reservoirs." In *Transactions of the American Society of Civil Engineers*, 1951.

Hurst, H. E., R. P. Black, and Y. M. Simaika. *Long Term Storage: An Experimental Study*. London: Constable, 1965.

Jacobs, B., and K. Levy. "The Complexity of the Stock Market." *Journal of Portfolia Management* 16, 1989.

Jacobs, R. A., M. I. Jordan, S. J. Nowlan, and G. E. Hinton. "Adaptive Mixtures of Local Experts." *Neural Computation* 3, 1991.

Jang, G. S., F. Lai, B. W. Jiang, and L. H. Chien. "An Intelligent Trend Prediction and Reversal Recognition System Using Dual-Module Neural Networks." In *Proceedings of the 1st International Conference on Artificial Intelligence Applications on Wall Street* (New York), 1991.

Jang, G. S., and F. Lai. "Intelligent Stock Market Prediction System Using Dual Adaptive-Structure Neural Networks." In *Proceedings of the 2nd International Conference on Artificial Intelligence Applications on Wall Street* (New York), 1993.

Jordan, M., "New Learning Algorithms." *Tutorial at International Joint Conference on Neural Networks* (Baltimore), 1992.

Kamijo, K., and T. Tanagawa. "Stock Price Pattern Recognition. A Recurrent Neural Network Approach." In *Proceedings of the International Joint Conference on Neural Networks* (San Diego), 1990.

Kandel, A. *Fuzzy Mathematical Techniques with Applications*. Reading, MA: Addison-Wesley, 1986.

Kasko, B. *Neural Networks and Fuzzy Systems*. Englewood Cliffs, NJ: Prentice-Hall, 1992.

Kaufman, P. J. *The New Commodity Trading Systems and Methods.* New York: John Wiley & Sons, 1987.

Kennel, M. B., R. Brown, and H. D. I. Abarbanel. "Determining Embedding Dimension for Phase-Space Reconstruction Using a Geometrical Construction." *Physics Review A* 45, March 1992.

Kennel, M. B., and S. Isabelle. "Method to Distinguish Possible Chaos from Colored Noise and to Determine Embedding Parameters." *Physical Review A* 46, Sept. 1992.

Klimasauskas, C. C. "Hybrid Technologies: More Power for the Future." *Advanced Technologies for Developers*, May 1992.

Klimasauskas, C. "Accuracy and Profit in Trading Systems." *Advanced Technology for Developers*, June 1992.

Klimasauskas, C. "Genetic Function Optimization for Time Series Prediction." *Advanced Technology for Developers*, July 1992.

Klimasauskas, C. C. "Hybrid Fuzzy Encodings for Improved Backpropagation." *Advanced Technology for Developers*, Sept. 1992.

Klimasauskas, C. "Hybrid Neuro-Genetic Approach to Trading Algorithms." *Advanced Technology for Developers*, Nov. 1992; and "An Excel Macro for Genetic Optimization of a Portfolio." *Advanced Technology for Developers*, Dec. 1992.

Kimoto, T., K. Asakawa, M. Yoda, and M. Takeoka. "Stock Market Prediction System with Modular Neural Networks." In *Proceedings of the International Joint Conference on Neural Networks* (San Diego), 1990.

Kosko, B. *Neural Networks and Fuzzy Systems.* Englewood Cliffs, NJ: Prentice-Hall, 1992.

Kostelich, E.J. and J.A. Yorke. "Noise Reduction: Finding the Simplest Dynamical System Consistent with the Data." *Physica D* 41, 1990.

Koza, J. *Genetic Programming.* Cambridge, MA: The MIT Press, 1993.

Kravtsov, Y. A. "Randomness, Determinateness, and Predictability." *Usp. Fiz. Nauk* 158, May 1989.

Krugman, P., and M. Miller eds. *Exchange Rate Targets and Currency Bands.* Cambridge: Cambridge University Press, 1992.

Kurz, M. "Asset Prices with Rational Beliefs." Unpublished paper. Stanford, CA: Stanford University, June 1992.

Lane, G. C. *Trading Strategies.* Future Symposium International, 1984.

Lapedes, A., and R. Farber. "Nonlinear Signal Processing Using Neural Networks." In *Prediction and System Modeling.* Los Alamos, NM: Los Alamos National Laboratories, 1987.

Larrain, M. "Testing Chaos and Non-Linearities in T-Bill Rates." *Financial Analysts Journal*, Sept.-Oct. 1991.

Lau, C., ed. *Neural Networks: Theoretical Foundations and Analysis.* New York: IEEE Press, 1992.

Lee, T. C. *Structure Level Adaptation for Artificial Neural Networks.* Norwell, MA: Kluwer Academic Publishers, 1991.

LIFE (Laboratory for International Fuzzy Engineering). "Fuzzy Engineering toward Human Friendly Systems." In *Proceedings of the 1st International Fuzzy Engineering Symposium (Yokohama), 1991; Proceedings of the 2nd*

International Conference on Fuzzy Logic and Neural Networks (Tizuka), 1992.

Lo, A. W. "Long-Term Memory In Stock Market Price." *Econometrica* 59(5), Sept. 1991.

Makridakis, S., S. C. Wheelwright, and V. E. McGee. *Forecasting: Methods and Applications*. New York: John Wiley & Sons, 1983.

Mandelbrot, B. B. *The Fractal Geometry of Nature*. New York: W. H. Freeman and Company, 1983.

Marcus, C. M., and R. M. Westervelt. "Dynamics of Analog Neural Networks with Time Delay." In *Advances in Neural Information Processing Systems I*. San Mateo, CA: Morgan Kaufmann, 1989.

Maren, A. J., C. T. Harston, and R. M. Pap. *Handbook of Neural Computing Applications*. San Diego: Academic Press, 1990.

Mark, R. "Risk Management." *International Derivative Review*, March 1991; and "Units of Management." *Risk*, June 1991.

Markowitz, H. M. *"Portfolio Selection." Journal of Finance* 7, March 1952.

Matsuba, I., H. Masui, and S. Hebishima. "Optimizing Multilayer Neural Networks using Fractal Dimensions of Time-Series Data." in *Proceedings of the International Joint Conference on Neural Networks* (Baltimore), 1992.

Matsuba, I., H. Masui, and S. Hebishima. "Prediction of Chaotic Time-Series Data Using Optimized Neural Networks." In *Proceedings of the International Joint Conference for Neural Networks* (Beijing), 1992.

Miller, G. E., et al. "Designing Neural Networks Using Genetic Algorithms." In *Proceedings of the Third International Conference on GA* (George Mason University). San Mateo, CA: Morgan Kaufmann, 1989.

Miller, M. H. *Financial Innovations and Market Volatility*. Cambridge, MA: Blackwell, 1991.

Montana, D., and L. Davis. "Training Feedforward Neural Networks Using Genetic Algorithms." In *Proceedings of the 11th International Joint conference on Artificial Intelligence*, 1989.

Moon, F. C. *Chaotic Vibrations: An Introduction for Applied Scientists and Engineers*. New York: John Wiley & Sons, 1992.

Murase, K., Y. Matsunaga, and Y. Nakade. "A Back-Propagation Algorithm Which Automatically Determines the Number of Association Units." In *Proceedings IEEE International Joint Conference on Neural Networks*, 1991.

Murphy, J. J. *Technical Analysis of the Futures Markets, A Comprehensive Guide to Trading Methods and Applications*. New York: Institute of Finance, 1986.

NeuralWare. *Neural Computing Reference Manual*. Pittsburgh: Neural Ware Inc., 1993.

Pardo, R. *Design, Testing and Optimization of Trading Systems*. New York: John Wiley & Sons, 1992.

Parker, T. S., and L. O. Chua. "Chaos: A Tutorial for Engineers." *Proceedings of the IEEE* 75(8), Aug. 1987.

Parlitz, U. "Identification of True and Spurious Lyapunov Exponents from Time Series." *International Journal of Bifurcation and Chaos* 2(1), 1992.

Peters, E. "Fractal Structure in the Capital Markets." *Financial Analysts Journal*, July-Aug. 1989.

Peters, E. *Chaos and Order in the Capital Markets: A New View of Cycles, Prices and Market Volatility*. New York: John Wiley & Sons, 1991.

Peters, E. "A Chaotic Attractor for the S&P 500." *Financial Analysts Journal*, March-April 1991.

Peters, E. E. *Fractal Market Analysis*. New York: John Wiley & Sons, 1994.

Plummer, T. *Forecasting Financial Markets: Technical Analysis and the Dynamics of Price*. New York: John Wiley & Sons, 1991.

Plutowski, M., and H. White. *Selecting Concise Training Sets from Clean Data*. San Diego: University of California, 1992.

Pool, R. "Is It Chaos, or Is It Just Noise?" *Science: Research News* 243, 6 Jan. 1989.

Pool, R. "Is Something Strange About the Weather?" *Science: Research News* 243, 10 March 1989.

Roll, R. "A Critique of the Asset Pricing Theory's Tests. Part I. On Past and Potential Testability of the Theory." *Journal of Financial Economics* 4(2), 1977.

Rosser, J. B., Jr. *From Catastrophe to Chaos: A General Theory of Economic Discontinuities*. Norwell, MA: Kluwer Academic Publications, 1991.

Ruelle, D. *Chaotic Evolution and Strange Attractors*. Cambridge: Cambridge University Press, 1989.

Ruell, D. *Chance and Chaos*. Princeton, NJ: Princeton University Press, 1991.

Rumelhart, D., G. Hinton, and R. Williams. *Parallel Distributed Processing*. Cambridge, MA: The MIT Press, 1986.

Ruthen, R. "Adapting to Complexity." *Scientific American*, Jan. 1993.

Schaffer, J. D., W. Darrell, and L. J. Eshelman. "Combinations of Genetic Algorithms and Neural Networks: A Survey of the State of the Art." *IIEE Workshop Proceedings*, 1993.

Scheinkman, J. A., and B. LeBaron. "Nonlinear Dynamics and Stock Returns." *Journal of Business* 67(3), 19xx.

Schepers, H. E., J. H. G. M. vanBeek, and J. B. Bassingthwaighte. "Four Methods to Estimate the Fractal Dimension from Self-Affine Signals." *IEEE Engineering in Medicine and Biology* 11(2), June 1992.

Schiller, R. J. *Market Volatility*. Cambridge, MA: The MIT Press, 1990.

Schuster, H. G. *Deterministic Chaos: An Introduction*. Weinheim, Germany: VCH Verlagsgesellschaft mbH, 1989.

Sharda, R., and R. B. Patil. "Neural Networks as Forecasting Experts: An Empirical Test." In *Proceedings of the International Joint Conference on Neural Networks* (San Diego), 1990.

Sharp, W. "Asset Allocation: Management Style and Performance Measurement." *Journal of Portfolio Management*, Winter 1991.

Shaw, R. *The Dripping Faucet as a Model Chaotic System*. Santa Cruz, CA: Aerial Press, 1984.

Smith, M. *Neural Networks for Statistical Modeling*. New York: Van Nostrand Reinhold, 1993.

Smith, V. L., and A. W. Williams. "Experimental Market Economics." *Scientific American*, Dec. 1992.

Stein, R. "Selecting Data for Neural Networks." *AI Expert*, Feb. 1993.

Stewart, I. *Does God Play Dice? The Mathematics of Chaos.* Cambridge, MA: Blackwell, 1989.

Sugeno, M., et al. *Fuzzy Systems Theory and Its Applications.* Tokyo: Tokyo Institute of Technology, 1992.

Sugeno, M., and Y. Takahiro. "A Fuzzy-Logic-Based Approach to Qualitative Modeling." *IIEE Transactions on Fuzzy Systems* 1, Feb. 1993.

Sy, W. "Market Timing: Is It a Folly?" *Journal of Portfolio Management* 16(4), 1990.

Szu, H., Y. Sheng, and J. Chen. "Wavelet Transforms as a Bank of Matched Filters." *Applied Optics*, 10 June 1992.

Taber, R. "Fuzzy Entropy." *Advanced Technology for Developers*, Sept. 1992.

Takens, F. In *Dynamical Systems and Turbulence.* Warwick 1980. D. Rand and L. S. Young, eds. Lecture Notes in Mathematics 898. Berlin: Springer-Verlag, 1981.

Tan, P. Y., Lim, K. S. Chua, F. Wong, and S. Neo. "A Comparative Study Among Neural Networks, Radial Basis Functions and Regression Models" In *Proceedings of the 2nd International Conference on Automation, Robotics and Computer Vision* (Singapore), 1992.

Theiler, J. "Estimating the Fractal Dimension of Chaotic Time Series." *Lincoln Laboratory Journal* 3(1), 1990.

Thompson, J. M. T., and H. B. Stewart. *Nonlinear Dynamics and Chaos.* New York: John Wiley & Sons, 1986.

Treleaven, P., and S. Goonatilake. "Intelligent Financial Technologies." In *Proceedings of Parallel Problem Solving from Nature: Applications in Statistics and Economics.* EUROSTAT, 1992.

Vaga, T. "The Coherent Market Hypothesis." *Financial Analysts Journal*, Nov.-Dec. 1990.

Vergnes, M. "Neural Trader's Assistant." In *Proceedings of the 3rd European Seminar on Neural Computing.* London: The Marketplace, IBC Tech. Services, 1990.

Walter, J., J. H. Ritter, and K. Schulten. "Nonlinear Prediction with Self-Organizing Maps." In *Proceedings of the International Joint Conference on Neural Networks* (San Diego), 1990.

Wang, P. P. *Fuzzy Control Bibliography*, Durham, NC: Duke University, Dept. of Electric Engineering, Dec. 1991.

Wassermann, P. D. "A Combined Back-Propagation/Cauchy Machine Network." *Journal of Neural Network Computing*, 1990.

Welstead, S. "Financial Data Modeling with Genetically Optimized Fuzzy Systems." In *Proceedings of the 2nd Annual International Conference on Artificial Intelligence Applications on Wall Street* (New York), 1993.

White, H. "Economic Prediction Using Neural Networks." in *Proceedings of the IIEE International Conference on Neural Networks*, vol. 2, 1988.

Wiggins, S. *Introduction to Applied Nonlinear Dynamical Systems and Chaos.* New York: Springer-Verlag, 1990.

William, F. E. *Technical Analysis of Stocks, Options & Futures: Advanced Trading Systems and Techniques.* Chicago: Probus, 1988.

Williams, T. "Fuzzy Is Anything but Fuzzy." *Computer Design*, April 1992.

Wilshire Asset Management. *Style Portfolios: A New Approach to Equity Investment Management.* 1990.

Wilson, J. *The SimEarth Bible.* New York: McGraw Hill, 1991.

Wolf, A., J. Swift, J. Swinney, and J. Vastano. "Determining Lyapunov Exponents from a Time Series." *Physica D* 16, 1985.

Wong, F. "Time Series Forecasting Using BackPropagation Neural Networks." *Neurocomputing* 2, 1990–91.

Wong, F. "FastProp: A Selective Training Algorithm for Fast Error Propagation." In *Proceedings of the International Joint Conference on Neural Networks* (Singapore), 1991.

Wong, F., and D. Lee. "A Hybrid Neural Network For Stock Selection." In *Proceedings of the 2nd Annual International Conference on Artificial Intelligence Applications on Wall Street* (New York), 1993.

Wong, F., P. Tan, and X. Zhang. "Neural Networks, Genetic Algorithms and Fuzzy Logic for Forecasting." In *Proceedings of the 3rd International Conference on Advanced Trading Applications on Wall Street* (New York), 1992.

Wong, F. S., and P. Z. Wang. "A Fuzzy Neural Network for FOREX Forecasting." In *Proceedings of the International Fuzzy Engineering Symposium* (Yokohama), 1991.

Wong, F. S., P. Z. Wang, T. H. Goh, and B. K. Quek. "Fuzzy Neural Systems for Stock Selection." *Financial Analysts Journal*, Jan.-Feb. 1992.

Yuize, H., T. Yagyu, M. Yoneda, Y. Katoh, et al. "Decision Support System for Foreign Exchange Trading." in *Proceedings of the International Fuzzy Engineering Symposium* (Yokohama), 1991.

Zeidenberg, M. *Neural Network Models in Artificial Intelligence.* New York: Ellis Horwood, 1990.

Zhang, X., and F. Wong. "A Decision-Support Neural Network and Its Financial Applications." In *Technical Report.* Singapore: Institute of Systems Science, National University of Singapore, June 1992.

Zimmermann, H. "Applications of the Boltzman Machine in Economics." In *Methods of Operations Research*, vol. 60/61. Reider, ed. 14th Symposium on Operations Research. Verlag Anton Hain, 1990.

Index

Trading on the Edge: CD-ROM

Neural, Genetic and Fuzzy Systems for Chaotic Financial Markets.

Two CD-ROMs are available as a companion to *Trading on the Edge*, courtesy of the editor:

Trading on the Edge CD-ROM: Toolkit is available now. It contains all software programs discussed in the book, a vast number of demo programs supplied by vendors, and sample market data you can use to build financial applications.

Trading on the Edge CD-ROM: Investment Navigator will be available in early Summer of 1994. It will contain all information and programs from the Toolkit CD-ROM plus tutorials, vendor database, and royalty free images on trading technology.

Both CD-ROMs can be accessed from Macintosh or IBM PC computers.

- - - - - - - - - - - - - - - - CUT HERE - - - - - - - - - - - - - - - -

Please send me		Total
_____ copies of **Trading on the Edge CD-ROM: Toolkit**	@ $ 9.95 each	$ _______
_____ copies of **Trading on the Edge CD-ROM: Investment Navigator**	@ $39.95 each	$ _______
Shipping and Handling:	Subtotal	$ _______
6.5% sales tax (Minnesota residents only)		$ _______
U.S. and Canada: add $4.50 per order		$ _______
International adds $10 per order (includes air shipment)		$ _______
	Total U.S.	$ _______

Method of payment

☐ Check ☐ Visa ☐ MasterCard ☐ Discover ☐ American Express

Card # ______________________ Expiration Date ________

Signature ______________________________________

Ship information [PLEASE PRINT]

Name ______________________ Company ______________________

Street ______________________________________

City ______________ State/Zip ______________ Country ________

Phone ______________________ Fax ______________________

Send Orders to: **Wayzata Technology, CD-ROM Publisher**
2515 East Highway 2
Grand Rapids, Minnesota 55744-3271

Send FAX to: 218-326-0598 **Or Phone** 1-800-735-7321 or 218-326-0597